MATHEMATICS:

IS GOD SILENT?

James Nickel

Ross House Books
Vallecito, California 95251

Mathematics: Is God Silent?

Library of Congress Catalog Card Number: 00-109001
ISBN: 1-879998-22-x

Printed in the United States of America.

First printing, January 2001
Second printing, October 2001
Third printing (with errata corrections), August 2002
Fourth printing, April 2003
Fifth printing (minor corrections/updates), August 2005

2005-7

The works of the Lord are great,
Studied by all who have pleasure in them.
Psalm 111:2

Johannes Kepler (1571-1630)
Courtesy of *Johannes Kepler Gesammelte Werke* (Munich: C. H. Beck, 1937)

DEDICATED TO THE ETHOS OF
JOHANNES KEPLER

MANY THANKS TO LILA, DANIEL, JOY, AND MARGARET *who have put up with my research idiosyncrasies over the years and have loved me through it all. May your crowns be many in heaven.*

SPECIAL THANKS TO SHANE GIBBONS *for service rendered beyond the call of duty.*

OVERVIEW

THE DETAILS INSIDE

LIST OF FIGURES

FOREWARD

Having taken my seat in the witness box I had little idea about what was to come next. I had been called to appear in the Family Court of the Australian Capital Territory to answer some "questions" about the curriculum of the Christian school of which I was the Principal and main secondary teacher. There had been an unfortunate family separation and the non-Christian father did not want his child in the school chosen by the mother, a Christian. More unfortunately, the father's lawyer was about to begin what was to be a forty-five minute "grilling" in an attempt to highlight any deficiencies in our school.

As the father's advocate moved me from one area of the school's curriculum and administration to another, the magistrate interrupted. He had been scanning our documents during this examination, but now stopped it and asked me, "What does faith have to do with mathematics?" I believe that I gave a more than satisfactory answer and I largely credit this book and its author for the ability to do so. Having had the privilege of working alongside James Nickel for about three years and absorbing his scholarly and spiritually insightful approach to the study of mathematics, I knew that I was prepared "to give an answer to everyone who asks you to give the reason for the hope [that in mathematics, God is not silent] that you have" (I Peter 3:15).

This book is groundbreaking in that it presents a clearly understandable Christian worldview explanation of the meaning and purpose of mathematics, as well as giving guidelines for the practical outworking of its theological and historical conclusions. If there was one book I was asked to recommend as a Biblical explanation of mathematics, then it would be this one. The style is easy, the layout clear, and there is extension for the keener students with review and discussion questions, along with expansive footnotes and a full reading list.

Paul wrote in Romans 1:20:

> For since the creation of the world God's invisible qualities – His eternal power and divine nature – have been clearly seen, being understood from what has been made, so that men are without excuse.

James Nickel has shown how one can see both God at work *and* God in His nature, as evidenced in the mathematical patterns and structures of the universe. As well, God has given the mind of His creature man an ability to organise his thoughts mathematically to better understand the world, and to provide insights into its workings. I trust this new edition finds its way into the curricula of every Christian schooling institution and home. For teachers, it provides a rationale and blueprint for their program; for the student, the beginning of an understanding of why, in mathematics, God is not silent.

Peter Cain, L.L.B (Hon), B. Math (Hon), Dip. Ed.
Light Educational Ministries, Belconnen, ACT, Australia
August, Anno Domini 2000

PREFACE TO THE FIRST EDITION

Skeptic: "You say Christian education is different. How?"

Christian: "Everything is taught from a biblical worldview."

Skeptic: "Is that so? What about mathematics?"

Christian: (Dead silence)

How often do people engage in this type of conversation? How many Christian parents, students, and educators are there that can articulate a biblical approach to mathematics, or even affirm that there *is* such a view?

As a Christian educator, and more specifically – a mathematics teacher, the author has been confronted with such questions time and time again. As the result of almost a decade of research and teaching, the author has come to the conviction that there is a distinctive, biblical approach to mathematics; an approach that will not only convert any skeptic, but also provide a potent motivation to anyone involved in Christian education. The dumbfounded response of dead silence can be turned into a dynamic declaration: the voice of God is *not* silent in mathematics.

Most of us have negative impressions of mathematics. Why is this? In school, our teachers probably taught us just the pure mechanics of the subject. To most, the "chicken scratches" on the blackboard meant nothing more than just that – chicken scratches. How any of it had any meaning to anything else was, well, a mystery. This book attempts to unveil the mystery of those chicken scratches. In so doing, the author hopes to show the real reason why mathematics teaching has created this mystery of meaninglessness.

It may surprise many laymen to realize that professional mathematicians are in a quandary as to the ultimate foundations and meaning of mathematics. Many of these professionals are not only responsible for the goals of school curricula, but also write the textbooks. Since textbook writers are unsure of the ultimate meaning, then most teachers, many who are too busy to look into the matter in more detail, will suffer from the same malady. It does not take long for the student to catch the germ of this dilemma.

The author believes that this mathematical uncertainty is caused by a philosophical prejudice; an assumption that the biblical God is silent in the realm of mathematics. This book attempts to show not only the ramifications of this assumption, but also the difference in perspective and meaning that results from assuming that the voice of the biblical God *is* speaking in mathematics.

Understanding and teaching mathematics from a Christian perspective does make a difference. First, when anyone removes God from any discipline, they end up approaching the subject assuming, not the autonomy of God, but the autonomy of man's mind. Given this assumption in the discipline of mathematics, a fundamental question *cannot* be answered. Why does a mere product of man's autonomous mind accurately model the workings of the physical world? Why can you, with the aid of mathematics, figure the trajectories, velocities, and fuel needed in order to place a man on the moon with an unrivaled degree of accuracy? Humanistic mathematicians and scientists answer using terms like "incredible, unreasonably effective, and mysterious." For the

Christian, the answer to this "mystery" lies in the biblical doctrine of creation. Man's mathematical constructions and the workings of the physical world cohere because of a common Creator.

Second, approaching mathematics from a Christian perspective will make a difference in mathematics education. A biblical Christian teacher will not be content to teach students *just* the mechanics of mathematics. A vast gold mine of history, philosophy, and breathtaking revelations of the manifold wonders of God's creation lie behind the mathematical formulae. In order for the student to see these rich nuggets, the teacher must be skilled in "prospecting," so to speak. For example, the surface beauty of a rainbow is appreciated by all. That is given. But, buried in this beauty, and uncovered only by the industrious researches of the mathematician, lies a marvelous complexity and order. Before the eyes and mind of the student, the teacher must dig up these treasures and bring them to the surface.

The book is divided into two parts. Mathematics does not exist in an historical vacuum. Knowing both the men and the movements of mathematics are essential to a clear understanding of the subject. Hence, part one gives the reader an overview of mathematics history and philosophy. Out from this foundation flows part two, which deals with mathematics teaching. This section is not just for mathematics teachers; it will help remove the veil of mystery from mathematics and enable any reader to catch a glimpse of its purpose in God's world.

Who is this book written for? Everyone – students, teachers, and mathematical laymen. The book has been carefully designed as a multipurpose text. It can be used in both high school and university classrooms as a primary or auxiliary text in the history, philosophy, and pedagogy of mathematics. Chapter questions are included to aid a teacher in directing class discussions or in assigning homework and research projects. For the benefit of teachers and others who want to know more detail, extensive resource documentation has been included.

Above all, this book is for the average layman. Most people are not neutral in their feelings towards mathematics. Either they like it or they hate it. It is the hope of the author that anyone who reads this book will, perhaps for the first time, hear the voice of the living God speaking in and through the discipline of mathematics. May they also come to appreciate, in a fresh and exciting way, the relevancy of biblical revelation to every aspect of life. The Word of the Living God is authoritative in every aspect of life because it speaks to every aspect of life.

James Nickel
Shreveport, Louisiana
Anno Domini 1989

PREFACE TO THE SECOND EDITION

The author is grateful that the demand for this book has necessitated a second edition. Ten years have passed since the first edition and during these years the author has continued his research into the relationship between the worldview delineated in Scripture and the manifold facets of the mathematics enterprise. The author would also like to thank the many people who have interacted with him concerning the thrust of this book during this time. These people have come from both a variety of backgrounds and from every corner of the earth. The author has appreciated their communication via letter, telephone call, book review, personal contact at conferences, or the budding medium of electronic mail. Because of these conversations – and with the kind permission of the publisher – the second edition is a considerable enlargement of the first. Although the chapter titles and internal structure have remained the same, the enlargement includes correction of typographical errors, expansion of seminal ideas, new thoughts and analyses, clarifications and qualifications as per peer review, enhanced and additional graphics, timelines, an amplified resource section, an extended bibliography, and an exhaustive index.

For those who have read the first edition, the second will add considerably more flavors to the mathematical stew and the author hopes that you will enjoy and appreciate the extra meat and spices. If you are reading this work for the first time, may you come to appreciate, in a fresh and exciting way, the relevancy of biblical revelation, not only to mathematics, *but also to every aspect of life*. May our Lord Jesus Christ exponentially increase over time and in history the tribe of those who embrace and implement such a comprehensive worldview.

Soli Deo Gloria,
James Nickel
Shreveport, Louisiana
August, Anno Domini 2000

INTRODUCTION

A biblical Christian is convinced that God applies to *all* of life. There is a biblical worldview – a biblical view of every area of life and every discipline of study, including mathematics. Knowing the biblical worldview and the biblical view of a discipline or profession positions a Christian for dominion leadership.

Effective Christian leadership in every area of life results from biblical Christian education and scholarship, which have three basic objectives: The first objective is to learn and teach the biblical worldview. The second objective is to know all other worldviews fully and fairly. (It is not possible to know what one believes and why without knowing what one does *not* believe and why.) The final objective is to reinterpret everything on the basis of the biblical worldview and biblical presuppositions.

For many years I have been reinterpreting the disciplines of history and political science from a biblical Christian perspective. In conducting biblical Christian Leadership Seminars for Christian educators and leaders I have challenged participants by example and admonition to reinterpret their disciplines, professions, or areas of endeavor. James Nickel, responding to this challenge, determined to reinterpret one of the most difficult disciplines – mathematics – from a biblical framework. His resulting work is of trailblazing significance. In making substantial strides toward reinterpreting mathematics, Nickel has uncovered truths that have been hidden for centuries.

Mathematics Is God Silent? demonstrates in an interesting, convincing, and easily readable fashion that mathematics can be reinterpreted and that, as in every discipline of study, mathematical conclusions can be understood only by mastering the worldview and presuppositions which undergird them. *Mathematics: Is God Silent?* also establishes that there is a biblical view of mathematics and that the notion of "neutrality," even in mathematics, is mythological because all mathematical conclusions are determined by the presuppositions on which they are based.

This reinterpretive study is written from a highly useful and informative historical perspective. Nickel recognizes and simplifies key movements and issues in mathematics, science, philosophy, theology, and the history of Christianity, demonstrating their interrelationship. *Mathematics: Is God Silent?* also brings to light the critical function and value of mathematics whereby man can fulfill the God-given mandate of exercising dominion over creation under the sovereign God.

Mathematics: Is God Silent? is an excellent model for all Christians who wish to reinterpret their disciplines, professions, or any area of life from a biblical position. In seminars I am currently conducting I recommend *Mathematics: Is God Silent?* as a reinterpretive model, and it has proven to be useful in this vital endeavor. It is my conviction that *Mathematics: Is God Silent?* will encourage a significant number of Christians to begin reinterpreting.

xxii

The Christian community is indebted to James Nickel for the prodigious research and protracted labor that this study represents. It is my hope that his scholarly reinterpretation of mathematics will produce a multiplication of biblical Christian education and leadership. I commend James Nickel for this significant service to God and man. And I recommend this work to everyone interested in biblical Christian Reformation – that Christians may become ever more effective in bringing all of life into captivity unto God.

Glenn R. Martin, B.A., M.A., Ph.D. (1935-2004)
Chairman, Division of Social Science
Indiana Wesleyan University
Marion, Indiana
Anno Domini 1989

CHAPTER SUMMARIES

PART I: MATHEMATICS IN HISTORY

CHAPTER 1: DOES IT REALLY MATTER?

A brief introduction to the book laying the groundwork for what is to follow. Establishes the fact that, among mathematics professionals, the philosophical foundations of mathematics are uncertain. Explains the meaning of presuppositions, worldview, and the components of a worldview. Summarizes the worldview perspective of Scripture as a basis for giving certain answers to this philosophical quandary of modern mathematics.

CHAPTER 2: FROM ADAM TO CHRIST

Surveys the historical flow of mathematics, beginning with perspectives gleaned from the early chapters of Genesis and then to the early postdiluvian cultures, the Hebrew culture, the Classical Greek age, and ending with the mathematicians and scientists of the Museum in Alexandria, Egypt. Examines the development of mathematical thought and method in these cultures showing the intimate relationship between worldview perspectives and mathematics. Concerning the Greeks, discussion focuses on the relationship of Greek philosophy to the progress and decline of Greek mathematics.

CHAPTER 3: FROM CHRIST TO WYCLIFFE

Investigates the historical flow of mathematics from the time of the Greeks and Romans to the High Middle Ages. Presents a detailed study of the philosophical struggles and technological innovations of the "so-called" Christian Dark Ages that served as a womb for the birth of quantitative and empirical analysis with its concomitant dependence upon the methods of mathematics. Details the mathematical input of the Muslim and Hindu civilizations. Focuses upon the key role of the scholastics (Grosseteste, Roger Bacon, Aquinas, Buridan, and Oresme) in developing germinal thoughts that proved to be efficacious in engendering a viable birth of modern science and mathematics.

CHAPTER 4: FROM WYCLIFFE TO EULER

Covers the period of the Scientific Revolution with special emphasis upon the biblical motivations and epistemology of key men like Copernicus, Galileo, Kepler, Newton, etc. Shows the intimate and powerful relationship between applied mathematics and the study of the physical world. Details why the birth of modern science and the ensuing explosion of useful mathematical knowledge *could only have occurred in Europe.* Shows why all other civilizations failed to produce the ingredients for a viable birth and growth of science and mathematics. Analyzes the critical role of Puritan theology in 17th century English science. Ends with a delightful study of the mathematical accomplishments of the Bernoulli family and Leonhard Euler.

CHAPTER 5: FROM EULER TO GÖDEL

Details the root and fruit of mathematical rationalism. Puts mathematics in the context of idolatry when mathematical method is absolutized. Traces the development of the non-Euclidean geometries and other algebras that literally threw a "wrench in the works" of the assumed "perfect mechanism" of mathematics. Shows the reaction of mathematicians to this foundational conundrum and their ensuing attempt to rigorously "shore up" the mathematical understructure. Details how Cantor developed the theory of infinite sets in an attempt to establish mathematics on an indubitable base. Shows how Cantor's theory produced paradoxes and ultimately split mathematical philosophy into three divergent groups. Each group claims their method to be *the way* to certainty and completeness in mathematics. Details the work of Kurt Gödel who, in 1930, dropped a philosophical "atomic bomb" on the whole processes by showing that ultimate foundations in mathematics *cannot* be found within "pure rationalistic" procedures alone.

CHAPTER 6: WHY DOES MATHEMATICS WORK?

Explores the somewhat baffling and intriguing comments made by twentieth century mathematicians in their attempt to explain the answer to why mathematics relates to the physical world. Quotations range from the likes of Jeans, Eddington, Duhem, Weyl, Einstein, Planck, Schrödinger, Heisenberg, Kline, Wigner, etc. Shows clearly that a philosophical prejudice resides among mathematicians who refuse to entertain the idea that the biblical God could be the answer to their philosophical dilemma. Presents a clear and concise biblical answer to the question, "Why does mathematics work?"

PART II: HOW SHOULD WE THEN TEACH?

CHAPTER 7: OBJECTIVES

A liberal arts curriculum is a "race-course" training a student in liberty. But, what is liberty? Christianity and humanism define "liberty" in direct opposites. Hence, a Christian curriculum of mathematics would differ, especially in content perspectives, from a humanist curriculum of mathematics.

First, summarizes a biblical world view perspective of mathematics. Second, details the biblical objectives of a mathematics curriculum and places the whole of mathematical knowledge within the range of these objectives. Mathematics is (1) a tool that is used to describe the wonders of God's creation. Since creation reveals a Creator, then mathematics is (2) a tool that reveals the attributes of God. In accordance with the dominion mandate, mathematics is (3) a tool that enables man to take constructive and godly dominion over God's creation. Since man must be godly in order to take constructive dominion, then man must be presented with the claims of Christ, the Redeemer. Certain mathematical principles can be used as (4) a tool that aids God's people in fulfilling God's worldwide missionary mandate.

CHAPTER 8: PEDAGOGY AND RESOURCES

Presents a biblical view of mathematical pedagogy. 90-95% of mathematics instruction appears to be, to the student, just a collection of abstract "chicken scratches" on the blackboard. Much of mathematics teaching divorces itself from the concrete, physical world and there is a biblical reason for this. To teach mathematics in its true perspective, the abstract principles must be derived from concrete, physical, and scientific foundations. The power of mathematics is that its methods can be applied to these abstract principles and produce new insights that can be returned to the concrete in extremely fruitful applications.

An annotated list of resources is included in order to aid the reader in learning more about the foundations and uses of mathematics.

PART I: MATHEMATICS IN HISTORY

1: DOES IT REALLY MATTER?

The issue in mathematics today is root and branch a religious one.[1]

Rousas J. Rushdoony

In 1948, Hermann Weyl (1885-1955), one of the foremost mathematicians of the 20th century, used the following words to describe the state of mathematics:

> The questions of the ultimate foundations and the ultimate meaning in mathematics remain an open problem; we do not know in what direction it will find its solution nor even whether a final objective answer can be expected at all.[2]

In 1976, the faculty of a major American Christian university defeated a proposal that was designed to integrate the Christian faith with the academic disciplines. A professor of mathematics was a key figure in the argument against integration. He said:

> Integration is not possible in mathematics. In mathematics, God's revelation is silent ... as far as mathematics goes there ain't nothin' there.[3]

---

[1]Rousas J. Rushdoony, *The Philosophy of the Christian Curriculum* (Vallecito, CA: Ross House, 1981), p. 58.

[2]Hermann Weyl, *Philosophy of Mathematics and Natural Science* (Princeton: Princeton University Press, 1948), p. 219.

[3]Cited in Larry L. Zimmerman, "Mathematics: Is God Silent?" *The Biblical Educator*, 2:1(1980), 1. The author credits Larry Zimmerman, a veteran mathematics teacher, for the title of this book.

1.1 CASE CLOSED ... OR IS IT?

To this university professor, the sincere and searching questions posed by Dr. Weyl could not be answered. Does the biblical God have anything to do with mathematics? Is God's revelation silent in this realm? Does it really matter? A Hindu, a Buddhist, a Christian, or even an atheist would all agree that 2 + 2 = 4 in the base 10 decimal system. Therefore, the case is closed. It would appear that mathematics has nothing to do with God.

Before we close the book on the subject, our conclusion must be inspected for possible faults in logic. What is really being said in the statement that mathematics has nothing to do with God? This statement is similar to an expression popular in certain quarters today: There is no such thing as absolute truth. Both statements find common ground in that they claim to be saying something absolutely. Since man coined these statements, they can be absolutely true only if man has absolute or total knowledge. Using theological terms, anyone making these statements is making a claim to be omniscient. Can man really make this sort of claim?

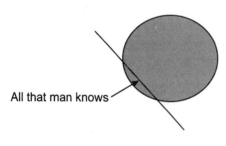

All that man knows

Figure 1: Man's limited knowledge

In Figure 1, the circle represents the totality of knowledge. All that man knows only takes up a portion of this knowledge. There is only one person in the entire universe who can rightly claim to know all and that person is the God of Scripture.

1.2 PENETRATING DECLARATIONS

The Bible informs us that the thoughts of God are above the thoughts of man (Isaiah 55:8-9). The God of Scripture is the God of all knowledge, a knowledge that is infinite in scope (I Samuel 2:3; Psalm 147:5). Given these biblical declarations, it is not a matter of uncertainty, but of *certainty* that mathematics lies within the realm of God's knowledge. Since His knowledge is infinite, He knows everything there is to know about mathematics (past, present, and future).

The Bible also proclaims that, in Jesus Christ, "all things consist [cohere, are held together – J.N.]" (Colossians 1:17). In context, the universal quantifier (i.e., all things) refers to things created, both visible and invisible. Since mathematics deals with things visible (the structure of the physical world) and things invisible (the structure of human thought), it would be reasonable and befitting to deduce that the person of Jesus Christ is the "cohesive" that holds the structure of mathematics together.[4]

1.3 LAYING FOUNDATIONS

A much-overworked phrase in Christian circles is, "Jesus is the answer." Many Christians piously proclaim Jesus to be the answer to every problem, but in actual practice they fail to apply the universal quantifier. The Lordship of Jesus Christ extends over all of life. He is Lord, not just within the confines of the Christian church, but also in heaven and on earth (Matthew 28:18). He is Lord over all knowledge, not just the so-called "spiritual" knowledge. God, through the Bible, speaks to our personal life, our home life, our church life, our business life, our civil life, and our educational life. Yes, God speaks even to the mathematical dilemma posed by Dr. Weyl.

> For no other foundation can anyone lay than that which is laid, which is Jesus Christ (I Corinthians 3:11).

Equating the above Scripture with mathematics seems to fly in the face of a cardinal hermeneutical[5] principle: All Scripture must be interpreted in its proper context. Indeed, the drift of this passage is not directed toward mathematics, but with the picture of building a stable substructure for the Christian life. The author does not minimize the importance of contextual interpretation, but he would also emphasize that, in this passage, a notable universal principle is revealed. This principle is that no structure or system can truly stand unless it is built upon the only true cornerstone – the person of Jesus Christ, God manifested in the flesh.

1.4 WORLDVIEW AND PRESUPPOSITIONS

If the cornerstone for every aspect of life is the person of the Lord Jesus Christ, then how does that premise interact with the "root and branch" of mathematics? It interacts with mathematics on what we will call the worldview level. Every person, whether he realizes it or not, understands the issues of life in a framework context. The man watching Sunday football with a beer in his hand sees life in a certain context. Try pulling the plug on the television set during the Super Bowl and you will see a reaction that is generated from a specific view of the world.

What specifically is a worldview? A worldview is a network of presuppositions not authenticated by the procedures of natural science, a perspective

---

[4]The argument of this paragraph is illustrative of a categorical syllogism, an organized procedure of reasoning consisting of three statements. The first two statements are called premises, and the last is the conclusion. Premise 1: All things visible and invisible hold together in Jesus Christ. Using symbols, All Q are S. Premise 2: Every branch of mathematics deals with either things visible or things invisible. Using symbols, All P are Q. Therefore, every branch of mathematics that deals with these things (the particulars of God's creation) holds together in Jesus Christ. Using symbols, all P are S.
[5]*Hermeneutics* is the science of biblical interpretation. The word is derived from the name of the ancient Greek god Hermes who functioned as "the interpreter or herald of the other gods."

through which everything in human experience is interpreted and human reason is guided.

- A worldview is a network ... it is a system or package of interconnected thinking.
- A worldview is a network of presuppositions ... it is a system of interconnected thinking governed by basic pre-commitments.
- A worldview is a network of presuppositions not authenticated by the procedures of natural science ... these basic pre-commitments cannot be proven by scientific procedure; they are basic and foundational faith commitments. They are an attempt to explain *why* science or mathematics works, not *how* science or mathematics works.
- A worldview is a network of presuppositions not authenticated by the procedures of natural science, a perspective through which everything in human experience is interpreted and human reason is guided ... these presuppositions are the filters through which every aspect of knowledge and the experiences of life are understood and interconnected.

Facts are always understood in the context of presuppositions. Philosopher and mathematician Alfred North Whitehead (1861-1947) said, "Theories are built upon facts; and conversely the reports upon facts are shot through and through with theoretical interpretation."[6] Evolutionary paleontologist Stephen Jay Gould (1941-) said, "Facts do not 'speak for themselves'; they are read in the light of theory."[7] In the context of discussing geocentricity[8] and heliocentricity,[9] astronomer Sir Fred Hoyle (1915-) said:

> Writers on scientific method usually tell us that scientific discoveries are made "inferentially," that is to say, from putting together many facts. But this is far from being correct. The facts by themselves are never sufficient to lead unequivocally to the really profound discoveries. Facts are always analyzed in terms of the prejudices of the investigator. The prejudices are of a deep kind, relating to our views on how the Universe "must" be constructed.[10]

Presuppositional commitments have consequences. Research metallurgist Ian T. Taylor (1931-) said "... presuppositions can not only make us see what does not exist but can also prevent us from seeing what does."[11] For example, in July of 1959, paleontologist Louis Leakey (1903-1972) found a bit of a skull bone and two teeth in Nairobi, Kenya. He said, "We knelt together to examine this treasure ... and cried with sheer joy. For years people had been telling us

---

[6]Alfred North Whitehead, *Adventures of Ideas* (New York: The Free Press, 1967), p. 3.
[7]Stephen J. Gould, *Ever Since Darwin* (New York: W. W. Norton, 1977), p. 161.
[8]Geocentricity, as a cosmology (how the universe works), means having the earth at the center of the solar system.
[9]Heliocentricity, as a cosmology, means having the sun at the center of the solar system.
[10]Fred Hoyle, *Highlights in Astronomy* (San Francisco: W. H. Freeman & Company, 1975), pp. 35-36.
[11]Ian T. Taylor, *In the Minds of Men* (Toronto: TFE Publishing, 1984), p. 194.

that we'd better stop looking, but I felt deep down that it had to be there. You must be patient about these things."[12] Note that Leakey knew what he found (i.e., the "missing link") *before* he examined it; he was finding proof for a theory already accepted. The Piltdown man forgery is another classic example of how a few men, blinded by their presuppositions, deceived and duped an entire generation (from 1912 until the hoax was uncovered in 1953). Concerning this, John Reader said, "When preconception is so clearly defined, so easily reproduced, so enthusiastically welcomed and so long accommodated as in the case of the Piltdown man, science reveals a disturbing predisposition towards belief before investigation."[13]

1.5 THE COMPONENTS OF A WORLDVIEW

There are three fundamental components of a worldview. The first deals with the nature of reality, the second with how we know what we know, and the third with how we live our lives. Using precise philosophical language, the three components of a worldview are:

1. Metaphysics[14] – the study of reality or the study of the nature, structure, and origin of what exists. It deals with the answer to the following questions: Is reality basically one or is reality basically many? Does reality consist of a unity of things or one concept or does reality consist of many things or many concepts? There are three sub-components of metaphysics. First, cosmology[15] deals with the nature and structure of what exists; i.e., how the universe works. Second, cosmogony[16] deals with how the universe came into being; i.e., the origin of what exists. Third, ontology[17] deals with the nature of existence and what kinds of things exist.

2. Epistemology[18] – the study of the nature and limits, the grounds, of human knowledge. It attempts to answer the following questions: What does it mean to know something? How can we know what we know? Do we

---

[12]John Pfeiffer, "Man-Through Time's Mists," *The Saturday Evening Post*, 25 (December 3, 1966), 41.

[13]John Reader, *Missing Links* (London: Collins, 1981), p. 81.

[14]Contrary to some widely shared medieval misconceptions, metaphysics is not the branch of study set up for determining the maximum number of angels that a pinhead can accommodate. *Meta* means "more comprehensive" or "what is behind or underneath." It deals with issues transcending the subject under discussion; i.e., ultimate reality. *Physics* deals with what exists; i.e., the study of the created reality and its workings. Metaphysics ultimately deals with what is behind or underneath, or what transcends or undergirds, that which exists. Since metaphysics undergirds physics, it cannot and should not be understood as something radically independent of physics (a prevalent sophism of modern scientism; e.g., positivism which states that the *only* valid knowledge is knowledge that is perceived through the senses). Metaphysics, while being a step *beyond* physics (or science), is *not* a step beyond the created reality.

[15]*Cosmos* is Greek for "world, universe, or enduring order." *Logos* is Greek for "reason, the word concerning, or study of."

[16]-gony comes from *gone* which is Greek for "origin."

[17]*Ont* is Greek for "being or existence."

[18]*Epistémé* is Greek for knowledge or understanding. It literally means "to cause to stand."

know things independent of experience *(a priori)*?[19] Or, do we know things based upon experience *(a posteriori)*?[20]

3. Ethics – the study of right and wrong attitudes and actions. It deals with human conduct. It attempts to answer the question: How shall we live?

1.6 THE BIBLICAL WORLDVIEW

The Biblical worldview is *unique* among all worldviews because it is founded upon the infallible, verbal revelation from an infinite, personal God; a revelation that is historical and progressive in character. Biblical Christianity addresses both time (this world) and eternity (the world to come). It tells us that God began history and time, superintends history and time, and, in fact, entered history redemptively in the fullness of time in the person of Jesus Christ (Galatians 4:4-5).[21] The Bible also informs us that history will have an end. The components of a worldview are found in Scripture and are briefly summarized in the following analysis.

1.6.1 METAPHYSICS

Genesis 1:1 speaks to one sub-component of metaphysics. In the beginning, the unbeginning and triune (one and the many; i.e., one God in three persons) God created all things (cosmogony). An infinite chasm separates the transcendent God from the finite creation. Creation reflects a personal origin but creation is not to be equated with God. The Creator God, the eternal one and the many, provides the only foundation for understanding the nature of created reality; i.e., the temporal one and the many.

As we have seen earlier, all things "hold together" in Christ (cosmology). God's created order reflects a lawfulness or wisdom or logic and this logic is centered in the person of Christ, the *logos* of God (John 1:1-14). This lawfulness makes the universe not only comprehensible and unified, but regular and predictable. Jeremiah 10:12 (cf. Jeremiah 51:15) states, "He has made the earth by His power, He has established the world by His wisdom, and has stretched out the heavens at His discretion." In Job 38:31-33 (cf. Jeremiah 31:35-36; Jeremiah 33:20-21), God challenges Job with cosmological questions, "Can you bind the cluster of the Pleiades, or loose the belt of Orion? Can you bring out Mazzaroth in its season?[22] Or can you guide the Great Bear with its cubs? Do you

---

[19]*a priori* is Latin for "from the previous." It means to offer a proof that does not depend upon, or is prior to observation. The proof is not based upon experience; rather, it is independent of experience. It is a rational proof with no need for authentication by experience.

[20]*a posteriori* is Latin for "from the subsequent." It means a proof that comes after observation; a proof that is derived from experience.

[21]God controls all the events of history. In Genesis 6, He judged the world via a flood. In Genesis 50:20, Joseph testified to his brothers that "you meant it for evil, but God meant it for good." The exhaustive sovereignty and providence of God lies behind this confession of Joseph. According to Ephesians 1:11, God works all things according to the counsel of His will, a will that is governed by infinite wisdom and goodness. God's eternal counsel and decree give meaning to history.

[22]This phrase refers to God bringing forth the nightly constellations in their seasonal parade.

know the ordinances of the heavens? Can you set their dominion over the earth?"

Acts 17:28 speaks to ontology stating that "in Him we live and move and have our being (or existence)." Revelation 4:11 states that all things base their ontological existence in their Creator, "You are worthy, O Lord, to receive glory and honor and power; for You created all things, and by Your will they existed and were created."

1.6.2 EPISTEMOLOGY

Psalm 36:9 states that "in Your light we see light." It is God and the revelation of light (understanding) from His Word that gives us our bearings in life. The Bible states that the fear of the Lord is the beginning of both wisdom and knowledge (see Proverbs 1:7; Proverbs 9:10). A biblical epistemology states that if you want to be wise and gain knowledge, you must first bow the knee to the Lord God of Scripture. The Hebrew word for beginning means "substructure or foundation." There can be no true knowledge about anything unless the Lord God of Scripture is first honored and respected. Our proximate knowledge of things, whether *a priori* or *a posteriori*, must first recognize God as the ultimate source of this knowledge. Any enlightenment that refuses to "give thanksgiving to God" is thereby pseudo-knowledge or the knowledge of fools (Romans 1:21-22).

1.6.3 ETHICS

Leviticus 20:7 states that "you shall be holy, for I am holy." We are to emulate God's ethical character in our lives learning to obey Him and His statutes from the heart.

All three components of a biblical worldview proclaim the autonomy and authority of God, not of man.[23] Man is not self-sufficient or independent in himself. God made man to live a life of dependence upon Him, the only independent One. *Ultimate* foundations and *ultimate* meaning (using the words of Hermann Weyl) in *any* area of life, including mathematics, can *only* be found in the revelation of the infinite, personal God of Scripture. You cannot understand any aspect of life rightly unless you understand it in a biblical Christian way.

In the ensuing pages, the author will show the reader that the structure of mathematics must be built upon a biblical worldview bedrock, the cornerstone of which is the person of the Lord Jesus Christ, or it will crumble into the dust of meaninglessness, mystery, and uncertainty.

---

[23]Autonomy means "self law." To say that man is autonomous is to say that man can analyze any aspect of life independent of any outside authority, including God and His revelation. It is to place man and his reasoning abilities as the final and ultimate reference point for all predication. It is to place man as the preeminent adjudicator in any issue under debate – metaphysical, epistemological, or ethical.

QUESTIONS FOR REVIEW, FURTHER RESEARCH, AND DISCUSSION

1. Analyze the logic of the statement: "Mathematics has nothing to do with God."
2. Present an argument that establishes the truth of the following statement: "The scope of biblical revelation applies to the whole of life, including mathematics."
3. Define worldview.
4. Critique the following statement: "The facts speak for themselves."
5. Define the three basic components of a worldview.
6. Explain the biblical worldview and its components.

2: FROM ADAM TO CHRIST

For since the creation of the world, His invisible attributes are clearly seen, being understood by the things that are made, even His eternal power and Godhead, so that they are without excuse (Romans 1:20).

Paul, the Apostle

In order to provide the proper groundwork for proving the thesis that mathematics finds its ultimate foundation in the biblical God, surveying the historical flow of mathematical thought becomes necessary. The reader should take note that this survey is not intended to be exhaustive (i.e., covering every mathematician or scientist, every mathematical concept, etc.). Though not exhaustive, this history will be sufficiently comprehensive to both stretch and challenge the reader's cognition of ideas, people, cultures, and events. To help the reader peg these elements into their historical context, detailed timelines for each chapter are located in back of the book. Those who think that their mathematical skill set is substandard will be glad to know that erudite knowledge of mathematics will not be a prerequisite for following its progression through the ages. Discussion of mathematical technique and method will be necessary at some points to illustrate certain themes, but as a whole it will be kept to a minimum. Implicit to all discussions will be a world and life view based upon biblical revelation.

2.1 ORIGINS

When it comes to the mathematics of ancient peoples, evolutionary presuppositions form the root of almost every history of mathematics or science. In these accounts of the early history of man, we first encounter him in the proverbial cave, grunting, groaning, and finger counting.[1] Biblical revelation is at one hundred and eighty-degree odds with this viewpoint. According to Scripture, we first meet man in a garden. He is highly intelligent and capable of intimate communication.

The early chapters of Genesis reveal God making and designing the heavens and fitting the earth as an optimal habitation for the crown of His creation – man.[2] God made man in His image as a dominion bearer giving him the responsibility to rule over the creation. The prerequisite for wise rulership is an intimate, worshipful, and dependent relationship with the infinite, personal, Creator God.

An example of intelligent dominion would be Adam's naming the animals. Note that God presented the animals before Adam in order "to see what he would call them" (Genesis 2:19). Although God named the stars (Psalm 147:4; Isaiah 40:26), he gave Adam the charge of naming the animals. He gave Adam the freedom to create names for the animals; freedom to observe them and to choose an appropriate name that aptly described their characteristics. The biblical understanding of naming is two-fold: (1) A name of an object reflects the characteristics of that object, and (2) a person cannot name an object unless he has previously been given dominion over it; hence, naming is a tool of dominion.

Adam was observing God's created order and classifying it for use. What does this have to do with mathematics? God's creation is replete with both numerical and spatial relationships. It is the duty of man, as a dominion bearer, to observe this reality and classify it for use. Man, in his mathematical and sci-

---

[1]Three examples of the evolutionary assumption concerning early man:

(1) "We can only speculate about the mental powers of our remote ancestors ... Any sense of number which they possessed was probably very elementary ... They were essentially food hunters living in quite small groups with at most a very limited form of speech." Donald Smeltzer, *Man and Number* (Emerson Books, 1958), p. 2.

(2) "Far away in the dim distances of the remote past we see it [the lantern of science – J.N.] emerging from lowly beginnings – possibly single-cell organisms on the sea shore – and gradually increasing in complexity until it culminates in the higher mammals of today, and in man, the most complicated form of life which has so far emerged from the workshop of nature." James Jeans, *Science and Music* (New York: Dover Publications, [1937] 1968), p. 1.

(3) "When the primitive savage began to develop number names, the process was a tedious one.... For a long time after the advent of man such simple numbers as two and three were sufficient for all purposes not met by nouns of multitude, like lot, heap, crowd, school (of fish), pack (of hounds), and flock." David E. Smith, *History of Mathematics* (New York: Dover Publications, [1923] 1958), 1:6.

Those historians who approach history from a non-biblical viewpoint assume that the arithmetic of the "stone-age" tribes of the modern times directly parallels the arithmetic of early, "primitive" man. Actually, what we observe today with regards to "stone-age" arithmetic is a degenerate version of an arithmetic that used to be on a higher plane. Every history of mathematics is an expression of a worldview. All study of history is governed by basic cosmogonic faith commitments.

[2]In this sense, the Bible teaches geocentricism (earth as center), not as a cosmology (as opposed to heliocentrism), but as the center of God's affairs with humankind. For more detail on the enormous degree of specificity (also known as the anthropic cosmological principle) in God's creation, see Stanley L. Jaki, *Cosmos and Creator* (Edinburgh: Scottish Academic Press, 1980).

entific endeavors, works with what God has created (i.e., the external world) and is, in a God-given sense, free to choose a name for it – free to create names that aptly describe its characteristics (i.e., scientific laws using mathematical formulae). Because man is made in the image of God, he is gifted with the ability to observe the physical creation and formulate relationships and consequences that both explain and predict. Throughout the history of mathematics, we find man doing just that.

In the mathematical realm, God's purpose is that man formulate these relationships out from a primary relationship with his Creator. Man's pursuit of mathematics is to be an act of worship to God, the only autonomous one. This response of worship would not be forced upon man, for God created him free to choose. The tragedy of Genesis shows us that man eventually chose, not to worship God his Creator, but himself the creation. In the historic, space-time fall, man became his own god (autonomous). According to Romans 8:20-21, the fall had cosmic repercussions in that the creation was subject to futility and decay. Note carefully the use of words by Paul, the Apostle. He did not say that, because of the fall, the creation became unordered or chaotic. According to Romans 1:18-23, creation after the fall is revelatory of the invisible attributes, the eternal power, and the Godhead of the Creator. Chaos and disorder cannot reveal the faithfully consistent God of Scripture. Since the fall affected the whole of man (mind, emotions, will, and body), he suppresses the knowledge of God known external to him and internal within him; i.e., in creation and in conscience. Sin has extinguished his ability to see and understand the creation rightly (as reflective of the Creator God). But the fall did *not* destroy the investigability of a universe that is both rationally ordered and dependent ontologically upon the Creator's sustaining word of power (Hebrews 1:3). Unredeemed man now investigates and categorizes on his own terms, not God's. This self-acclaimed autonomy will, to some extent, generate personal, systematic (including mathematics), and cultural fragmentation and schizophrenia (Romans 1:24-32).[3]

Although man is now in rebellion, he still bears the image of God, albeit a shattered likeness. He can still take dominion over the earth using the tools of science and mathematics. He can still observe and categorize the patterns in creation, but he does so autonomously and in a blinded sense. As Rousas J. Rushdoony (1916-2001), reformed theologian and educator, notes, "He is rational, but his rationality is spiritually blind, emotionally distorted, and out of kilter in terms of its created purpose, i.e., to function analogically, to think God's thoughts after Him and to interpret and experience life in terms of the will of God."[4] Rushdoony continues his superb analysis, "Man's fall was his attempt to become the original interpreter rather than the re-interpreter, to be the

---

[3]See Rousas J. Rushdoony, *Intellectual Schizophrenia: Culture, Crisis and Education* (Phillipsburg: Presbyterian and Reformed, [1961] 1980).

[4]Rousas J. Rushdoony, *By What Standard?* (Tyler, TX: Thoburn Press, [1958] 1983), pp. 40-41.

ultimate instead of the proximate source of knowledge."[5] The ultimate result is decay and chaos although due to God's goodness to all and His sovereign and wise counsel, He does not allow man's depravity absolute dominion in human affairs. Man needs redemption through grace in Christ in order to become an effective and constructive dominion bearer and to know facts in their true condition as God-created and God-dependent facts.

Throughout the history of mathematics, we find man understanding and using this dominion tool either autonomously or in relationship to the biblical God. The author will attempt to document and analyze the ensuing fruits (epistemological, metaphysical, and ethical) of these two faiths.

2.2 EARLY CIVILIZATIONS

According to Rousas J. Rushdoony, "History after the Flood shows that man rapidly reproduced great civilizations and then declined from them."[6] We will observe this fact as we document the status of mathematics in these civilizations.

We know from archaeological findings that mathematics played a prominent role in the early post-diluvian civilizations.[7] Concerning the extent of mathematical method, we find procedures such as tables of squares and square roots, tables of cubes and cube roots, solutions to cubic equations, the theory of compound interest (exponential functions), and even the solution of quadratic equations.[8] Mathematics served not only as a practical tool in the construction of buildings, but also in agriculture and commerce. Mathematical method and application surfaced in painting, architecture, religion, and the study of the heavenly bodies; i.e., astronomy.

A. Seidenberg has developed an interesting thesis that posits the origin of mathematical method in the religious ritual of these civilizations.[9] In the early chapters of Exodus we find such people as wizards, sorcerers, and magicians performing the religious rituals of Egypt (Exodus 7:11; Exodus 8:7, 18-19). As priests, they mediated deity to the Egyptians. But, the Egyptians worshipped, not God the Creator, but elements of God's creation, from the river Nile to the sun itself.

Egyptian mathematics was elementary, but not child's play.[10] The pyramids, irrigation canal systems, granaries, and business transactions reveal the

---

[5]*Ibid.*, p. 55.

[6]Rousas J. Rushdoony, *World History Notes* (Fairfax: Thoburn Press, 1974), p. 21.

[7]Post-diluvian means "after the worldwide flood" of Genesis 6-8.

[8]Otto Neugebauer, *The Exact Sciences in Antiquity* (New York: Dover Publications, [1957] 1969), pp. 34, 42. Using modern symbolic algebra, the general form of a cubic equation is $ax^3 + bx^2 + cx + d = 0$ and the general form of a quadratic equation is $ax^2 + bx + c = 0$.

[9]See A. Seidenberg, "The Ritual Origin of Geometry," *Archive for the History of Exact Sciences*, 1 (1960-1962), 488-527 and A. Seidenberg, "The Ritual Origin of Counting," *Archive for the History of Exact Sciences*, 2 (1962-1966), 1-40.

[10]See Richard Gillings, *Mathematics in the Time of the Pharaohs* (New York: Dover Publications, [1972] 1982). See also James R. Newman, "The Rhind Papyrus," *Mathematics: An Introduction to Its Spirit and Use*, ed. Morris Kline (San Francisco: W. H. Freeman, 1979), pp. 10-14.

rudiments of algebra, geometry, and trigonometry. Historians have noted that in Egypt, mathematics did not progress; it stagnated. Otto Neugebauer (1899-1990), world renowned expert on science in ancient civilizations, documents that Egyptian mathematics acted as a "retarding force upon numerical procedures."[11] In reference to Egyptian astronomy, he states that it "remained through all its history on an extremely crude level."[12] Howard Eves (1911-2004), mathematics historian, indicates that "ancient Egyptian sources of more recent dates ... show no appreciable gain in either mathematical knowledge or mathematical techniques. In fact, there are instances showing definite regressions."[13]

Babylonian astronomy has bequeathed to us the sexagesimal (base 60) system.[14] It endures today in: (1) degree measurement where a full circle has 360 degrees (6 x 60), each degree divisible in 60 minutes of arc, each minute divisible in 60 seconds of arc and (2) the hours, minutes, and seconds recorded on the timepiece strapped on our wrists. Concerning Babylonian mathematics, Neugebauer states that its "contents remained profoundly elementary."[15] He reaches the same conclusions as Rushdoony, "All historically well known periods of great mathematical discoveries have reached their climax after one or two centuries of rapid progress following upon, and followed by, many centuries of relative stagnation."[16]

Figure 2: Otto Neugebauer

Why this stagnation? Animistic in nature, each ancient culture worshipped the creation rather than the Creator. The flow of time occurred in cycles, not in terms of a linear beginning and end. Hence, history repeated itself again and again in a never-ending circle of time. With a treadmill view of history, no meaningful advance could be confidently made in any aspect of human endeavor.[17] When a culture rejects basic biblical truths, fragmentation tends to manifest itself in every area of life. As history shows us, this cultural fragmentation can be pandemic or it can impact only particular spheres (for example, in ancient Greece, as we shall see, mathematics did blossom for a time).

---

[11]Neugebauer, p. 30.

[12]*Ibid.*

[13]Howard Eves, *An Introduction to the History of Mathematics* (New York: Holt, Rhinehart and Winston, [1953, 1964, 1969] 1976), p. 40.

[14]It is likely that Babylonian astronomers worked with this number system because they assumed a year consisted of 360 days.

[15]Neugebauer, p. 48.

[16]*Ibid.*, p. 30.

[17]For detailed analysis of these ancient cultures in terms of their treadmill view of time and history, see Stanley L. Jaki, *Science and Creation: From Eternal Cycles to an Oscillating Universe* (Edinburgh: Scottish Academic Press, 1974), pp. 1-101.

2.3 CLASSICAL GREECE

Figure 3: The Stage of Greek Mathematics

The formulation of mathematical structure and theory as we know it today finds its roots in the age of Classical Greece (600-300 BC). The Greeks systematized the mathematics and science bequeathed to them by antiquity. In so doing, they also developed some original concepts that greatly enhanced its effectiveness.

Almost all of the famous Greek philosophers, including Plato and Aristotle, were knowledgeable of mathematics. In fact, mathematics played a key role in their philosophical development. What was the catalyst in Greek culture that produced this union?

2.4 REASON AND NATURE ALONE

The Greek culture is unique in that, for the first time, a people attempted to seek answers to the basic questions of life (What is real? How do we know? How should we live?) in the power of human reason alone. Given this conviction, they believed that they could not only solve the so-called mystery of man's origin, value, and destiny but also resolve these answers in the context of a completely autonomous and naturalistic explanation. Note Rousas J. Rushdoony's analysis of the proper context in which to understand the nature of Greek philosophy and science:

> It appears now what constituted Greek "philosophy" and "science." Earlier cultures, in their legal codes, mathematics, and often remarkable calcu-

lation, indicated their high order of intelligence and rationality, but they were not "scientific" because they were not naturalistic.[18]

2.5 HEBREW REVELATION

It is of utmost significance to note that within their own boundaries there existed a people who possessed a book that contained the true supernaturalistic answer to these questions and this mystery. The God of Scripture had chosen the people of Israel to be His voice to all peoples, a lighthouse shining the rays of truth to the then known world. Throughout the history of the Old Testament, the Israelites were used, more involuntarily than voluntarily, as a channel of God's blessings. For example, one of the greatest blessings that ancient Egypt ever received from God was the ten plagues. Each plague shattered one of their cherished gods (Numbers 33:4).[19] In the end, there was no god left but one, the true God of Israel. In the 8th century BC, God had to throw Jonah, the reluctant prophet, in the belly of a great fish in order to make him cry out against the wickedness of Ninevah, the capital of Assyria.

The high point of Hebrew culture, lasting only 80 years, was the reign of David and Solomon (approximately 1000 to 920 BC). Solomon's distinguished wisdom was encyclopedic. According to Scripture:

> He spoke three thousand proverbs, and his songs were one thousand and five. Also he spoke of trees, from the cedar tree of Lebanon even to the hyssop that springs out of the wall; he spoke also of animals, of birds, of creeping things, and of fish. And men of all nations, from all the kings of the earth who had heard of his wisdom, came to hear the wisdom of Solomon (I Kings 4:32-34).

A small subset of that wisdom is found in the book of Proverbs. The book answers, in a multi-faceted way, the ethical question, "How should we live?" Concerning epistemology, only by an antecedent respect and honor of the Lord Creator can true wisdom and knowledge be found (Proverbs 1:7; Proverbs 9:10). It also speaks of wisdom in a personified manner. Concerning the meta-

---

[18]Rousas J. Rushdoony, *The One and the Many: Studies in the Philosophy of Order and Ultimacy* (Fairfax: Thoburn Press, 1978), p. 66.

[19]The ten plagues were (along with the Egyptian god judged):

(1) Nile turned to blood (the Nile was the sacred giver of life).
(2) Swarm of frogs (Egyptians saw the frog as sacred).
(3) Plague of lice (no Egyptian could approach a sacred altar with lice).
(4) Plague of flies (Beelzebub, the lord of the flies, was worshipped to remove flies).
(5) Diseased cattle (Egyptians worshipped the whole animal creation).
(6) Plague of boils (started with ashes from the furnace of the Egyptian god Typhon - this furnace was a place of human sacrifice).
(7) Thunder and hail (Egyptians worshipped all the forces of nature).
(8) Plague of locusts brought with the wind (wind was another force of nature worshipped).
(9) The sign of darkness (the Egyptian sun god, Ra, is shown to be powerless).
(10) Death of the firstborn (this included the firstborn of Pharaoh who was worshipped as god).

physical nature of reality, this personified wisdom made all things (Proverbs 8:12-31).

Another wisdom book of the Hebrews, now a part of the apocrypha, reflects the quantitative nature of God's creation, "He arranged everything according to measure, and number, and weight" (Wisdom of Solomon 11:20-21). Although the Protestant wing of the Church does not consider Wisdom of Solomon to be a part of the official canon of Scripture, both Roman Catholics and Protestants agree on the canonicity of the books of Isaiah and Job. These books also conceptualize the quantitative nature of God's creation:

> Who has measured the waters in the hollow of His hand, measured heaven with a span and calculated the dust of the earth in a measure? Weighed the mountains in scales and the hills in a balance (Isaiah 40:12)?

> For He looks to the ends of the earth, and sees under the whole heavens, to establish a weight for the wind, and apportion the waters by measure, when He made a law for the rain, and a path for the thunderbolt, then He saw wisdom and declared it; He prepared it, indeed, He searched it out, and to man He said, 'Behold, the fear of the Lord, that is wisdom, and to depart from evil is understanding' (Job 28:24-26).

We must recognize that God's measurement, like His thoughts and ways, are higher than man's as the heavens are higher than the earth (cf. Isaiah 55:8-9). Man, being finite, cannot think as God's thinks. He can only "think God's thoughts after Him." In the same way, man cannot measure in kind or in the same way as God. He can only measure in degree after Him. Using philosophical terminology, man can never think or measure *univocally;* i.e., to think or measure as though the Creator/Creature distinction was irrelevant. Man can only think or measure *analogously;* i.e., to think or measure in terms of the light and understanding given by God's revelation. Stanley L. Jaki (1924-), nuclear physicist, Roman Catholic theologian, and science historian, notes the nature of the image of God in man, "Obviously, man could not be a mirror image, that is, a univocal replica of God. Nor could the phrase 'image of God' be a sheer play upon words, a meaningless equivocation. Between the two extremes was the realm of analogy."[20]

The light and understanding that the above texts provide is that it is possible to measure creation in quantifiable terms. As we will see in sections 3.10 through 3.17, the extension of mathematics to the whole of physical science (in principle if not in fact) using quantitative measurement was one of the principle and original contributions during the Christian Middle Ages to the viable birth and nourishment of natural science.[21]

---

[20]Stanley L. Jaki, *The Road of Science and the Ways to God* (Edinburgh: Scottish Academic Press, 1978), p. 54.

[21]For more analysis, see the chapter entitled "The Beacon of the Covenant" in Jaki, *Science and Creation: From Eternal Cycles to an Oscillating Universe,* pp. 138-162.

The people of Israel unfortunately tended to absolutize their status of being God's chosen. From the beginning of the nation politic in the reign of David (and even before), we also see them continuously imbibing the idolatrous practices of the surrounding cultures.[22] Because of these two factors, the voice of God was silenced among the nations. God called many prophets to stand against, not only the sins of Israel, but also the powerful gods (and false worldview conceptions) of Egypt, Babylon, and Assyria. In judgment, God sent Israel to these nations in what is known as the captivity. A subjugated and syncretic people can make very little progress in science and mathematics. Out of their land and away from their temple, the Jews created a new focal point for their worship, the synagogue. Here, Greek "God-fearers" could sit "in the back pew" and listen to the Word of God. In Scripture, we find the answers to the questions regarding the metaphysical nature of reality and to the place of man in that reality:

- Where he came from (cosmogony),
- Who he is (ontology),
- What, if anything, should be his ultimate value in life (axiology),
- What his problem is (sin) and what the solution to that problem is (redemption in Christ; i.e., soteriology),
- How he can know (epistemology),
- How he is to live (ethics), and
- Where he is going (teleology).

2.6 REVELATION REJECTED

Synagogues dotted the Greek Empire established by Alexander the Great. In the 3rd century BC, Jewish scholars in Alexandria, Egypt, translated the Old Testament into the Greek language.[23] Any Greek could visit the "House of the Book" and listen to the words of Scripture in his own language. Yet, the Greek culture, as a whole, rejected the message of the Book.

The Greeks, like other civilizations before them, turned away from God's revelation of Himself, not only in Scripture, but also in creation and conscience. When man runs from God, he ends up suppressing the truth revealed by God (Romans 1:18-20). Instead of seeing creation as God's handiwork, man either views it as divine, or as completely independent in itself.

Worship is endemic to the human race. Every civilization testifies to that fact. Either man worships the creation or the Creator. In the case of the Greeks, we see them deifying human reason and nature.

2.7 THE IONIAN SCHOOL

Thales of Miletus (636-546 BC), one of the seven wise men of antiquity, is believed to be the first of the Greek philosophers to take special interest in

---

[22]This compromise with pagan worship is called syncretism.
[23]This translation is called the Septuagint.

mathematics and science. Miletus was a thriving commercial and trading center and, in his younger years, Thales traveled extensively as a merchant. By this means, he may have accumulated enough wealth to live independently as a philosopher and mentor in his later years.[24] During one business trip to Egypt, he is reputed to have determined the heights of the pyramids by using an angle and shadow technique. Greek historians also report that he used the theory of similar triangles to calculate the distance of a ship from shore.[25] These same historians credit him with logically proving six basic propositions in elementary geometry (all of which are intuitively obvious). According to mathematics historian David E. Smith (1860-1944), with Thales, "we first meet with the idea of a logical proof as applied to geometry…."[26]

2.7.1 THE SIX PROPOSITIONS OF THALES

1. Any circle is bisected by its diameter.
2. The angles at the base of an isosceles triangle are equal.
3. The vertical angles of two intersecting lines are equal.
4. An angle in a semicircle is a right angle.
5. The sides of similar triangles are proportional.
6. Two triangles are congruent if they have two angles and a side respectively equal.

Figure 4: Thales of Miletus

As a philosopher, Thales reduced all reality to water (a monistic metaphysic[27]) and believed that everything is filled with gods. By deity he meant an active power or agent in the material things of the universe. His metaphysical theorizing reflected a firm commitment to empiricism.[28]

Anaximander (611-547 BC), a student of Thales, rejected his mentor's metaphysical and epistemological positions. Instead of everything being reduced to water, Anaximander posited that the basic essence of reality cannot be defined. Instead, it is boundless.[29] Concerning the nature of time, he followed the ancient cosmic destruction and regeneration theory of cycles. He said, "From the infinite eternity comes the destruction, in the same way as does generation issued from it long

---

[24]He founded the Ionian school of philosophy.
[25]Two triangles are similar if they have the same shape but different sizes. Two triangles are congruent if the have exactly the same shape and size.
[26]Smith, 1:68.
[27]Monism is the view that reality consists of only one kind of substance.
[28]Empiricism, as an epistemology, is *a posteriori* in that all knowledge is dependent upon observation, experience, or sense perception.
[29]He used the Greek word *apeiron* to denote this boundless concept.

before: all these generations and destructions reproduce themselves in a cyclic manner."[30]

During his tenure as leader of the Ionian school, Anaximander had little use for mathematics but he did make use of an imported device called a gnomon for determining noon, the solstices, and the equinoxes.[31] He also predated Charles Darwin's (1809-1882) theory of evolution by asserting cosmic matter to be like a spinning mixing bowl the contents of which, by chance, spills out and becomes the particular things of our experience. He believed that man evolved from fish.

In Thales and Anaximander, we see the two basic epistemologies that govern non-biblical thinking. One is the empirical, the "look and see," or the *a posteriori* way. A radical empiricist tends to exclusively rely upon observations believing that all knowledge is dependent on them. The other is rational, the "stop and think," or the *a priori* way. An extreme rationalist tends to exclusively rely upon the mind's ability to conceptualize truth believing that some (or most) knowledge is independent of observation, experience, or sense perception.

Anaximenes (588-526 BC) was the last member of the Ionian school. He rejected the "boundless essence" metaphysics of Anaximander and proposed, in its place, that the unity of all things is air. He noted that air has density (it can be heavy or light) and that it can change *quantitatively* (you can compress or expand it). He surmised the possibility that these quantitative changes could account for the qualitative (hot, cold, hard, soft) changes in the world. He may be the first person in history to attempt to reduce all of reality to quantitative terms, which is the basic thrust of modern science. Like his mentors before him, he also embraced the eternality of the world and a cyclic view of time. Concerning the world, he said "it is not the same world that exists forever; the one which exists is now one world and another later; its generation takes place anew after certain periods of time."[32]

2.8 THE RISE AND FALL OF PYTHAGORAS

Pythagoras (572-492 BC), a student of Thales, is perhaps one of the most famous mathematicians of all time. He based his metaphysic, not on matter as the Ionian school did, but upon non-material number. He observed with wonder that there are reliable and predictable mathematical patterns in the universe. He also noted that the world is in constant flux, but the geometric and numerical relationships do not change. Because of these patterns, he posited that ultimate reality consists of numbers and geometric shapes. Since physical phenomena are determined by the mathematical relationships of shape and proportion,

---

[30]Cited in Jaki, *Science and Creation: From Eternal Cycles to an Oscillating Universe*, p. 105. Jaki cites his source as H. Diels, *Die Fragmente der Vorsokratiker griechisch und detsch*, ed. W. Kranz (Dublin/Zurich: Weidmann, 1968), 12A-10.

[31]A gnomon is an object like a sundial that projects a shadow and is used as an indicator.

[32]Cited in Jaki, *Science and Creation: From Eternal Cycles to an Oscillating Universe*, pp. 105-106.

then number effectively "rules the universe."[33] He believed that the whole universe could be explained within the context of the counting numbers (1, 2, 3,...). In fact, he and his followers worshipped number as the full intelligibility and the generating source of all things. This veneration is revealed in one of the Pythagorean creedal confessions, "Bless us divine number, thou who generatest gods and men."[34]

Pythagoras' discovery of counting number ratios in musical notes served to intensify his conviction that "all is number." According to David E. Smith, "Pythagoras is said to have discovered the fifth and the octave of a note that can be produced on the same string by stopping at ⅔ and ½ of its length, respectively."[35] He also believed that musical proportions governed the motion of the planets and that "the heavenly bodies in their motion through space gave out harmonious sounds: hence the phrase the harmony of the spheres."[36]

Figure 5: Pythagoras

He attracted students to him and eventually formed a mystical mathematical cult in Crotona, a wealthy Greek seaport in southern Italy. There a closely-knit brotherhood bound by secret rites and observances furthered the study and worship of number.[37]

In addition to believing number to be transcendent and the driving force of the universe, the Pythagoreans religiously adhered to its mystical qualities. Carl Boyer (1906-1976) and Uta Merzbach, mathematics historians, remark on the Pythagorean worship of number:

> Many early civilizations shared various aspects of numerology, but the Pythagoreans carried number worship to its extreme, basing their philosophy and their way of life upon it. The number one, they argued, is the generator of numbers and the number of reason; the number two is the first even or female number, the number of opinion; three is the first true male number, the number of harmony, being composed of unity and diversity; four is the number of justice or retribution, indicating the squaring of accounts; five is

---

[33]"Number rules the universe" was the popular motto of the followers of Pythagoras.

[34]Cited in Stanley L. Jaki, *The Relevance of Physics* (Edinburgh: Scottish Academic Press, [1966] 1992), p. 136. Contra Pythagoras, numbers do not *generate* the particularity of things; numbers *report* on the particularity of things.

[35]Smith, 1:75. Some historians believe that this harmony gave rise to the "harmonic proportion," since $1 : \frac{1}{2} = 1 - \frac{2}{3} : \frac{2}{3} - \frac{1}{2}$.

[36]W. W. Rouse Ball, *A Short Account of the History of Mathematics* (New York: Dover Publications, [1908] 1960), p. 27.

[37]His followers were divided into two groups: the hearers and the mathematicians. In order to become a full-fledged mathematician, you had to prove yourself worthy during a proscribed period of probation as a hearer. This brotherhood has ever since served as a model for secret societies in Western civilization.

the number of marriage, the union of the first male and female numbers, and six is the number of creation. Each number had its peculiar attribute. The holiest of all was the number ten, or the tetractys, for it represented the number of the universe, including the sum of all possible dimensions. A single point is the generator of dimensions, two points determine a line of dimension one, three points (not on a line) determine a triangle with area of dimension two, and four points (not in a plane) determine a tetrahedron with volume of dimension three; the sum of the numbers representing all dimensions, therefore, is … ten. It is a tribute to the abstraction of Pythagorean mathematics that the veneration of the number ten evidently was not dictated by anatomy of the human hand or foot.[38]

Due to his mathematical metaphysic, Pythagoras believed that the good life is the contemplative life. To properly exercise the brain, one must learn to live an ascetic life; a life emancipated from material comforts, overindulgence, and bodily desire.[39] Therefore, an odious "sinful" habit eschewed by any good Pythagorean was the eating of beans!

The wonder of mathematical patterns that Pythagoras observed in the world around him is reflective of the creative and sustaining order of the infinite, personal, Creator God of Scripture. Masking this truth in biblical impiety, Pythagoras ended up abstaining from the world in order to become like it intellectually. Instead of appreciating a patterned universe of God's making and obeying the Creator with all of his heart, soul, mind, and strength, Pythagoras bowed to the idol of intellect (mathematics) and ascetic purification. As the Apostle Paul states in Romans 1:25, Pythagoras "exchanged the truth of God for the lie, and worshipped and served the creature rather than the Creator." He worshipped the reflection of God (mathematical order), not the source of that reflection. All too quickly, and as we shall see next, his idolatrous mathematical metaphysic collapsed.

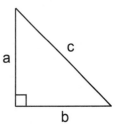

Figure 6: Pythagorean Theorem

Pythagoras is perhaps best known for the famous theorem named in his honor. In the right triangle pictured in Figure 6, $a^2 + b^2 = c^2$. Needless to say, he was thrilled to see his beloved counting numbers fit so beautifully into this equation.[40] But his joy was short-lived. If $a = 1$ and $b = 1$, then $c = \sqrt{2}$! $\sqrt{2}$ is not a counting number and to the despair of Pythagoras, it cannot even be writ-

---

[38]Carl Boyer, *A History of Mathematics*, rev. Uta C. Merzbach (New York: John Wiley & Sons, [1968] 1991), p. 53.

[39]Pythagoras also firmly held to the Hindu belief in the transmigration of souls. Because of this and his commitment to contemplative intellectualism, many scholars believe that Pythagoras had some interaction with Hinduism during his life.

[40]Numbers that satisfy this equation are called Pythagorean triples; e.g., {3,4,5}, {5,12,13}, {7,24,25}, and {17,144,145}.

ten as the ratio of two counting numbers.[41] The Pythagoreans denoted $\sqrt{2}$ as an inexpressible number.[42] This discovery is attributed to the Pythagorean Hippasus of Metapontum. Tradition states that the Pythagoreans were at sea and they threw Hippasus overboard for producing an element in the universe that countered the Pythagorean doctrine that all phenomena can be reduced to whole numbers or whole number ratios. According to Eric Temple Bell (1883-1960), professor of mathematics for many years at the California Institute of Technology, "One obstinate mathematical discrepancy demolished Pythagoras' discrete philosophy, mathematics, and metaphysics."[43]

2.9 HERACLITUS, THE OBSCURE

The metaphysic of Heraclitus (544-484 BC), an Ephesian, was not number, but fire. Everything is and will be eternally living fire (used metaphorically) which he called *logos*. He understood *logos* to mean reason. Thus, reason is the main feature of everything. In this light, he was perhaps the originator of the *natural law* concept; i.e., the idea that a common wisdom or order pervades the entire universe. The *logos* that is the reason behind and in everything is non-moral, neutral, and indifferent to notions of justice, righteousness and compassion; it is within nature and discoverable by man's mind. In contrast, the Apostle John used the same Greek word, *logos*, to describe the Word that was with God in the beginning, is God, and through whom all things were made. Biblically, the *logos* is not an immanent force that brings order to the world; it is a transcendent person of creative and sustaining power (John 1:1-4).

Heraclitus believed that all reality is constantly changing, although it does not seem that way to our senses. He made famous the remark, quoted by physician Hippocrates (ca. 460-ca. 377 BC), "You could not step twice into the same rivers; for other waters are ever flowing on to you."[44] The waters may look the same, but they are not. Hence, you cannot trust your senses; things are not as they seem. Continuing this logic, he said that the sun was new *every* moment and that its width was one foot. Why? He trusted his senses. But, he said that you cannot trust your senses. Here is an example of the conundrum that afflicts many a non-biblical philosopher. Obscuring all of his thinking was his conception of cyclical time. Using the fire metaphor, he proposed that the universe regularly "ignites and extinguishes itself" every 10,800 years.[45] Note Stanley L. Jaki's analysis of Heraclitus:

---

[41]The Pythagoreans called the ratio of two counting numbers *commensurable* ratios.

[42]The Pythagoreans used the Greek word *alogos* (literally means "no reason") to express this number. The term *arratos* (without ratio) was also used. In modern understanding, an irrational number is a number that cannot be written as the *ratio* of two counting numbers. A ratio that involves an irrational number in the numerator or denominator is called an *incommensurable* ratio.

[43]Eric Temple Bell, *Men of Mathematics* (New York: Simon and Schuster, 1937), p. 21.

[44]Hippocrates, *On the Universe*, aph. 41. Cited in Robert Andrews, *The Columbia Dictionary of Quotations* (Columbia: Columbia University Press, 1993), p. 129.

[45]Jaki, *Science and Creation: From Eternal Cycles to an Oscillating Universe*, p. 107.

His thinking is indeed a classic example of the sad unbalance which is ultimately imposed on one's thinking by the acceptance of the endless, cyclic recurrence as the basic pattern of existence. Within that framework one was ultimately left with no consistency in reasoning and observation ... Clearly, in a philosophy of nature steeped in the idea of perennial cycles there remained ultimately no room except for disconnected sense perceptions.[46]

2.10 THE ELEATIC SCHOOL

Parmenides (515-450 BC) lived in the southern Italian city of Elea. He countered the Heraclitean flux with the idea that change or motion is impossible. He agreed with Heraclitus that your senses are not to be trusted, but he came to an entirely opposite conclusion: it appears that things change, but they really do not. His metaphysic was radically monistic; reality for him was a single, permanent substance that is uncreated, indestructible, and unchangeable. According to Stanley L. Jaki, the philosophy of Parmenides blended the eternity and uncreatedness of nature with its endless conflagrations.[47]

Figure 7: Zeno of Elea

Zeno of Elea (495-430 BC) proposed four paradoxes on time and space to support the teachings of Parmenides. The paradoxes were Dichotomy, Achilles, Arrow, and Stadium. Through these intellectual puzzles, he attempted to discredit trust in the senses. Two paradoxes will be explained to illustrate his thinking.

In Dichotomy, Zeno proved that motion was impossible by the following reasoning. Imagine that you are going to run a race. As you approach the starting line, you think, "Before I get to the finish line, I must pass the halfway mark. Before I reach the halfway mark, I must pass the quarter mark. Before I reach the quarter mark, I must pass the one-eighth mark, and so on, indefinitely. Hence, I can *never* start running!"

In Achilles, Zeno continued this same theme, but in a different context. The Tortoise challenged Achilles to a race. All the Tortoise asked for was a 100m head start.[48] Achilles, whose speed was legendary, could run ten times faster than the Tortoise. It seemed like a sure win for Achilles even with the handicap. But, reasoned Zeno, Achilles *could never win*. Why? Zeno said in the words of the Tortoise, "Achilles, after you run 100m, I will have run 10m. After

---

[46]*Ibid.*

[47]*Ibid.*, p. 166.

[48]Using the metric system (m = meters) will make the calculations more accessible to the modern reader.

you run the next 10m, I will have run an extra 1m. After you run that 1m, I will have run an extra 0.1m. After you run 0.1m, I will have run an extra 0.01m and so on, indefinitely. You will never overtake me!" No matter how long the race continues, the Tortoise would continue to preserve his lead, no matter how infinitesimally small!

Zeno showed that the current mathematical treatment of space and time required that they be broken up into infinite sets of points and instants. Reason says that motion is impossible, but perception says that motion is possible. Both Zeno and Parmenides said, "Trust reason." Hence, motion is impossible.

Where is the "catch"? The Greek methods of logical thought failed to solve this problem, but we can solve the Achilles paradox using algebra (a method unknown to Zeno). First, at what distance will Achilles "take the lead"? We will let x = this distance. Second, since distance = rate multiplied by time (d = rt), then r = d/t and t = d/r. The rate of the Tortoise, r_t = 1/unit of time. The rate of Achilles, r_a = 10/unit of time. Third, Since t = d/r, we need to solve the following equation:

$$\frac{x}{1} = \frac{x+100}{10}. \ Cross \ multiplying, \ 10x = x+100, \ 9x = 100, \ x = 11\frac{1}{9} \ m.$$

Therefore, Achilles passes the tortoise after he runs 111 1/9 m or the Tortoise is passed after he runs 11 1/9m. Why were problems like this a "paradox" to these Greek mathematicians? First, they had no knowledge of Algebra. Second, they had no knowledge of ratio and velocity. Third, they did not have the use of the Hindu-Arabic decimal place-value system.[49]

This paradox can be viewed from a different perspective. If you add progressively larger quantities of anything to a pile, the pile grows with ever increasing speed as long as you persist in adding more. Why doesn't the same thing happen when *ever-decreasing* amounts are added indefinitely? It seems that you should *not* reach a limit. The problem facing Zeno was to write an expression for the distances traveled by the Tortoise. Using the decimal system:

$$10+1+\frac{1}{10}+\frac{1}{100}+\frac{1}{1000}+...or$$

$$10+1+0.1+0.01+0.001+...=11.1111...or \ 11.\overline{1}$$

Note that $11.\overline{1}$ *will never be as great as* 11.2

Therefore, it is possible to add *ever-decreasing* amounts to a pile *forever* without accumulating much of a pile at all!

Since $\frac{1}{9} = 0.111...or \ 0.\overline{1}, then \ 11\frac{1}{9} = 11.111...or \ 11.\overline{1}$.

---

[49] Using Roman numerals, $10+1+\frac{1}{10}+\frac{1}{100}+\frac{1}{1000}+...$ becomes a very cumbersome expression: $X+I+\frac{I}{X}+\frac{I}{C}+\frac{I}{M}+...$

The tortoise is found to run the same distance whether calculated by algebra or determined by the reasoning above. It is possible to add *ever-decreasing* quantities to a pile indefinitely until a point is reached at which the pile ceases to grow. Modern mathematicians call this the convergence of an infinite series to a limiting value.

2.11 GEOMETRY TO THE RESCUE

We break momentarily from our historical survey of Greek philosophical and mathematical thought for a parenthetical commentary. The Greeks had great difficulty in confronting concepts that baffled the human mind. In arithmetic, the Pythagorean discovery of irrational numbers proved to be an "unreasonable" hindrance. In geometry, according to the Greek historian Thomas L. Heath, no Greek mathematician made any use in proof of the idea of infinity.[50] This was probably due to the paradoxes of Zeno. Since they could not put their hands on these concepts, they tended to sweep them under the proverbial rug. Eventually, they shied away from the rudiments of arithmetic and algebra and turned to geometry instead. As Morris Kline (1908-1992), mathematics historian, notes, "in turning to geometry to handle irrational numbers, the classical Greeks abandoned algebra and irrational numbers as such."[51] They still worked with arithmetic and algebra, but now these topics were understood in the context of geometry. Even today, we carry vestiges of this shift in our mathematical verbiage. A square with dimensions 3 by 3 can be used to illustrate the number 3^2, expressed as "three squared." A cube with dimensions 3 by 3 by 3 is likewise used to illustrate the number 3^3, expressed as "three cubed."

2.12 PLURALISTIC METAPHYSICS

A group of Greek philosophers rejected the monistic metaphysical premise of Parmenides that there is one underlying substance of all things. Instead, they posited that there is a plurality of underlying substances.

Empedocles (ca. 490-430 BC), born in Sicily, was a Pythagorean and a disciple of Parmenides. He posited that reality is made up of a plurality of particular things. Specifically, all things are composed of four primal elements (all of which are eternal and unchanging): earth, air, fire, and water.[52] Two active and opposing forces, love and strife, act upon these elements, respectively combining and separating them into varied forms. To him, everything oscillated between these two extreme principles, a process he equated with pure reason and

---

[50]Thomas L. Heath, *A History of Greek Mathematics* (New York: Dover Publications, [1921] 1981), 1:272.

[51]Morris Kline, *Mathematical Thought from Ancient to Modern Times* (New York: Oxford University Press, 1972), p. 49.

[52]The Greek word for elements is *stoicheia*. This word is used in the New Testament in three places (Galatians 4:3-8; II Peter 3:10; Colossians 2:8). The writers probably meant either the elements of Judaism (as contrasted to the New Covenant) or the governing princes (angels) over the nations which were made obsolete by the definitive and mediatorial work of the Prince of Life, Jesus Christ.

which together he defined as deity. He noted that not all of these combinations that come about through love and strife survive. In this notion, he predates the "survival of the fittest" doctrine postulated by Charles Darwin. Empedocles also adhered to the cyclic rejuvenation theory.[53]

Born near Clazomenae (near modern Izmir, Turkey), Anaxagoras (ca. 500-428 BC) was the first Greek philosopher to reside in Athens. Socrates (ca. 470-ca. 399 BC) was probably one of his students. His mathematical interests centered upon the quadrature of the circle[54] and on perspective. His metaphysic asserted that matter existed originally as infinitely numerous, infinitesimally small, and qualitatively different seeds that had been present from all eternity. His universe was purely and only mechanistic. Order was produced out of this chaos by an eternal intelligence, called the *nous* (Greek for mind). His prime postulate was that "reason rules the world." Note the progression from Thales, who said "everything was full of gods," to Anaxagoras who said "nothing is full of gods" (i.e., everything is pure mind). According to Stanley L. Jaki:

> In the universe of Anaxagoras ... there was no room ... for considering what the gods represented, namely, will, purpose, and personal determination. In short, in the book in which Anaxagoras ascribed everything to the mind there were regularities, but no mind which regulated, planned, and did so for a purpose.[55]

Democritus (ca. 460-ca. 370 BC) was born near Abdera, Trace. Known as the "Laughing Philosopher," he inherited great wealth, spent it all on travel, was remarkably diligent in study, and died in poverty. According to Archimedes (ca. 287-212 BC), Democritus was the first to show the relationship between the volume of a cone and that of a cylinder of equal base and equal height, and similarly for the pyramid and prism.

Democritus posited that there is no spiritual reality behind the processes of the world. His metaphysic, like Anaxagoras', was purely and only mechanistic. Reality is composed of an infinite number of qualitatively *identical* and self-propelling atoms.[56] These atoms have both primary and secondary qualities. Their primary qualities are shape, size, and speed. Secondary qualities result from the interaction of the material object with a sense organ (e.g., color, taste). The difference between objects is determined by the mechanical arrangement of atoms. Democritus reasoned to these conclusions on a purely *a priori* basis. He had no external (*a posteriori*) confirmation to support his claims and yet his assertions anticipated, in a very elementary way, the atomic structure profiled by modern physics.

Note the conundrum that afflicts the thinking of Democritus: If everything is a purely mechanical concoction of atoms, then on what can man base his

---

[53]Jaki, *Science and Creation: From Eternal Cycles to an Oscillating Universe*, p. 108.
[54]The process of constructing a square equal in area to a given circle.
[55]Jaki, *The Road of Science and the Ways to God*, pp. 19-20.
[56]From the Greek, *atomos*, meaning an irreducible, indestructible material unit.

ability to reason and discover knowledge? In other words, how can reason, which is immaterial, account for the material world? Democritus answered by positing that reason is a divine activity accounted for by "refined fire atoms." His answer unveils an internal tension: he must resort to irrational fudging (a "pick a straw out of the hat" explanation) to account for the immaterial nature of man.

Note how the ethics of Democritus depended upon his metaphysic. Like Pythagoras before him, Democritus contended that true happiness is *not* found through sense experiences (he called this *coarse* atomic activity), but through knowledge of the truth; a life of contemplation and intellect that produces tranquility and moderation. He said, "One must keep one's mind on what is attainable, and be content with what one has, paying little heed to things envied and admired, and not dwelling on them in one's mind."[57] A life devoted to tranquillity consists of passively accepting the circumstances around us. In this context, as Stanley L. Jaki reflects, "he did not consider life an opportunity to change man's material conditions."[58] Ideas have consequences. The tendency of the Greek philosophers to absolutize intellectual pursuits resulted in their general scorn of manual crafts, what they called the *illiberal* arts.[59]

2.13 THE IDEAL WORLD OF PLATO

Born in Athens and mentored by Socrates, Plato (427-347 BC) did not want any student "destitute of geometry" to enter his classroom.[60] In *Republic* (Book 7, section 527b), he said that "geometry is the knowledge of the eternally existent … it would tend to draw the soul to truth, and would be productive of a philosophical attitude of mind."[61] What "eternally exists" to Plato was a world of the impersonal idea developed in the recesses of the human mind. This was his "god who ever geometrizes."

Plato's thought was influenced by the Persian religion of Zarathustra.[62] Because of this, we find in Plato the roots of two philosophical concepts: dualism and realism.[63] From Parmenides, he learned that knowledge requires a permanent, unchanging, stable object. In order to know something, the object of your attention cannot be in constant flux. But, our natural experience informs us that everything that we know is changing (as per Heraclitus). Hence, the permanent,

---

[57]Cited in Jaki, *Science and Creation: From Eternal Cycles to an Oscillating Universe*, p. 129. Jaki cites as his source Kathleen Freeman, trans., ancilla to the *Pre-Socratic Philosophers* (Cambridge: Cambridge University Press, 1962), p. 109.

[58]*Ibid.*

[59]To the Greeks, artisan activity (skilled manual work) was something ill-bred, vulgar, and lacking in true culture.

[60]He founded and conducted a famous school of philosophy, the Academy, the first school in history with an entrance requirement: *Let no one ignorant of geometry enter here.*

[61]Plato, *The Collected Dialogues of Plato*, ed. Edith Hamilton and Huntington Cairns (Princeton: Princeton University Press, 1961), p. 759.

[62]Neugebauer, p. 151.

[63]For example, this Persian religion was a dualistic faith reflecting a cosmic struggle between two equal powers: good and evil.

unchanging, and stable object of knowledge must lie in another world, a world not of time and space.[64] By this logic, Plato embraced metaphysical and epistemological dualism. Our senses tell us about one realm, our reason another. Dualism is a view of life that separates the invisible from the visible. The invisible, the world of ideas (also known as ideals or forms), is that which is real, perfect, and unchanging while the visible, the world of matter or natural phenomena, is that which is imperfect, changing, and not to be trusted. In this context, Plato unveiled his unflinching commitment to rationalism and intellectualism:

Ideals

Time and Space

Figure 8: Plato's two realms

> When anyone by dialectic attempts through discourse of reason and apart from all perceptions of sense to find his way to the very essence of each thing and does not desist till he apprehends by thought itself the nature of the good in itself, he arrives at the limit of the intelligible.[65]

To Plato, the material part of reality has a "drag" of imperfection to it. The perfect, ideal world is accessed by intellectual contemplation. According to Vern S. Poythress, theologian and mathematician, "Plato was against the body [or matter – J.N.] and its 'messy' corruption of the pure vision of the abstract ideal. Plato was against creation, in fact."[66] According to Rousas J. Rushdoony, "The goal of man was to be an incarnation of the idea, the universal, and hence the study of geometry, of abstract forms, was more religious than practical, or, more accurately, was practical because religious."[67] Mathematical objects (e.g., triangles, circles, etc.) were a part of Plato's impersonal world of abstract and perfect ideas and therefore fused with his religious philosophy.

How does man come to knowledge of the ideal? Plato presented his famous cave allegory to answer this question. You can only see images from the real world as shadows inside the cave. Man, in his sense experiences of the world, sees only shadows of reality. When he sees the shadow, he recollects the ideal behind the shadow. What stands for the shadows in our actual experience are the physical objects that we see and these objects are merely an instantiation of the ideal. According to Stanley L. Jaki, "The physics and cosmology that could be had on such a basis was as elusive as shadows are."[68]

---

[64]This idea probably came from his teacher, Socrates, who said that there could be no life unless there was an absolutely perfect, or immortal, life.

[65]Plato, p. 764 (cited in *Republic*, Book 7, section 532a).

[66]Vern S. Poythress, "Science as Allegory-Mathematics as Rhyme," *A Third Conference on Mathematics from a Christian Perspective*, ed. Robert L. Brabenec (Wheaton: Wheaton College, 1981), p. 38.

[67]Rousas J. Rushdoony, *The Philosophy of the Christian Curriculum* (Vallecito, CA: Ross House, 1981), p. 5.

[68]Jaki, *The Road of Science and the Ways to God*, p. 21.

Figure 9: Plato

It is important to note that the learning process is a reminding or recollection process. This process of reminiscence reflects upon Plato's unswerving commitment to cyclical time and his view of deity. Plato did not know the answer to the question of why the ideal is "impressed" on physical matter, so he resorted to creating a fiction called the demiurge.[69] Plato understood the demiurge as a supernatural force that only fashions (not creates) physical matter out of chaos; it unites ideals and forms. This force was limited, finite, and unknowable. He associated the demiurge with the maker and father of the universe, but this "god" was surely "past finding out, and even if we found him, to tell of him to all men would be impossible."[70] To Plato, the world only briefly remembers the demiurge because of the compounding and darkening effect that ever-recurring cycles of time had upon the recollection process.[71] According to Stanley L. Jaki:

> Plato subjects man's comprehension of God to the pessimistic logic of eternal recurrence in which each cosmic cycle is largely dominated by increasing dissolution. For much of each Great Year neither man nor the world remembers God's instructions because the world is destined by and large to "travel on without God."[72]

Biblical theology cannot be equated with Plato's philosophy of the demiurge, an impersonal force that exists outside of the physical world. Since Plato did not want to bow the knee to the infinite, personal, Creator God of Scripture, he was left to his own devices and developed an autonomous philosophy. Plato erred by not going beyond the physical world *far enough*. Since the biblical God created the physical world and declared it good, it is not to be shunned or considered irrelevant to life (in spite of sin's impact on man and creation).[73] The real world is not a shadowy dream of the ideal world. Ideas have consequences. Because Plato saw the physical world in terms of shadows, his few applications of geometry to the real world were merely playful gestures, fanciful pastimes, and intellectual cogitations. He did not encourage any serious quantitative and empirical study of the physical world (the essence of true science). As Carl Boyer summarizes:

---

[69]From the Greek word *demiourgos* which originally meant "public contractor or artisan."
[70]Plato, p. 1162 (cited in *Timaeus*, section 28c).
[71]*Ibid.,* p. 1038f (cited in *Statesman*, section 273-274).
[72]Jaki, *The Road of Science and the Ways to God*, p. 340.
[73]The incarnation (God made flesh in the person of Jesus Christ) is devastating to Platonic metaphysics.

The Pythagoreans and Plato noted that the conclusions they reached deductively agreed to a remarkable extent with the results of observation and inductive inference. Unable to account otherwise for this agreement, they were led to regard mathematics as the study of ultimate, eternal reality, immanent in nature and the universe, rather than as a branch of logic or a tool of science and technology.[74]

2.14 ARISTOTELIAN METAPHYSICS

Figure 10: Aristotle

Aristotle (384-322 BC) of Stageira, one of Plato's students and a teacher of Alexander the Great (356-323 BC), shifted the emphasis from the invisible ideal to the visible world of natural phenomena. He rejected Plato's dualistic metaphysic while keeping the form concept, but with a different theoretical definition. Instead of the ideal or form existing outside the physical world, the ideal or forms are embodied within matter; i.e., forms coincide with the visible world. His metaphysic was the immanent transcendence of the particular. Man can distinguish form from matter but he cannot separate form from matter (except in the mind via the process of abstraction).

For Aristotle, reality consists of individual particulars. Each particular is a substance that can be analyzed in two ways: by its form and by its matter. To Aristotle, form deals with the class of an object; i.e., what it is (e.g., a tree). Matter deals with the specific existence of an object; i.e., this tree versus that tree. Everything that exists is *formed matter* and the form that an object has always serves a function or purpose, called its *telos*.[75] Aristotle was able to account for the process of change with these definitions, a concept that Plato disregarded since it dealt with the shadows of time and space. For example, consider the process of an acorn growing into an oak tree. At the acorn stage, the form of an acorn represents its actuality. The matter represents the oak tree; i.e., the potentiality of the form. The particular therefore keeps its identity through time. The form changes as the acorn matures, but matter (oak tree) remains the same.

In his *Metaphysics* (1075a), the universe bespoke of law, order, and coherence.[76] In 1075b, he remarked that this order could be compared, not to the order of an army that is due to its commander, but to the order of a house that is

---

[74]Carl Boyer, *The History of the Calculus and its Conceptual Development* (New York: Dover Publications, [1949] 1959), p. 1.
[75]*telos* is Greek for "end result or aim."
[76]Aristotle, *The Metaphysics*, trans. John H. McMahon (Amherst: Prometheus Books, 1991), pp. 266-268.

due merely to its being a household.[77] Yes, the universe displayed law and order, but not a lawgiver.

Aristotle's cosmology began with common sense observations. For example, it was obvious that a spherical universe rotates around a stationary earth. But, from then on, his discourse on the cosmos was entirely *a priori*. He presupposed a finite universe consisting of a series of concentric and moving circles. The celestial circular motion reflected the perfection of the heavens. The planets and stars hung on these circular astronomical bands like lamps.[78] In contrast, linear motion (either straight up or straight down) governed the imperfect and constantly changing terrestrial arena, made up of the four basic elements of earth, air, fire, and water.

Since the heavens consisted of different matter than earth, different laws also governed celestial and terrestrial motion. In the terrestrial realm, one of his laws of motion stated that projectiles move across the air because the air keeps closing behind them. Celestially, the movement of the most outward band caused friction on its adjacent inward band causing it to move. The friction from this band caused the movement of the next band, etc. What initiates the movement of the most outward band? Aristotle developed the concept of the "Unmoved mover" as the source of this eternal circular motion. This "god" was eternal, perfect, impersonal, and unmovable. What caused the "Unmoved" to move? Since Aristotle believed that motion arises from loving desire and that "thought" is the most intelligible essence to be desired, then "thought thinking thought" causes the motion of the universe. He also believed that circular movement reflects this perfect thought. Since this "thought" knows nothing about imperfection or change, then this "Unmoved mover" has no interest in the changing world; it only knows itself. "Pure form" or "thought thinking thought" determined the boundaries of Aristotle's conception of deity, a conception that excluded the attributes of personality and providence; e.g., care for the world. The Greek word "physis" meant world soul; the moving, animating, and living principle of the universe. This physis, or nature, was a living organism, unchangeable, eternal, uncreated, self-generating, and, above all, rational or intelligible.[79]

Aristotle also embraced the doctrine of eternal returns. Everything, including the sciences, had repeatedly reached their perfection in former cycles. In *Metaphysics* (982b), he said that the mathematics, science, the arts, and craftsmanship of his day had already risen to their highest point of achievement.[80] According to Stanley L. Jaki, "This statement was a sad prophecy for Greek science. Euclid, Archimedes, Diophantes [or Diophantus – J.N.], and Ptolemy

---

[77] *Ibid.*, p. 267.
[78] These bands were made up of a fifth element or essence, the *quinta essentia* (also called ether).
[79] Reijer Hooykaas, *Religion and the Rise of Modern Science* (Grand Rapids: Eerdmans, 1972), pp. 6, 11.
[80] Aristotle, pp. 14-16.

were unable to raise it to a level much higher than where it stood when Aristotle parted from life."[81]

To Aristotle, mathematical objects were abstracted from the physical world. The mind, also being rational, could determine basic truths about the physical world without exerting much observational or experimental effort.[82] He did embrace empiricism, but only to a point (in the realm of biology, his favorite subject). In the realm of mathematics, he placed strong emphasis on the power of human reason, in the form of deduction, to discover all truth. He also completely ignored empirical epistemology (testing theories with experiments) in the realm of astronomy and physics (e.g., the laws of motion). His systematization of syllogistic logic laid the groundwork for the great deductive system of Euclid.

2.15 THE "ELEMENTS OF EUCLID"

Around 300 BC, in Alexandria, Egypt, a man by the name of Euclid (ca. 330-ca. 275 BC) compiled the previous three centuries of Greek mathematical achievements in a massive, thirteen-book edifice.[83] According to David E. Smith, "He is the only man to whom there ever came or ever can come again the glory of having successfully incorporated in his own writings all the essential parts of the accumulated mathematical knowledge of his time."[84] In his "Stoicheia (the *Elements*)," he applied deductive reasoning in a manner that has since become the "bread and butter" of mathematical method.

In deductive reasoning, certain general facts are assumed to be true. These facts are called postulates or axioms.[85] Given these basic premises or hypotheses, logical reasoning, called syllogistic thinking, is applied to discover new facts, called theorems.[86] For example,

> **Defining the terms of proof**
> Axiom: think worthy
> Postulate: a thing demanded
> Hypothesis: a thing laid down under
> Syllogism: to reason
> Theorem: a subject of contemplation

---

[81]Jaki, *The Road of Science and the Ways to God*, p. 24.

[82]*Ibid.*, p. 25.

[83]Euclid drew from approximately sixty known Greek geometers who contributed to the discipline in the preceding four to five generations. See Jaki, *Science and Creation: From Eternal Cycles to an Oscillating Universe*, p. 103.

[84]Smith, 1:104.

[85]The Greeks made a fine distinction between axioms and postulates. An axiom is an assumption that is intuitively obvious and acceptable while a postulate is an assumption that is neither necessarily obvious nor necessarily acceptable. Also, an axiom is an assumption common to all sciences, while a postulate is an assumption specific to the science being studied.

[86]Sometimes you encounter the Greek word *lemma* (meaning "a thing taken") which stands for an ancillary theorem whose result is not the target of the proof. After a theorem was proved Euclid wrote οπερ εδει δειξαι meaning, loosely translated, "we have proved the proposition we have set out to prove." In Latin, this phrase becomes *quod erat demonstrandum* meaning "which was to be demonstrated" and abbreviated QED.

let us assume as a basic general fact that all cats have whiskers. Felix is a cat. Therefore, the new fact observed is that if Felix is a cat, then he has whiskers.

Beginning with about twenty-three basic definitions, five axioms, and five postulates, Euclid went on to logically prove four hundred and sixty-five new theorems (also called propositions) in plane and solid geometry, plus other topics.[87] Howard Eves comments about the impact of this work:

> As soon as the work appeared, it was accorded the highest respect, and from Euclid's successors on up to modern times the mere citation of Euclid's book and proposition numbers was regarded as sufficient to identify a particular theorem or construction. No work, except the Bible, has been more widely used, edited, or studied, and probably no work has exercised a greater influence on scientific thinking. Over a thousand editions of Euclid's *Elements* have appeared since the first one printed in 1482, and for more than two millennia this work has dominated all teaching in geometry.[88]

There are two remarkable facts about these theorems that Euclid deductively proved. First, in the 17th century, a brilliant French mathematician named Blaise Pascal (1623-1662) discovered, with no one teaching him, many of the same conclusions that Euclid formulated.[89] Note that Pascal did not copy what Euclid had done. He did his work completely independent of the *Elements*. Second, although each one of Euclid's theorems could have been worked out in an "ivory tower with blindfolded eyes," when they were applied to physical phenomena they described the situation perfectly.

Figure 11: Euclid

To the Greeks, these theorems were not only the essence of reality, but also eternal and unchangeable. Why? Human reasoning, the touchstone for them of all truth, had determined them to be so.

2.15.1 EUCLID'S FIVE AXIOMS AND FIVE POSTULATES

In modern terminology.

A1. Things being equal to the same thing are also equal to each other.

A2. If equals are added to equals, the wholes are equal.

A3. If equals are subtracted from equals, the remainders are equal.

A4. Things which coincide with each other are equal to each other.

---

[87]See Thomas L. Heath, *Euclid: The Thirteen Books of the Elements* (New York: Dover Publications, [1925] 1956), 3 vols.

[88]Eves, p. 115.

[89]*Ibid.*, p. 261.

A5. The whole is greater than the part.

P1. Any two points, A and B, determine exactly one straight line.

P2. A finite straight line can be extended continuously as a straight line.

P3. A circle can be completely described given only its center, O, and its radius measure r.

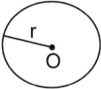

P4. All right angles are equal to each other.

P5. Through a point not on a line, there is only one line parallel to the line.

2.15.2 APPLIED GEOMETRY

An example of the application of basic geometry to physical problems is the use of the properties of similar triangles in order to determine the height of an object. Here are two approaches that can be readily submitted to the test of an experiment.

1. To do this experiment, you need a mirror and a measuring tape. Point M represents the position of the mirror so that the light from the top of the tree travels along TM and ME before reaching your eyes. Using one of Euclid's theorems, the two triangles, ΔTBM and ΔEFM are similar because $\angle B = \angle F$ and $\angle 1 = \angle 2$ (by the law of reflection of light). Hence, if you know EF, MF, and BM, you can find TB, the height of the tree. Why? Corresponding sides of two similar triangles are proportional; i.e., in this case:

$$\frac{TB}{EF} = \frac{BM}{MF} \Rightarrow TB = \frac{(BM)(EF)}{MF}$$

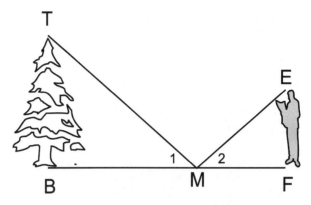

Figure 12: First example of Applied Geometry

2. In this experiment, all you need is a sunny day, a measuring tape, and a friend to assist you. Walk away from a tree along its shadow until your head is in line with the top of the tree and the tip of the shadow. ΔHFS is now similar to ΔTBS. Why? If you know BF, HF, and FS, you can find TB, the height of the tree; i.e.,

$$\frac{TB}{HF} = \frac{BF + FS}{FS} \Rightarrow TB = \frac{HF(BF + FS)}{FS}$$

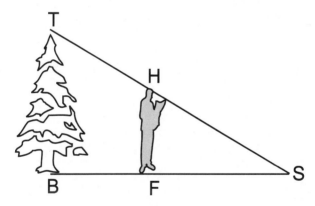

Figure 13: Second example of Applied Geometry

2.15.3 GEOMETRY AND LATITUDE

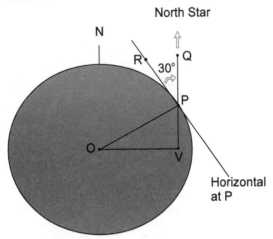

Figure 14: Geometry and Latitude

To calculate one's latitude by sighting the North or Pole star (Polaris or alpha Ursae Minoris), the observer at P measures the angle between the horizontal at P and the direction of the North Star. Suppose this angle is 30°. The latitude ($\angle$POV) is determined as follows:

1. Since the horizontal at P is tangent to the circle O, $\angle$RPO = 90°.
2. Since supplemental angles equal 180°, $\angle$QPR + $\angle$RPO + $\angle$VPO = 180° or $\angle$VPO = 180° - (30° + 90°) = 60°.
3. Since $\triangle$OPV is a right triangle, $\angle$POV (latitude) = 180° - ($\angle$PVO + $\angle$VPO) = 180° - (90° + 60°) = 30°.

2.16 ACCOMPLISHMENTS IN ALEXANDRIA

The conquests of Alexander the Great led to the demise of the classical Greek age. He envisioned an international empire unified by the Greek language. After his death in 323 BC, his generals fought each other for control. No one could gain the upper hand so the empire was split into four sections. Egypt, ruled by the Greek Ptolemy dynasty, became the heir of the mathematics of Classical Greece. It was no accident that Euclid did his work in Alexandria Egypt, for here Ptolemy Soter built an educational base where scholars of all fields could study and teach.[90] This base became known as the Museum and next to it was built a massive library containing upwards of about 750,000 vol-

---

[90]Alexandria, Egypt was one of the largest cities in the ancient world with a population of about 150,000 people. The Pharos Lighthouse, considered one of the Seven Wonders of the ancient world was also built in this city in the 3rd century BC. It towered four hundred feet and its signal of blazing fire could be seen for miles offshore. Here also stood the Serapeum, cult center of the worship of Apis, the sacred bull of ancient Egypt.

umes (including Aristotle's personal library).[91] In fact, Alexandria became the magnetic north of the book-copying trade due to the readily available Egyptian papyrus. In addition to these scrolls, the Museum housed vast collections of scientific instruments and biological specimens. A maze of lecture halls was built and generous stipends provided for scholars so that they could devote themselves entirely to research and writing.

The character of Alexandrian Greek mathematics differed from that of its classical ancestors. Corresponding to the ingestion of Platonic thought, the mathematics of Classical Greece tended toward the absolutization of abstract thought. In Alexandria, these abstractions were mostly devoted to useful and practical results partly due to the ancient Egyptian and Babylonian influence. Some of the mechanical devices created are astonishing. Let Morris Kline provide the list:

> Pumps to bring up water from wells and cisterns, pulleys, wedges, tackles, systems of gears, and a mileage-measuring device no different from what may be found in the modern automobile were used commonly. Steam power was employed to drive a vehicle along the city streets in the annual religious parade. Water or air heated by fire in secret vessels of temple altars was used to make statues move. The awe-struck audience observed gods who raised their hands to bless the worshipers, gods shedding tears, and statues pouring out libations. Water power operated a musical organ and made figures on a fountain move automatically while compressed air was used to operate a gun. New mechanical instruments, including an improved sundial, were invented to refine astronomical measurements.[92]

Also, these mathematicians did not shy away from irrational numbers. Although they had no logical foundation for their use, they did apply these numbers extensively in their calculations. The result was the development of trigonometry. Some of the more famous mathematicians from the period 300 BC to the birth of Christ were Aristarchus, Archimedes, Eratosthenes, Apollonius, and Hipparchus. In addition to these men, we must note Sosigenes, an Alexandrian astronomer, who assisted Julius Caesar (100-44 BC) with his historic calendar reform (45 BC).

Aristarchus (ca. 310-230 BC) of the island of Samos was the first to postulate the heliocentric theory of planetary motion; i.e., the planets, including the earth, revolve in perfect circles around a fixed sun. However, this idea was too radical for most Greeks.[93] He also modified a sundial (called a *skaphe*) that enabled him to measure the height and direction of the sun.

Archimedes (ca. 287-212 BC), the son of an astronomer, made discoveries in two important fields, mechanics and hydrostatics, which are as useful and

---

[91]Kline, *Mathematical Thought from Ancient to Modern Times*, p. 102.
[92]*Ibid.*, p. 103.
[93]See Thomas L. Heath, *Aristarchus of Samos: The Ancient Copernicus* (New York: Dover Publications, [1913] 1981).

practical today as they were over two thousand years ago. The Roman encyclopedist Pliny (ca. 23-79 AD) called him "the god of mathematics." He made many mechanical instruments and resulting observations, but only because his hometown, the Greek city-state of Syracuse, was in danger of attack by the Romans. Legend states that he set fire to the besieging ships in the harbor of Syracuse with the aid of parabolic mirrors that focused and intensified the rays of the sun on the enemy warships.

In the area of hydrostatics, another story from the city of Syracuse illustrates the achievements of Archimedes. King Hiero of Syracuse had ordered a crown of pure gold to be made. He suspected the finished product to be a mixture of silver and gold. He asked Archimedes, who knew the basic principles of hydrostatics, to prove his suspicions by doing some detective work. Archimedes knew that, given the same volume, gold weighs twice as much as silver (more accurately, gold weighs 19.3 g/cm³ while silver weighs 10.5 g/cm³). Hence, a given weight of silver has twice as much volume as the same weight of gold. That is, for example, one kilogram of gold has a volume of 50 cm³ while one kilogram of silver has a volume of 100 cm³. Ar-

Figure 15: Archimedes

chimedes placed the crown in a basin of water filled to the brim and measured the volume of the water that overflowed. The overflow represents the volume of the gold. He then weighed the crown and compared its weight with its volume. He discovered that the volume of the crown, if pure gold, should be less than the actual volume of displaced water. The gold, therefore, was not pure, but a mixture. His reasoning can be illustrated using modern measurement. Assume that the volume of the water that overflowed measured 140 cm³ and that the crown weighed 2 kg. From above, 2 kg of pure gold should have a volume of 100 cm³, not 140 cm³. All that glitters is not always pure gold!

In the area of mechanics, he found the centers of gravity of many plane and solid figures and developed the principles of mechanical advantage in pulleys and levers. He boasted that, given a long enough lever and a fulcrum strong enough, any given weight could be moved, even the earth! He also developed a spiral mechanical screw, called the Archimedean screw, that pumped water uphill. It was used in Egypt to drain the fields after the inundation of the Nile and was frequently applied to take water out of the holds of ships.

In mathematics, he solved what we now know as cubic equations, developed formulae for the volumes of solids, perfected a technique for determining the square roots of numbers, and obtained a very close approximation of π so that circular areas could be calculated. He crudely anticipated the basic propositions of the infinitesimal calculus. He could go no farther than presentiment because the static form (the study of objects at rest) of Greek geometry constrained his conceptions. The infinitesimal calculus requires a conceptual break from statics to dynamics (the physics of motion), a break that Greek mathematicians failed to accomplish. Stanley L. Jaki explains why, " … ancient Greek scientific thought fell prey to the lure of sweeping generalizations that sidetracked the cultivation of physics for two thousand years."[94]

It is crucial to note that Archimedes, notably the greatest mechanical engineer in antiquity, did not write a handbook on the subject. Plutarch (ca. 46-ca. 120 AD), Greek biographer and essayist, wrote that Archimedes looked upon "the work of an engineer and every art that ministers to the needs of life as ignoble and vulgar."[95] We have noted this before – the tendency of the Greeks to generally scorn manual crafts, the *illiberal* arts, because of their association with the world of Platonic shadows. In this context, note Stanley L. Jaki's remarks about the mechanical wonders of Alexandrian period, "The fact that the best results of ancient engineering were used mostly for purposes of warfare or as devices of deception and magic in temples also provides a vivid illustration of a passive attitude toward nature."[96] The combination of the Archimedean disdain for practical arts and the previously mentioned inability to conceptualize a dynamic geometry dealt a death blow to a genuine formulation of the rudiments of the infinitesimal calculus.

The death of Archimedes symbolized what was to eventually happen to the entire Greek world. Rome was engaged in warfare (Second Punic war) with Carthage. As the Romans attacked Syracuse, Archimedes was drawing mathematical figures in the sand. A Roman soldier drew near to him and ordered him to surrender. Archimedes, lost in a train of mathematical thought, did not respond and the soldier speared him to death in spite of a specific order from his commander, Marcellus, that he be captured unharmed.

Eratosthenes (ca. 275-194 BC), a custodian of the Museum and nicknamed "the second Plato" by his admirers, accurately measured the circumference of the earth using basic propositions derived by Euclid. He also measured, within a tenth of a degree, the tilting of the earth's axis of rotation. His contribution to arithmetic was the famous "sieve of Eratosthenes," a method of sifting out the composite numbers in a natural series, leaving only the prime numbers. He was also a mapmaker and was known to dabble in poetry.

2.16.1 ERATOSTHENES MEASURES THE CIRCUMFERENCE OF THE EARTH

The earliest known measurement of earth's circumference was made in Egypt by Eratosthenes about 240 BC. In summary, he based his calcula-

---

[94]Jaki, *Science and Creation: From Eternal Cycles to an Oscillating Universe*, p. 103. This lure was the result of Aristotle's linking biology to physics. This link, according to Jaki, is the reason for the Greek failure in physics.

[95]Cited in *Ibid.*, p. 130. Jaki cites *Plutarch's Lives*, trans. B. Perrin (London: W. Heinemann, 1955), 5:479.

[96]*Ibid.*

tions on (1) the angular height of the sun and (2) the linear distance between Alexandria, the great center of ancient learning, and Syene. This distance was easy to calculate because both cities were located on the river Nile, the only long river on the earth whose course is essentially south to north. Note that Eratosthenes had to assume that the earth was round in order to make his calculations. Of course, the book of Isaiah, written five centuries *before* Eratosthenes, also described the earth as spherical (Isaiah 40:22).

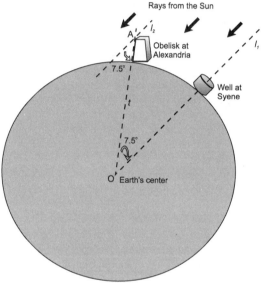

Figure 16: Eratosthenes' calculation of the circumference of the earth

At noon, during the summer solstice, Eratosthenes observed that the sun shone down a well at Syene indicating that it was at Syene's zenith. Syene is today known as Aswan and this city was conveniently located very close to the Tropic of Cancer (latitude of 23.5°N). Some 5,000 stades (approximately 500 miles) north, he measured a 7.5° angle between the sun's rays and an obelisk at Alexandria. Of course, Eratosthenes had countless occasions to note this angle at equinoctial noon in Alexandria (he did not have to travel the 500 miles from Syene to Alexandria on the same day for confirmation). By assuming that the sun's rays reaching earth are parallel and employing simple theorems of Euclidean geometry, he calculated the circumference of earth as follows:

1. $\angle O = \angle A$ because corresponding angles formed by the parallel lines l_1 and l_2 and the transversal t are equal.

2. Since 7.5° is 1/48th of 360°, then the distance from Alexandria to Syene is 1/48th of earth's circumference.

3. Therefore, the circumference of Earth equals $(48)(500) = 24,000$ miles, a remarkably accurate estimate.[97]

2.16.2 THE SIEVE OF ERATOSTHENES

3	5	7	9	11	13	~~15~~	17	19
~~21~~	23	~~25~~	~~27~~	29	31	~~33~~	~~35~~	37	~~39~~
41	43	~~45~~	47	~~49~~	~~51~~	53	~~55~~	~~57~~	59
61	~~63~~	~~65~~	67	~~69~~	71	73	~~75~~	~~77~~	79
~~81~~	83	~~85~~	~~87~~	89	~~91~~	~~93~~	~~95~~	97	~~99~~

A prime number is a number that is divisible *only* by itself and one. A number is composite if it has other factors. The number one is neither prime nor composite. Knowing that all even numbers greater than two are composite, Eratosthenes listed the odd numbers starting from three and then canceled out the successive multiples of each. What is left from this sieve procedure are prime numbers (**in bold**).

Apollonius of Perga (ca. 262-ca. 190 BC), a friend of Archimedes, was known as the "great geometer" because of his creative work on conic sections.[98] These sections were probably discovered by Menaechmus (ca. 350 BC), tutor of Alexander the Great.[99] Otto Neugebauer notes the ancient and concrete origin of the conic sections, "In antiquity the conic sections are needed for the theory of sundials and I have conjectured that the study of these curves originated from this very problem."[100] The conic sections consist of the circle, the ellipse, the parabola, and the hyperbola, the last three named by Apollonius. The Greek literal meanings of ellipse (ελλειπειν – elleipein), parabola (παραβαλλειν – paraballein), and hyperbola (υπερβολη – hyperbola) are "less than or falling short," "the same as," and "exceeding" respectively.[101]

---

[97]The Greek philosopher Posidonius (ca. 100 BC), the teacher of the Roman orator Marcus Tullius Cicero (106-43 BC), used another method to calculate the circumference of the earth and, by a variety of geometrical errors, *shortened* Eratosthenes' calculation by three-quarters. This error was propagated to posterity through the works of the Greek geographer Strabo (ca. 63 BC-ca 24 AD). In the 15th century, although he knew of the calculations of Eratosthenes, Christopher Columbus (1451-1506) fastidiously used Strabo's work and Posidonius' calculations to convince King Ferdinand V (1452-1516) and Queen Isabella I (1451-1504) of the practicality of his proposed voyage to the East Indies by sailing west. No other geometrical error has had such profound consequences in human history. For details of the calculations of Posidonius, see Heath, *A History of Greek Mathematics*, 2:219-220.

[98]Apollonius was nicknamed *epsilon* (ε), the fifth letter of the Greek alphabet, and Eratosthenes *beta* (β), the second letter of the Greek alphabet. Since the lecture halls at Alexandria were numbered, some mathematics historians have concluded that the two friends always taught in the rooms numbered *epsilon* (5) and *beta* (2) respectively.

[99]Smith, 1:92. The conic sections were long called Menaechimian triads.

[100]Neugebauer, p. 218, 226.

[101]For the geometric context as taught by Apollonius, see Alistair Macintosh Wilson, *The Infinite in the Finite* (New York: Oxford, 1995), pp. 324-365. See also Heath, *Euclid: The Thirteen Books of the Elements*, 1:343-345.

Three figures of speech – ellipsis, parable, and hyperbole – are derived from the same source.

Technically, the circle is defined as the locus[102] of all points at a given distance from a given fixed point. The ellipse is the locus of points for which the sum of the distances from each point to two fixed points is equal. The parabola is the locus of points equidistant from a fixed line and a fixed point not on the line. The hyperbola is the locus of points for which the difference of the distances from two given points is a constant.

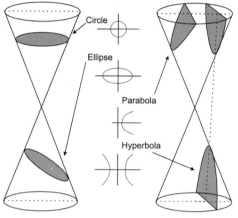

Picture cutting through an upside-down ice cream cone with a hack saw (see Figure 17). Given a cone, if you cut through it parallel to the base, the intersection of the cut with the cone is a circle. If you make an oblique cut, the intersection is the ellipse, identified as the path of a planet around the sun by Johannes Kepler in the 17th century. If you make a cut parallel to a line on the cone (make the slice in the same direction as the side of the cone), the intersection is the parabola, iden-

Figure 17: Conic Sections

tified in the 17th century as the path of a projectile. If the slice is tilted away from the direction of the side of the cone and toward the vertical, the cut is the hyperbola, known today as the "shock wave" curve. When an airplane flies faster than the speed of sound, it creates a shock wave heard on the ground as a "sonic boom." The shock wave has a shape of a cone with its apex located at the front of the airplane. This wave intersects the ground in the shape of the hyperbola.

With the works of Apollonius, Greek mathematics reached its zenith. According to Morris Kline, "Mathematical activity in general, and in geometry in particular, declined in Alexandria from about the beginning of the Christian era."[103]

The astronomer Hipparchus (ca. 180-125 BC) of Nicea developed trigonometry to help solve astronomical problems, perfected the theory of stereographic projection (for representing the projection of the celestial sphere upon the plane of the equator), measured the length of the solar year as 365 days, 5 hours, and 55 minutes (six minutes too long), catalogued 850 fixed stars, and

---

[102]Locus is defined as the set of all points whose coordinates satisfy a single equation or one or more algebraic conditions.

[103]Kline, *Mathematical Thought from Ancient to Modern Times*, p. 126.

discovered the precession of equinoxes.[104] His catalogue of stars laid the astronomical groundwork for the geocentric system of Claudius Ptolemy as detailed in his *Almagest* (see section 3.1).

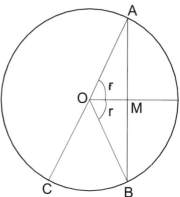

Figure 18: The Method of Hipparchus

To be able to do his calculations, he needed a table of trigonometric ratios. He began with the circle and its circumference of 360⁰ as per the ancient Babylonians (see Figure 18). He then divided the diameter of the circle into 120 equal parts. Each part of the circumference and diameter was further divided into 60 parts (minutes) and each of these into 60 more parts (seconds) using the Babylonian sexagesimal fractions. Thus, for a given arc AB (in degrees), Hipparchus gave the number of units in the corresponding chord AB. The number of units in the chord AB corresponding to the number of degrees in arc AB is equivalent to the modern sine function. In the figure, sin r = AM/OA (the ratio of the length of the side opposite angle r to the length of the hypotenuse, the side opposite the right angle). Since Hipparchus worked with the diameter instead of the radius (the measure of OA), then sin r = AB/AC. If 2r is the measure of the central angle of the arc AB, then the chord length of AB can be denoted as "CL 2r." Thus, AB = CL 2r and AC = 120 units. So we have this equation defining the trigonometric sine ratio:

$$\sin r = \frac{AM}{OA} = \frac{AB}{AC} = \frac{CL\ 2r}{120}$$

Based upon the work of Hipparchus, Ptolemy produced a "table of chords" in 15-minute intervals from 0⁰ to 90⁰.

The etymology of the word "sine" has a fascinating history. We start our journey in India. The Hindu mathematician and astronomer Aryabhata the Elder (476-ca. 550) called it *ardha-jya* (meaning "half-chord"). It was later abbreviated to *jya*. Note that *jya* could also mean "bowstring." This makes sense since arc AB resembles a bowstring. Now we travel to the Islamic culture. Arab translators turned this phonetically into *jiba* (without meaning in Arabic) and according to the Arabic practice of omitting the vowels in writing (similar to Hebrew), wrote it *jb*. Our journey now ends in Western Europe. European Arabic-to-Latin translators, having no knowledge of the Sanskrit (ancient Indic) origin, assumed *jb* to be an abbreviation of *jaib* (Arabic for "cove", "bay", "bulge", "bosom"). This also makes sense, since arc AB looks like a curve or a

---

[104]The precession of equinoxes is the slow shift westward of the equinoctial points against the stars. Isaac Newton (1642-1727) determined that this shift was caused by a very subtle gravitational tug of the moon and the sun on the earth.

bulge. When Gerard of Cremona (ca. 1114-1187) translated Ptolemy's *Almagest* in the 12th century, he translated *jaib* into the Latin equivalent *sinus* (from which we derive the English sine).

Figure 19: Hipparchus

With a readily available measurement on the surface of the earth, you can, with trigonometry and a good sextant, literally calculate, in succession, the height of a mountain, the radius of the earth, the distance to the moon, the distance to the sun, any planet, and the stars.[105] Hipparchus applied a similar procedure and with his measuring tools his calculations were remarkably accurate. Not only were they accurate, these "astronomical" distances staggered the Alexandrian mind. If they knew the book of Job, they would have echoed the sentiments of Eliphaz the Temanite:

> Is not God in the height of heaven? And the highest stars, how lofty they are! (Job 22:12)

Or, they could have sung with the Psalmist David:

> When I consider Your heavens, the work of Your fingers, the moon and the stars, which You have ordained, What is man, that You are mindful of him, and the son of man, that You visit [care for] him? For You have made him a little lower than the angels, and You have crowned him with glory and honor! (Psalm 8:3-5)

But they did not. In the words of Rousas J. Rushdoony, the Alexandrian mind was "learned but ignorant, and given to masses of detail without a focus."[106]

2.16.3 RUDIMENTARY TRIGONOMETRIC RATIOS

$\sin A = \dfrac{a}{h}$; *side opposite ∠A over the hypotenuse*

(*literally means "to stretch under"*)

$\cos A = \dfrac{b}{h}$; *side adjacent ∠A over the hypotenuse*

$\tan A = \dfrac{a}{b}$; *side opposite ∠A over the side adjacent ∠A*

---

[105]The principles of trigonometry are simple and have a varied scope of application. Today, surveyors, navigators, and mapmakers employ trigonometric principles constantly.

[106]Rushdoony, *The One and the Many: Studies in the Philosophy of Order and Ultimacy*, p. 18.

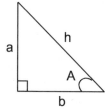

Figure 20: The sides of a right-angle triangle

Mnemonic memory aid:

- SOH (**S**ine of an angle equals the side **O**pposite over the **H**ypotenuse).
- CAH (**C**osine of an angle equals the side **A**djacent over the **H**ypotenuse).
- TOA (**T**angent of an angle equals the side **O**pposite over the side **A**djacent).

2.16.4 FINDING THE HEIGHT OF A TREE

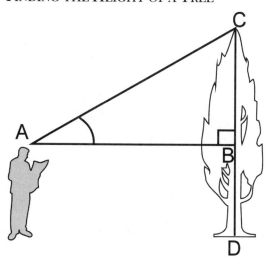

Figure 21: Finding the height of a tree

Given: $\angle A, BD,$ *and* AB

Find: $DC = CB + BD$

$$\tan A = \frac{CB}{AB}$$

Hence, $CB = AB \cdot \tan A$ *and*

$$DC = CB + BD$$

2.16.5 WHAT HAPPENS IF A MOUNTAIN GETS IN THE WAY?

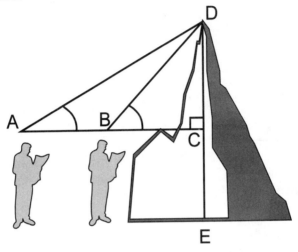

Figure 22: Finding the height of a mountain

Given: $\angle A$, $\angle B$, AB, *and* CE

Find: $DE = DC + CE$

$$\tan A = \frac{DC}{AB + BC} \text{ or } DC = (AB + BC)\tan A$$

$$\tan B = \frac{DC}{BC} \text{ or } BC = \frac{DC}{\tan B}$$

$\therefore$ *(the mathematical symbol for therefore)* $DC = \left(AB + \dfrac{DC}{\tan B} \right)\tan A$

$$DC = AB \cdot \tan A + DC\frac{\tan A}{\tan B} \Rightarrow DC - DC\frac{\tan A}{\tan B} = AB \cdot \tan A \Rightarrow$$

$$DC\left(1 - \frac{\tan A}{\tan B} \right) = AB \cdot \tan A \Rightarrow DC = \frac{AB \cdot \tan A}{1 - \dfrac{\tan A}{\tan B}} \Rightarrow$$

$$DC = \frac{AB \cdot \tan A}{\dfrac{\tan B - \tan A}{\tan B}} \Rightarrow DC = \frac{AB \cdot \tan A \cdot \tan B}{\tan B - \tan A}$$

Hence, $DE = DC + CE$.

2.16.6 CALCULATING THE RADIUS OF THE EARTH

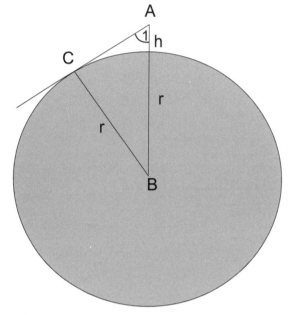

Figure 23: Calculating the radius of the earth

Given that the radius of a circle to the point of tangency, C, is perpendicular to that tangent, then ∆ACB is a right triangle.

$$Hence, \sin \angle 1 = \frac{r}{r+h}$$

$$(r + h)\sin \angle 1 = r$$

$$r \cdot \sin \angle 1 + h \cdot \sin \angle 1 = r$$

$$r - r \cdot \sin \angle 1 = h \cdot \sin \angle 1$$

$$r(1 - \sin \angle 1) = h \cdot \sin \angle 1$$

$$r = \frac{h \cdot \sin \angle 1}{1 - \sin \angle 1}$$

If h (height of a mountain) = 3 miles, then ∠1 =87°46′ (modern value) and r = 3944 miles. If the reader found the algebra too difficult, tiresome, or boring, then he should remember that the method described is an alternative to tunneling down to the center of the earth and then measuring the radius by applying a yardstick from the center all the way to the surface.

2.16.7 DETERMINING THE DISTANCE BETWEEN ANY TWO POINTS ON THE EARTH

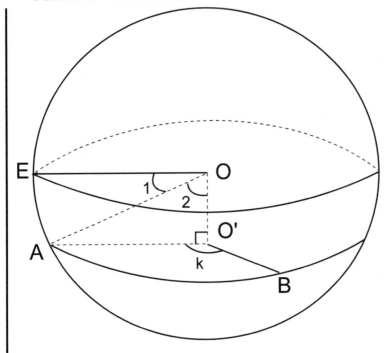

Figure 24: Determining the distance between any two points on the earth

Find: measure of AB or mAB

Since ΔEOO' is a right triangle,

then $\angle 2 = 90° - \angle 1$ *(the latitude of A).*

Hence, $\sin \angle 2 = \dfrac{O'A}{OA}$ *or* $O'A = OA \cdot \sin \angle 2$

where OA = radius of the earth.

Let k = difference in longitude from A to B.

Hence, $mAB = \dfrac{k}{360} \cdot 2\pi \cdot O'A.$

2.16.8 Determining the Distance from the Earth to the Moon

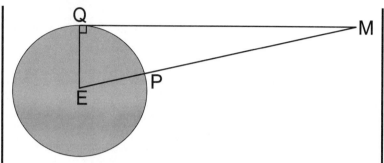

Figure 25: Determining the distance from the earth to the moon

Let P and Q be two points on the equator. Assume the moon is directly overhead at P. Regard the moon as a point M. Assume the moon to just be visible from Q; i.e., the line segment QM is tangent to the earth. QE = radius of the earth. ∠E = difference in degrees longitude from P to Q (modern value is 89°4').

Hence, $\cos E = \dfrac{QE}{EM}$ *or* $EM = \dfrac{QE}{\cos E}$ *and PM = EM – EP.*

∴ PM = 242,126 - 3,944 = 238,182 miles.

2.16.9 Determining the Radius of the Moon

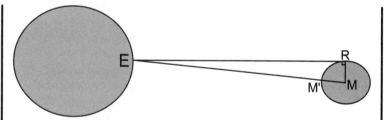

Figure 26: Determining the radius of the moon

At E, measure the angle between line segment EM and line segment ER (modern value is 15'). Let EM = PM (the distance from the earth to the moon).

Hence, $\sin E = \dfrac{MR}{EM}$ *and MR = EM · sin E.*

∴ MR = 238,182 sin 15' = 1,039 miles.

2.17 SOVEREIGN GOD AT WORK

Undoubtedly, Greek culture produced some significant developments in both mathematics and science. In fact, the Greek establishment of the correlation between mathematics and the study of the design of nature has since become the very basis of modern scientific work.

From a Christian standpoint, how are these accomplishments to be viewed? This explosion of genuinely useful knowledge must be seen in its proper historical perspective, especially as it relates to God's purposes. The close proximity of Greek discoveries with the birth of Jesus Christ is not just a coincidence of history. In reference to the Alexandrian Greek period, Morris Kline states that "the long sea voyages of the Alexandrians called for far better knowledge of geography, methods of telling time, and navigational techniques...."[107]

Why was God moving upon these men at this time in history to explore the possibilities of long sea voyages? Why did the calculations of Eratosthenes in reference to the circumference of the earth turn out to be so accurate? Why did Alexander the Great install the Greek language as the universal trade language of his empire? When the Romans came along, why did they construct a universal road system? In these developments, we see the hand of the sovereign God of Scripture directing the flow of history preparing practical tools for His people to use in preaching the liberating Gospel of the Lord Jesus Christ, not only throughout the Roman Empire, but globally to all the peoples of the earth.

2.18 SANDY FOUNDATIONS

The temple of Greek mathematics, even though it outwardly looks strong, is built upon a sandy foundation. Although their geometric developments were profound, they failed to incorporate key concepts like infinity, the theory of perspective, and the capacity to generalize.[108] Art curator William M. Ivins, Jr. (1881-1961) observes that "perhaps the most interesting thing about it [geometry – J.N.] is that just at the precise stage in its development when it was about to become the most important and exciting the Greeks lost interest in it."[109] According to Stanley L. Jaki, "Greek science had already lost its creativity by the time the Museum opened in Alexandria."[110] Ivins goes on to say that "within another hundred years [after Euclid – J.N.] ... the great imaginative work of the Greek geometers had for all practical purposes come to an end."[111] Concerning Greek philosophy, he states that "at the end, Greek thought, following its own nature and methods, wound up hopeless in a blind alley of its

---

[107]Kline, *Mathematical Thought from Ancient to Modern Times*, p. 102.
[108]William M. Ivins, Jr., *Art and Geometry: A Study in Space Intuitions* (New York: Dover Publications, 1964), p. 95.
[109]*Ibid.*, p. 38.
[110]Jaki, *The Road of Science and the Ways to God*, p. 32.
[111]Ivins, p. 50.

own making."[112] The glory that was Greece was "not only short but immediately and necessarily followed by ages of exhaustion and sterility."[113]

Why, near the time of the birth of Christ, was Greek mathematics and science dying and decaying? According to Pierre Duhem (1861-1916), French science historian, the failure of Greek science was due to the influence of such theological doctrines as the divinity of the heavens and the cyclical view of time.[114] Like the ancient civilizations before them, their view of history trapped them into a treadmill of endless repetitions. Stanley L. Jaki states that "the problem of the failure of ancient Greek science is largely the failure of the Greeks of old to go resolutely one step beyond the prime heavens to a prime mover absolutely superior to it."[115] He continues, "Needless to say, a world not governed by its divine pilot is a largely irrational world which discourages natural theology and science by the same logic."[116] Francis Schaeffer (1912-1984), theologian and philosopher, observes that the gods of the Greeks, no matter what form they took, were "simply inadequate for they were not big enough."[117] As a result, "they had no sufficient reference point intellectually."[118] Greek mathematics and science stagnated because the theological ponderings that undergirded these disciplines were not biblical.

Greek theology did affect the stability of the timbers of the Greek mathematical framework. Contrary to biblical thought, the Greeks viewed the flow of history in terms of eternal, recurring cycles. In addition, they absolutized both nature and reason. The universe could be understood by reason and was understood *only* in terms of itself, not in terms of the personal, transcendent, Creator revealed by Scripture. Greek philosophers always understood this reasoning capacity of man in an autonomous fashion. In the words of Rousas J. Rushdoony, in Greek philosophy "the human mind is capable of knowing all finite facts without any reference to God and is itself a neutral agent capable of weighing and evaluating facts without prejudice. This, of course, runs counter to the Biblical faith that man is a sinner, fallen in all his being, and incapable of neutrality."[119] In accordance with biblical thought, the Greeks understood nature to reveal design and order. It must be noted, though, that this design and order existed as a reflection of the Platonic ideal determined by human reason or the Aristotelian form embodied in matter concept, not by an eternal and personal lawgiver. Since nature is rational and man is rational, the common denominator of the two is mathematics. The Greeks saw in their mathematical structures a

---

[112]*Ibid.*, p. 55.

[113]*Ibid.*, pp. 55-56.

[114]Pierre Duhem, *Le Systéme du monde: Histoire Des Doctrines Cosmologiques* (Paris: Librairie Scientifique A. Hermann et fils, 1913-1959), 1:65-85.

[115]Jaki, *The Road of Science and the Ways to God*, p. 320.

[116]*Ibid.*, p. 340.

[117]Francis Schaeffer, *He is There and He is Not Silent* (Wheaton: Tyndale House Publishers, 1972), p. 40.

[118]Francis Schaeffer, *How Should We Then Live? The Rise and Decline of Western Thought and Culture* (Old Tappan: Revell, 1976), p. 21.

[119]Rushdoony, *World History Notes*, p. 42.

perfect and complete description of the order of a universe eternal and unchangeable.

To the Greeks, Euclid's geometry exemplified the glories of deductive reasoning. They viewed syllogistic thinking as the sole pathway to all truth since this method of thought compels the truthfulness of a conclusion based upon accepted axioms. This overemphasis on deductive geometry impeded the development of mathematics as Richard Courant (1888-1972) and Harold Robbins point out, "For almost two thousand years, the weight of Greek geometrical tradition retarded the inevitable evolution of the number concept and of algebraic manipulation, which later formed the basis of modern science."[120]

The sandy foundation of Greek mathematics is the denial of the autonomy of the biblical God, creator and sustainer of both the universe and man. In the place of the autonomy of God, the Greeks substituted the autonomy of man's mind. When the reason of man is not submitted to a higher law, the God of Scripture, his grandiose structural systems will soon stagnate and retard. Later, they will collapse. And that is just what eventually happened to the temple of Greek mathematics.

QUESTIONS FOR REVIEW, FURTHER RESEARCH, AND DISCUSSION

1. Explain how the biblical mandate of dominion relates to mathematics.
2. Document the evidences and explain the reasons why mathematics stagnated in ancient civilizations.
3. Detail fully the components of the biblical worldview from the Book of Proverbs.
4. Give reasons why science and mathematics did not flourish to a great extent in Hebrew culture in spite of the worldview revelation of Scripture. What seeds did this worldview perspective plant that eventually sprouted during the Christian Middle Ages?
5. Explain the epistemology and metaphysics of the key thinkers of the Ionian school.
6. What was the Pythagorean doctrine concerning the essence of reality? How did this doctrine impact the Pythagorean view of ethics?
7. Explain the metaphysical and epistemological response of the Eleatic school to the thinking of Heraclitus.
8. Detail the pluralistic metaphysic of Empedocles, Anaxagoras, and Democritus.
9. Compare and contrast the worldview components (metaphysics and epistemology) of Plato with the worldview components of Aristotle.
10. Explain the link between Greek philosophy and Greek mathematics.
11. What mathematical concept turned out to be the "fly in the ointment" of Greek mathematics? Why?

---

[120]Richard Courant and Harold Robbins, *What is Mathematics?* rev. Ian Stewart (London: Oxford University Press, [1941] 1996), p. xvi.

12. Characterize deductive reasoning. Can it *always* be applied to prove a desired statement? Why or why not?
13. Relate the accomplishments of the mathematicians of Alexandria with the sovereign hand of God in history.
14. What was the Greek method for establishing truth?
15. Analyze both the strengths and weaknesses of Greek mathematics.

3: FROM CHRIST TO WYCLIFFE

Be fruitful and multiply, fill the earth and subdue it ... (Genesis 1:28).

The Living God

Essentially, the Greeks viewed reason, in its deductive form, as the pathway to absolute truth. Originally, Greek scientists observed the workings of nature and then derived the physical laws describing its revealed order. For example, Aristotle observed that a stone falls faster than a feather; therefore, he concluded that the speed of a falling object is proportional to its weight. In spite of his emphasis on the importance of physical phenomena, he developed a view of science based upon deductive reasoning that eventually negated the value of experimental, or inductive, verification. Alfred North Whitehead comments about the resulting influence of Aristotelian thought:

> The popularity of Aristotelian Logic retarded the advance of physical science throughout the Middle Ages.... Mathematics, as a formative element in the development of philosophy, never, during this long period, recovered from its deposition at the hands of Aristotle.[1]

3.1 PTOLEMAIC THEORY

In astronomy, the Greeks observed that the earth appeared to be motionless and the sun encircled it daily. It was Claudius Ptolemy (ca. 85-ca. 165), another Alexandrian mathematician, who popularized a cosmology that is now known as the geocentric theory of the heavens, a theory that held sway in as-

---

[1]Alfred North Whitehead, *Science and the Modern World* (London: Free Association Books, [1926] 1985), pp. 37-38.

tronomy until the Scientific Revolution of the 16th century. To review, not only was the earth motionless, it was also the pivot point of the universe. In contrast to an irregular, corrupt, and imperfect earth, the perfect circular orbits of the heavens displayed eternal changelessness.[2] Ptolemy compiled his conclusions in a book originally titled *Syntaxis mathematica* ("Mathematical collection"). In 827, the Arabs, viewing this work as almost divine, translated it and gave it a new title, *Almagest*, which means "the greatest work."[3]

Figure 27: Claudius Ptolemy

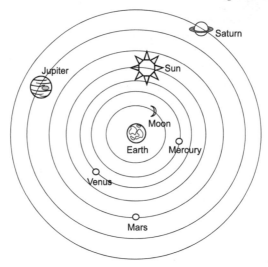

Figure 28: Ptolemaic cosmology

Ptolemy admitted that his theories were just mathematical models. The actual explanation for the first cause of these motions remained a mystery to him. In Book 9, he said, "After all, generally speaking, the cause of the first principles is either nothing or hard to interpret in its nature."[4] To him, the deity was invisible, motionless, and unknown.[5] He characterized the heliocentric views of

---

[2] Ptolemy had to "save the phenomena" by introducing a complex set of epicycles in order to account for the retrograde motion of the planets. An epicycle is a small circle, the center of which moves on the circumference of a larger circle at whose center is the planet earth and the circumference of which describes the orbit of one of the planets around earth. Retrograde motion refers to the brief, regularly occurring, apparently backward movement of a planetary body in its orbit as viewed against the fixed stars, caused by the differing orbital velocities of the earth and the body observed.

[3] Kline, *Mathematical Thought from Ancient to Modern Times*, p. 191.

[4] Claudius Ptolemy, *Almagest*, trans. G. J. Tooner (London: Gerald Duckworth, 1984), p. 423.

[5] *Ibid.*, pp. 35-36.

Aristarchus as "absurd and ridiculous" and explained, with full mathematical demonstrations, the motions of the universe. Stanley L. Jaki comments, "Instead of prompting scientific growth, Ptolemy's work enshrined an already old stagnation of science among the Greeks."[6]

3.2 PRAGMATIC ROME

As God had foretold through the prophet Daniel, the juggernaut of Rome eventually swallowed up the Greek empire of Alexander the Great (Daniel 2, 7). Highly pragmatic and only slightly creative, the Romans basically copied and organized whatever the Greeks had done. Marcus Tullius Cicero (106-43 BC), statesman and orator, said that "the Greeks held the geometer in the highest honor, accordingly, nothing made more brilliant progress among them than mathematics. But we have established as the limit of this art its usefulness in measuring and counting."[7] Because of this general practical and earthbound viewpoint, advancement in mathematics halted. Moreover, the Romans acted as an agent in furthering the decline of Greek culture by destroying their Alexandrian citadel of learning. In 47 BC, the great Museum was accidentally set aflame by the armies of Julius Caesar during a nighttime sortie. Many of its 750,000 volumes containing works by Greek authors and others dating back to antiquity went up in smoke.

In addition to Platonism and Aristotelianism, three other philosophical views became popular in Roman times. They were Skepticism, Epicureanism, and Stoicism. The Apostle Paul encountered the last two philosophies at the Aeropagus in Athens (Acts 17:16-34). The philosophers of these traditions primarily concerned themselves with ethics; metaphysics and epistemology were important, but secondary issues.

The Greek advocate of Skepticism was Pyrrho (ca. 360-ca. 272 BC). The basic epistemological presupposition of the Skeptics was relativism. According to this view, no one knows anything for sure. Even if you could know something, you could not communicate it to anyone. Hence, very little written documentation survives of individual skeptics. We base our information on them from Sextus Empiricus (ca. 200) who reportedly wrote down what they said. Since everyone does what is right in his own eyes, the Skeptics embraced a relativistic or eclectic approach to ethics.

The Epicureans, named after their founder Epicurus (341-270 BC), embraced a metaphysic of atomistic materialism and an ethic of refined hedonism. They defined the pursuit of pleasure, not in sensual "eat, drink, and be merry" terms, but in conservative intellectual terms. They sought to live the "good life" of culture and refinement. A famous Roman Epicurean was Lucretius (94-55 BC) who wrote *On the Nature of Things*. In the text, he repeatedly denounced the

---

[6]Jaki, *The Road of Science and the Ways to God*, p. 10.
[7]Cicero, *Tusculan Disputations*, trans. J. E. King (Cambridge, MA: Harvard University Press, 1951), p. 7.

Greek gods and falsely concluded that since some evil deeds had been specifically prompted by religion, then all religions should be abandoned.

Founded by Zeno of Citium (ca. 335-ca. 263 BC), the Stoics[8] were a perfect fit for the Roman juggernaut with their emphasis on duty as the essence of the good life and the utilitarian character of mathematics and science. The Romans made particular use of this ethic by applying it in the context of the Empire: a good citizen does his duty to the State.[9] They posited that knowledge is based upon perceptive sensations. These sensations are not like the form of Plato or Aristotle. They are simply a mental construct. Only matter was real. They equated matter with a *logos* (reason or virtue) that permeates and controls everything that happens. Man must strive to live in harmony with nature.

3.3 THE ADVENT OF CHRIST

During the pinnacle of Roman power and influence, God manifested Himself in the person of Jesus Christ. In His death, resurrection, and ascension, He established the kingdom of God as foretold by the prophets of old. King Jesus would rule in the midst of His enemies and sooner or later His followers would have to confront the philosophy of the Greeks (Psalm 110:1-2; Hebrews 10:12-13). He predicted that His kingdom would slowly, gradually, and almost imperceptibly grow in influence like a mustard seed (where the beginning is wholly inadequate to explain the result) or like leaven working invisibly in a loaf of bread (Matthew 13:31-33).

Not many secular historians notice that Jesus spoke to worldview issues. They relegate His life and work to the realm of irrelevant religion without realizing the dynamic that He brought to every aspect of life. Rousas J. Rushdoony gives a penetrating critique of this prejudicial perspective:

> His [Jesus – J.N.] birth and life are events which most historians, being in the Hellenic tradition, are content to mention briefly at best and then to ignore. They choose to ignore the Biblical record because it does not jibe with their conception of history, which leaves no room for the supernatural. By virtue of their naturalistic prejudice they refuse to consider anything to the contrary and choose to act as if it were not there.[10]

To illustrate the impact that the person of Jesus Christ had on philosophy (and by implication, on mathematics and science), let us analyze what Jesus said about Himself and what one of His followers said about Him.

> Jesus said to him, "I am the way, the truth, and the life. No one comes to the Father except through Me" (John 14:6).

---

[8]From the Greek word *stoa* meaning "porch." Stoic philosophers originally discussed their ideas around a porch.
[9]A famous Roman stoic was the Emperor Marcus Aurelius (121-180) who wrote *Meditations*.
[10]Rushdoony, *World History Notes*, p. 48.

In the beginning was the Word, and the Word was with God, and the Word was God. He was in the beginning with God. All things were made through Him, and without Him nothing was made that was made. In Him was life, and the life was the light of men.... That was the true Light which gives light to every man coming into the world.... And the Word became flesh and dwelt among us, and we beheld His glory, the glory as of the only begotten of the Father, full of grace and truth (John 1:1-4, 9, 14).

First, Jesus proclaimed that He is the way. Way implies a pathway to walk upon. Christ presented before man an ethic, a lifestyle to follow, and He provided the moral power through His redemptive work to walk in it. Man, at root, is a sinner and sin must be dealt with. According to Ephesians 4:17-19, sin produces futility in the mind, darkens the understanding, and generates an ignorant and blind heart. This epistemological corruption of mind and heart results in a behavioral lifestyle committed to lewdness, uncleanness, and greediness. According to Romans 1:18-22, man outside of Christ "suppresses the truth" of the power and person of the Godhead revealed by creation "in unrighteousness." Because of this willful suppression, man's thoughts turn to futility, his heart is engulfed by a shadow of darkness, and his wisdom becomes a mere profession of foolishness. Christ came to earth to redeem sinners ethically and epistemologically. In this context, note that no Greek philosopher saw a need for redemption of the heart and mind. Why? Man does not need any renewal because there is nothing inherently wrong with man's heart and mind. The autonomy of man is the presuppositional starting point not only for Greek, but all unbelieving philosophy. Because of the heritage of Greek culture, most people think of philosophers as "ivory tower" thinkers who immerse themselves in the neutral study of abstract truth. In contrast, biblical philosophy is intensely personal. It requires obedience of the whole man to the word of a higher authority. Philosophy means "love of wisdom." Since Christ is wisdom and knowledge personified (Colossians 2:3), then a true philosopher is an obedient lover of Christ. *Redemption in and through Christ is the key to philosophy.*

Second, it is clear from Ephesians 4 and Romans 1 that man's problem is not just behavioral; it is epistemological. Jesus said that He is the truth and that He enlightens every man that comes into the world. This enlightenment means that Christ gifts all men with the ability to know what they know.[11] It is light in the person of Christ that makes any knowledge possible. Note that the Greeks, especially Pythagoras, marveled at the mathematical order of the world. Their ability to note that order is a direct gift of Christ. Due to sin, they misinterpreted the source and purpose of that order. If men will not see the world as Christ reveals it, then they will not be able to see the world in truth. As Jesus said, "If you abide in My word, you are My disciples indeed. And you will know

---

[11]This enlightenment does not confer salvation upon all men; it is one aspect of God's common grace to all men. Another aspect is that "He makes His sun rise on the evil and on the good, and sends rain on the just and on the unjust" (Matthew 5:45).

the truth, and the truth shall make you free" (John 8:31-32). The truth that is in Jesus brings ethical and epistemological freedom.

Third, Jesus is the life. In Him, there is true life (metaphysical reality). Through Him, the true *logos*, all things were made. And it is in the *logos* of God that all things consist and it is the word (Greek: *rhema*[12]) of His power that upholds all things (Colossians 1:17; Hebrews 1:3). The biblical Christian metaphysic is that a personal *logos* has made the world giving life to all things. Not only has the *logos* made the world, the *logos* has also entered the world taking on human flesh, the glory as of the only begotten (Greek: *monogenes*) of the Father in order to save it (John 1:14, John 3:17). As we have noted, the Greeks also used the *logos* concept. Heraclitus defined the *logos* as the principle of order in the changing world. But, the *logos* of Greek philosophy was a component of the natural order; it was an immanent *logos*. The Bible reveals a transcendent *logos* (an eternal, yet personal Creator). The created order of cause and effect reflects the way wisdom in the person of Christ made it (Proverbs 8:12-36). The Greeks also employed the *monogenes* concept. The Greeks called the universe *monogenes* (or *unigenitus* in Latin). In order for an educated Greek to embrace Christianity, he was faced with one of two alternatives: either Christ was the "only begotten" or the universe was the "only begotten." In order to follow Christ, the only begotten of the Father, an educated Greek had to renounce the culturally pervasive concept of the universe as the "Supreme Being" with its concomitant emanationist system.[13]

The good news of the Gospel is that the *logos* of God is the way of (ethics), the truth (epistemology), and the life (metaphysics). The bad news is that this way, truth, and life is exclusive – no one can come to the Father except through Christ. A worldview is not only a network of presuppositions concerning how to live, how to know, and what is real; it is an *antithetical* network of presuppositions. There is one way to a true worldview and that way is in the person of the Lord Jesus Christ and His redemption.

3.4 JERUSALEM CONFRONTS ATHENS

First century Christianity had little time for mathematics. Converts included men and women in high places and low, educated and uneducated. Numbering maybe close to half a million by 70 AD, it was still a scattered and oppressed group. Faced with intense persecution first from the Jews and then from the Romans, survival was the name of the game, not the pursuit of mathematics.

---

[12]*rhema* means the express statements or utterance of a living voice.
[13]The emanation concept was first used to describe divine procreation in Hellenistic Jewish works of the 2nd and 1st centuries BC. The systematic application of the concept of emanation was the product of Gnosticism and the neo-platonism popularized by Plotinus in the 3rd century AD. Under these systems, a succession of emanations caused a diminishing of the divine essence. One consequence of this was the eventual creation of the material world.

It did not take long until those educated in Greek philosophy converted to the faith. Their ensuing reaction to Greek thought was mixed. Justin the Martyr (100-165) tried to show that Christianity was contained in Greek philosophy. In his *Apologies*, he went so far as to announce that Greek philosophers, being enlightened by the divine *logos*, where themselves Christians (a misapplication of John 1:9). Clement of Alexandria (ca. 150-ca. 220) argued that Greek philosophy found its fulfillment in Christianity. Origen (185-254), who was born in Alexandria but later moved to Caesarea, attempted to prove the truth of Christianity using logical arguments that found their source in Greek thought. To understand why many of the early Church fathers perceived that "common ground" existed between Greek philosophy and Christianity, we must remember that a belief in "God the Father almighty, maker of heaven and earth" dominated all their thinking.[14] In this context, note the analysis of Rousas J. Rushdoony:

> ... what Plato and Aristotle actually said was very different from what Plato and Aristotle were thought to mean when read by Christians who assumed the reality of the creator God and His handiwork, the universe, and then approached the Greek philosophers as though they had presupposed a like order. This error is still with us.[15]

When compared with Christianity, our detailed analysis of Greek philosophy (sections 2.3 through 2.18) has shown that an antithesis (not a synthesis) exists. Among the many contrasts, three need to be noted and emphasized:
1. The Greeks founded their thinking on the neutral autonomy of man's mind. For the Christian, man's mind is not neutral. It has been infected by sin and needs renewal via God's grace and God's word.
2. The Greeks embraced a cyclical view of time and history. The Christian view of time and history is linear. Time and history had a beginning; time and history will have an end.
3. The Greeks believed in the divinity of nature. They also viewed it as a living organism.[16] It revealed order due to an inherent, imminent, and impersonal *logos* (reason or natural law). Christianity embraces a metaphysical dualism: a distinction and division between the Creator and the creation. The creation reflects harmonious order due to the decrees of a transcendent and personal Lawgiver who upholds and sustains all things by the word of His power.

One exception to the "common ground" thought of most Church fathers was Tertullian of Carthage (150-230). He expressed his absolute abhorrence of anything originating from "Athens." He so detested the philosophy of Athens that he gravitated toward the rejection of the validity of reasoning altogether. His apologetic was counterproductive in that he tried to prove the veracity of

---

[14]This is the first confession of the famous "Apostles' Creed."

[15]Rousas J. Rushdoony, "The Quest for Common Ground," *Foundations of Christian Scholarship*, ed. Gary North (Vallecito, CA: Ross House Books, 1976), p. 28.

[16]See Jaki, *The Relevance of Physics*, pp. 3-51.

the Christian faith on the basis of its rational *absurdity*. He did not understand that it was the teaching of Athens that was presuppositionally absurd (foolish to use the biblical terminology of I Corinthians 1:20-25) and that Christianity only *appeared* to be absurd or foolish to the unbeliever.

3.5 THE BISHOP OF HIPPO

The bishop of Hippo, Aurelius Augustinus (354-430), known better as Augustine, deserves special notice because it was his theology that provided much of the groundwork for medieval civilization. In his pre-conversion days he drank deep from the well of Manichaeism and neo-platonism. Manichaeism, founded by the Persian prophet Mani (ca. 216-ca. 276), was a dualistic philosophy dividing the world between good and evil principles and regarded matter as intrinsically evil and mind as intrinsically good. Neo-platonism, pioneered by Plotinus (204-270), sought to revise Plato in its search for a mystical vision of the ideal. After his conversion, self-portrayed in *Confessions*, Augustine grew in his understanding of Scripture and eventually came to reject these views as anti-biblical.

Figure 29: Aurelius Augustinus

He said, "Let every good and true Christian understand that wherever truth may be found, it belongs to his Master."[17] He encouraged Christians to recognize and acknowledge truth, but to reject "the figments of superstition."[18]

He acknowledged value in the methods of mathematics. To him, "numbers … have fixed laws which were not made by man, but which the acuteness of ingenious men brought to light."[19] He did have trouble, though, with many mathematicians. In his time, they were called *mathematici*, a code name for astrologers![20] Augustine rejected their speculations and superstitions and warned Christians to keep clear from their contaminating influence.[21]

---

[17]Aurelius Augustinus, "On Christian Doctrine," *Great Books of the Western World: Augustine*, ed. R. M. Hutchins (Chicago: Encyclopaedia Britannica, 1952), 18:646.

[18]*Ibid.*

[19]*Ibid.*, 18:654.

[20]The root word of mathematics, *mathesis*, means "learning in general or disciple." In its historical context, when it has the long penultima (next to last syllable), the word signifies the figments of divination or magic (e.g., astrology). A short penultima means the word refers to a department of philosophy.

[21]Aurelius Augustine, "On Christian Doctrine," p. 647f. Augustine's condemnation, along with the tie of mathematics to astrology, may be the reason why some of the early Christians strongly opposed mathematics.

In his great work, *City of God*, he attacked the pantheistic and cyclical thought of Greek philosophers. He said, "Let us therefore keep to the straight path, which is Christ and with Him as our Guide and Saviour, let us turn away in heart and mind from the unreal and futile cycles of the godless."[22] For him, in the beginning, God created all things, and as the Lord of history, He will bring history to an end with His personal coming in glory. Stanley L. Jaki details how Augustine's thought eventually laid the cornerstone for the building called modern science:

> Augustine's *City of God* molded more than any other book by a Christian author the spirit of the Middle Ages. Its pages were as many wellsprings of information and inspiration for the emerging new world of Europe about the meaning of mankind's journey through time. He declared that the physical universe and human history both had their origin in the sovereign creative act of God, which also established a most specific course and destiny for both.... This book became the intellectual vehicle for a confidence which centuries later made possible the emergence for the first time of a culture with a built-in force of self-sustaining progress.[23]

Concerning epistemology, Augustine emulated his mentor in the faith, Ambrose (ca. 340-397), by emphasizing that faith precedes understanding.[24] To him, the reasoning abilities of man are safeguarded by a previous faith commitment. Man must begin with what God has revealed and believe it. Then man's reason can function properly. In essence, a man of faith "thinks God's thoughts after Him."[25] Cornelius Van Til (1895-1987), reformed theologian, emphasizes Augustine's epistemology as he compares the thought of Plato with the thought of Augustine:

> Plato first tried to interpret reality in terms of the sense world. Then he tried to interpret reality in terms of the Ideal world. Finally he tried to interpret reality in terms of a mixture of temporal and eternal categories. In this way Plato exhausted the antitheistic possibilities. Plato assumed that the human mind can function independently of God; Augustine held that man's thought is a thinking of God's thoughts after Him. Accordingly, Augustine did not seek to interpret reality by any of the three Platonic methods.[26]

---

[22]Aurelius Augustine, *City of God*, trans. Marcus Dods (New York: Modern Library, 1950), p. 404.

[23]Jaki, *Science and Creation: From Eternal Cycles to an Oscillating Universe*, pp. 177-178.

[24]This epistemological emphasis reflects the principle of Hebrews 11:3, "*By faith we understand* that the worlds were framed by the word of God, so that the things which are seen were not made of things which are visible" (emphasis added).

[25]In review, this type of thinking is philosophically called analogous thinking. In the 17th century, Johannes Kepler consciously employed this type of thinking in his scientific work.

[26]Cornelius Van Til, *A Survey of Christian Epistemology* (Phillipsburg: Presbyterian and Reformed, 1969), p. vi.

3.6 THE ESSENCE OF THE CONFLICT

The basic debate between Greek and Christian thought structures focused on the correct purveyor of truth. To the Greeks, the autonomy of man's mind, in the form of deductive reasoning, provided the means. To the Christians, the autonomy of God, as revealed by faith in biblical revelation, showed the way. In the early days of Christianity, this conflict was such an issue that many Greek works were burned.[27]

The Roman Empire had tried to stamp out Christianity as early as the reign of the Roman Emperor Nero (54-68), but persecution had an uncanny way of both multiplying and purifying the Christian movement. As a whole, this early strong and robust Christian faith could handle Greek scholarship without compromising.

3.7 EVENTUAL COMPROMISE

But this "no compromise" stance did not last for long. The beginning of the wane can be traced back to the reign of the Roman Emperor Constantine (306-337). He saw that Christianity was too strong of a force to fight against any more, so he sincerely joined ranks with the faith. Persecution stopped and the political machine was set in motion to eventually establish the Christian faith as the one legal religion of the empire. As a result, the unconverted began to flood the front doors of every church and gradually pushed the vibrant, purified faith out the windows.

As time passed, Christianity slowly began to mix with the humanistic ideas of Greece and Rome. As Rousas J. Rushdoony comments, "Humanism was in constant warfare with Christianity and often successfully infiltrated it."[28] Eventually, after many *centuries* of battle, the authority of human reason replaced the authority of Scripture. A robust faith can handle autonomous reasoning, but a weak faith will fall prey to it.

In reference to mathematics and science, this accommodation can be seen in two specific examples. First, as a result of Augustine's writings, Anicius Manlius Severinus Boethius (ca. 475-524), a Roman citizen, laid the groundwork of a school curriculum by mixing Christianity with Platonic education.[29] Greek educators directed the training of children toward encompassing the whole of life. The Greek word *paideia* (train) reflected this worldview emphasis.[30] Paul, the Apostle, used the same Greek word in his command to parents to *train* their children in the admonition of the Lord (Ephesians 6:4). According

---

[27]In 392, the Roman Emperor Theodosius (ca. 346-395) banned pagan religions. In the ensuing milieu, thousands of Greek books were burned. Christians in Alexandria, Egypt destroyed the Serapeum temple that still housed the only extensive collection of Greek works (about 300,000 manuscripts were destroyed).

[28]Rushdoony, *World History Notes*, p. 150.

[29]In this mix, Boethius fused Greek rationalization with his Christian faith. He tried to defend Christianity on purely rational grounds and embraced a purely rational theology.

[30]For documentation, see the three-volume work by Werner Jaeger, *Paideia: The Ideals of Greek Culture*. trans. Gilbert Highet (New York: Oxford University Press, [1943] 1971).

to science historian David C. Lindberg, "… the majority of the early church fathers valued their own classical education and, while recognizing its deficiencies and dangers, could conceive of no viable alternative to it; consequently, instead of repudiating the classical culture of the schools, they endeavored to appropriate it and build upon it."[31]

The curriculum propagated by Boethius became known as the Seven Liberal Arts.[32] This educational system, used throughout the Middle Ages, was divided into the Trivium (three roads) and the Quadrivium (four roads). The Trivium dealt with the "tools of learning" by studying the structure of language, thought, and speech.[33] The initial thrust of most of the Church fathers who used the Trivium was to provide the tools, or establish a model, for *biblical* worldview training without amalgamating it with the vain philosophy of the Greeks (see Colossians 2:8).[34] Where this admixture occurred due to a compromised and synthesized faith (e.g., in some of the scholastics of the late Middle Ages), the door was opened that eventually produced the humanistic Western civilization of the Renaissance and the Enlightenment.[35]

3.7.1 THE SEVEN LIBERAL ARTS

Trivium

1. Grammar: The structure of language
2. Logic: The structure of thought
3. Rhetoric: The structure of speech

Quadrivium

Number (The discrete)
4. Arithmetic: The absolute – the structure of number
5. Music: The relative – applied number
Space (The continued)
6. Geometry: The stable – the structure of static forms
7. Astronomy: The moving – the structure of moving forms

To Pythagoras we owe the basic mathematical structure of the Quadrivium. The school that he founded concentrated on the study of four subjects: (1) *arithmetica* (number theory), (2) *harmonia* (music), (3) *geometria* (geometry), and (4) *astrologia* (astronomy).[36] The findings of Pythagoras formed the

---

[31]David C. Lindberg, *The Beginnings of Western Science* (Chicago: University of Chicago Press, 1992), p. 153.

[32]In Latin, the Seven Liberal Arts are enumerated as follows: *Lingua, tropus, ratio; numberus, tonus, angulus, astra.*

[33]For a popular and insightful analysis of the Trivium and its impact on education, see Dorothy Sayers, *The Lost Tools of Learning* (Canberra: Light Educational Ministries, [1947] 1996).

[34]A strong biblical faith could and can do this.

[35]The basic premise of the Renaissance was the preeminence of reason over Scripture.

[36]During the Middle Ages, arithmetic was used to keep accounts, music for church services, geometry for land surveying, and astronomy was necessary in order to calculate feast and fast days.

theory of arithmetic and music and the medieval source material for geometry and astronomy came from Euclid and Ptolemy respectively.

Second, since Scripture does not present a detailed cosmological system, Christian thinkers during the Middle Ages, when confronting the Greek philosophy of science, had a difficult time distinguishing between the wheat and the tares.[37] The first chapter of Genesis declared man to be the crown of God's creation. Man was created to fellowship with his Maker and the earth was the stage upon which the drama of this fellowship would take place. Greek science had stated that the earth was the apex of the universe in a *physical* sense, and the institutional church read this interpretation of the physical universe *into* Scripture. Following the scheme of Ptolemy, beneath the earth was hell as evidenced by the fiery fumes that belched from volcanoes. Above were seven spheres in which the sun and the planets encircled the earth. The eighth sphere was an incorruptible, unmovable dome upon which the stars hung like lamps. The ninth, or crystalline sphere, was the abode of the saints and at the top, sphere number ten, was the dwelling place of Almighty God, called paradise or the empyrean.[38] It must be emphasized that this viewpoint was an attempt to understand God's word in the light of a self-stated *theory* of cosmology; a theory that tended to be more and more accepted as fact.

3.8 MATHEMATICIANS OF THE EARLY CHRISTIAN ERA

A few mathematicians/scientific commentators of note during the early Christian era were Heron (or Hero), Menelaus, Diophantus, Pappus, Theon, and John Philoponus. We have no evidence as to what extent Christianity influenced any of their work (except for Philoponus). Most, if not all, lived in Alexandria, a city that eventually had a thriving Christian presence.

Heron or Hero (ca. 100 BC or 100 AD), known as *mechanikos* (the "machine man"), was an Alexandrian scientist who invented many water-driven and steam-driven machines. He also developed a formula for determining the area of a triangle known as Heron's formula.[39] Given a triangle with sides of length a, b, and c where s is half the perimeter, then A, the area of the triangle, is determined by: $A = \sqrt{s(s-a)(s-b)(s-c)}$.

Menelaus (ca. 100) of Alexandria wrote a six-book treatise on the theory of chords in a circle (which has not survived). The Arabs translated his three-book work on spherical trigonometry, entitled *Sphaerica* (the Latin title of the Latin translation). This work sheds considerable light on the development of Greek

---

[37]Scripture speaks of the decreed ordinances (laws) of the heavens, but these laws are never quantitatively detailed. The only detail given is that God is the qualitative author and sustainer of these laws (see Job 38:33; Psalm 148:6; Jeremiah 31:35-36). It is left up to man, as part of his dominion calling, to approximate (name) these laws using the tool of mathematics combined with empirical confirmation.

[38]Empyrean means "belonging to the sky; celestial." For the Greeks, this sphere contained the pure element of fire or light.

[39]The formula is credited to him, but it was probably first formulated by Archimedes.

trigonometry. He is reported to have made astronomical calculations in Rome in 98. The theorem of Menelaus states that if a transversal intersects the sides BC, CA, AB of a triangle ABC in the points L, M, N, respectively, then L, M, and N are collinear if (NB)(LC)(MA)=(AN)(BL)(CM). Because of the six segments involved, this proposition was known in the Middle Ages as the *regula sex quantitatum.*

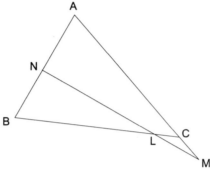

Figure 30: The Theorem of Menelaus

All that is known of the life of Diophantus (ca. 250) of Alexandria is a curious riddle requiring algebra to solve. The problem, found in an anthology called the *Palatine*, assembled by the grammarian Metrodorus (ca. 500), states: His boyhood lasted 1/6 of his life, his beard grew after 1/12 more, and after 1/7 more he married. Five years later his son was born and the son lived to half his father's age. The father died four years after his son. How old was Diophantus when he died? Let x = his age at death and let the reader do the algebra!

$$x = \frac{x}{6} + \frac{x}{12} + \frac{x}{7} + 5 + \frac{x}{2} + 4$$

$$x = 84$$

His work on indeterminate equations, one equation with two unknowns (e.g., 5x + 2y = 20), anticipated by several centuries the progress of algebra.[40] Before his time, all algebra was rhetorical, in which the solution to a problem was written as a pure argument and without symbols. The algebra of Diophantus was syncopated, or cut short, in which some abbreviations were adapted for oft-recurring quantities and operations. This was still very crude and difficult to work with given the fact that Greek letters were used as numerals. According to Thomas L. Heath, Diophantus used the first two letters of the Greek word *arithmos*, α (alpha) and ρ (rho), as the abbreviation for unknowns.[41] Symbolic algebra, what we find in our elementary algebra textbooks – the systematic use of

---

[40]The branch of modern algebra devoted to the solution of indeterminate equations is called Diophantine analysis in his honor.
[41]Arithmetic comes from the Greek *arithmetike*. *Arithmos* means "number" and *techne* means "science."

letters to represent coefficients and unknown quantities, did not come into use until the work of the French mathematician François Viète (1540-1603).

We owe much of our knowledge of Greek geometry to the eight-book work of Pappus (ca. 300) of Alexandria entitled *Mathematical Collection* (six of the eight books are extant). The work is a gold mine of numerous and original theorems, improvements, extensions, and historical commentary.

Figure 31: Hypatia

Theon of Alexandria (ca. 390) lived in the turbulent death throes of the Roman Empire. He authored a commentary of Ptolemy's *Almagest* and the modern versions of Euclid's *Elements* are based upon his revised work. His daughter, Hypatia (370-415), is reported to have written commentaries of the works of Apollonius and Diophantus. She also developed astrolabe designs.[42] A group of zealous Christian monks shamefully murdered her in a riot engendered by political tensions between the Christian and pagan populace of Alexandria.[43]

The death of Hypatia at the hands of a Christian mob is often characterized in the popular literature of modernity as the end of classical learning and free inquiry and the commencement of the dogmatic and often brutal New World order (i.e., the Dark Ages). This cliché pays little attention to the fact that Greek science had already collapsed of his own inert weight sometime between 300 BC and 100 AD.[44]

John Philoponus (ca. 500 AD), a Monophysite theologian, spent his adult life in Alexandria.[45] After being trained in the Museum, he wrote several commentaries on Aristotle. He converted to Chris-

Figure 32: Astrolabe

---

[42]An astrolabe is similar to a sextant in that it can be used to determine the altitude of the sun or other celestial bodies.

[43]See Wilson, *The Infinite in the Finite*, pp. 420-423. See also Maria Dzielska, *Hypatia of Alexandria* (New York: Harvard University Press, 1995). Dzielska notes that Theon practiced divination. She also documents that there was some evidence (whether contrived or real) to support Hypatia's practice of witchcraft and that through it, she beguiled many people of Alexandria (pp. 90-91). It is most likely this accusation, along with the power politics of that time, that guaranteed Hypatia's cruel death. For a study of some insight concerning the place of women in the history of science, see Margaret Alic, *Hypatia's Heritage: A History of Women in Science from Antiquity through the Nineteenth Century* (Boston: Beacon Press, 1986).

[44]Paul Tannery, *La géométrie grecque* (Paris: Gauthier-Villars, 1887), pp. 10-11. This downfall was due to its domination by Stoic philosophy, which, because of its essential utilitarian character, was hostile to science.

[45]A Monophysite is an adherent of the doctrine that in the person of Jesus there was but a single, divine nature. Historically, Egyptian (Coptic) and Syrian Christians generally professed this doctrine.

tianity in 517 shortly after he completed his critical commentary on Aristotle's *Physics*. Contrary to Aristotle, Philoponus made the following revolutionary resolutions:[46]

1. All bodies would move in a vacuum with the same speed regardless of their weight (or mass).
2. Bodies of differing weights would, falling from the same height, hit the ground at the same time.
3. Projectiles move across the air, not because the air keeps closing behind them, but because they were imparted with a "quantity of motion" (or momentum).

Philoponus also rejected Aristotelian cosmology seeing that "the alleged divinity of celestial matter and the eternity of motion in a pantheistic world could not be accommodated in the Christian interpretation of the cosmos."[47] It is important to note that these theological ponderings stimulated the first crucial step toward a new theory of matter and motion.

We shall shortly see that medieval scholars Jean Buridan (ca. 1295-1358) and Nicole Oresme (ca. 1323-1382) picked up on these ideas of Philoponus, especially the elementary impetus theory of motion, and provided the foundation for Galileo Galilei's (1564-1642) work in momentum and inertia and Isaac Newton's (1642-1727) formulation of the first law of motion.[48]

3.9 THE DARK AGES?

The period of European history from the fall of Rome to the Renaissance has been commonly labeled the Dark Ages. This is a misnomer, not based upon history, but upon presuppositional prejudice. Medieval historian Friedrich Heer (1916-1983) comments:

> It is sheer prejudice which condemns the entire Middle Ages as a Dark Age, based on the hackneyed assumption that the medieval mind was narrow, dominated by a fanatical clergy, strait-jacketed by a rigid set of dogmas. Reality presents quite a contrary picture.[49]

Another medieval historian, James Westphal Thompson, spoke of the Dark Ages as the age of pioneers:

> There was much evidence that the human spirit was alive. It was above all an age of pioneers who, because no education could be had outside the

---

[46]See Jaki, *Science and Creation: From Eternal Cycles to an Oscillating Universe*, pp. 185-187.
[47]*Ibid,*, p. 187.
[48]See Herbert Butterfield, *The Origins of Modern Science* (New York: The Macmillan Company, 1961), pp. 1-16. For documentation on the possible cultural transmission link connecting Philoponus with Buridan and Oresme, see Fritz Zimmermann, "Philoponus's Impetus Theory in the Arabic Tradition," *Philoponus and the Rejection of Aristotelian Science*, ed. Richard Sorabji (Cornell: Cornell University Press, 1987).
[49]Friedrich Heer, *The Medieval World*, trans. Janet Sondheimer (New York: New American Library, 1961), p. 20.

Church, and because there was no other civilizing institution, were for the most part churchmen. Popes and monks were forging the new instruments of a new culture.[50]

Medieval historian William Carroll Bark (1910-1997) echoes Thompson with his observations:

" ... the early medieval society was a pioneer society living on a frontier, both geographical and intellectual, and engaged in advancing it. It is remarkable that historians of the West should so long have failed to apprehend this absolutely vital truth about the origins of their own tradition."[51]

Note also that a pioneering spirit is characteristic of youth, not age. Rousas J. Rushdoony observes:

Moreover, the scholastic, as well as much of the medieval world, were marked by the eminence of *youthfulness* ... *Youthfulness* flourishes in a deeply rooted culture which has vitality and communicates it readily and early to its sons.[52]

William Carroll Bark continues with his characteristically insightful analysis:

... yet to Western man since the Renaissance, the historian as well as the philosopher, the artist, and the scientist, it has been all but unpalatable. Blinded by our prejudice in favor of classical "civilization" as contrasted with medieval "barbarism," we have grossly misinterpreted the creative character of what was taking place in late Roman and early medieval times. We have confused adjustment with decay, and failing to recognize what may be called a change of pace and direction, we have branded it as exclusively an ending ... What may seem today to have been quite simply retrogression and nothing else, may from another point of view be regarded as a cutting away of dead wood.[53]

Rousas J. Rushdoony observes that this "Frontier Age" was "by no means the unchanging, sterile era of most caricatures; it was an important nursing ground and battlefield of basic Christian liberties."[54]

Francis Schaeffer comments:

We must not think that everything prior to the Renaissance had been completely dark. This false concept grew from the prejudice of the humanists (of the Renaissance and the later Enlightenment) that all good things

---

[50]James Westphal Thompson and Edgar Nathaniel Johnson, *An Introduction to Medieval Europe: 300-1500* (London: George Allen and Unwin, 1938), p. 184.
[51]William Carroll Bark, *Origins of the Medieval World* (Stanford: Stanford University Press, 1958), pp. 27-28.
[52]Rushdoony, *The One and the Many: Studies in the Philosophy of Order and Ultimacy,* pp. 187-188.
[53]Bark, p. 65.
[54]Rushdoony, *World History Notes,* pp. 147-148.

began with the birth of modern humanism. Rather, the later Middle Ages was a period of slowly developing birth pangs.[55]

In addition to an anti-Christian presuppositional bias[56] for the use of the "Dark Ages" cliché, medieval historian Jean Gimpel (1918-1996) offers another reason for this misnomer:

> The reputation of the Middle Ages has never really recovered from the attack launched by the Renaissance upon the centuries that preceded humanism. Being passionately interested in the literature and poetry of classical civilization, people in the Renaissance were convinced that their forebears in the Middle Ages – later to be termed the Dark Ages – were altogether ignorant of or indifferent to ancient Greek and Roman authors, whereas in actual fact the medieval men were passionately interested – not so much in the literature and poetry of the classical world, as in its philosophical, scientific, and technological works.[57]

3.10 THE MEDIEVAL INDUSTRIAL REVOLUTION

James Burke (1936-), British science popularizer, in his witty and informative *Connections* – a book in which he examines the ideas, inventions, and "coincidences" that have culminated in the major technological achievements of today – denotes the period between the 10th and 14th centuries as the "Medieval Industrial Revolution."[58]

During this period of time, technology progressed considerably. Professor Lynn T. White (1907-1987) of the University of California at Los Angeles has done considerable research into this and observes:

> Indeed, the technical skill of classical times was not simply maintained: it was considerably improved. Our view of history has been too toplofty.... In technology, at least, the Dark Ages mark a steady and uninterrupted advance over the Roman Empire.[59]

In the masterful work by Frances and Joseph Gies, entitled *Cathedral, Forge, and Waterwheel: Technology and Invention in the Middle Ages*, the authors reflect on this technological innovation:

> Today we recognize that one of the great technological revolutions took place during the medieval millennium with the disappearance of mass slavery, the shift to water- and wind-power, the introduction of the open-field system of agriculture, and the importation, adaptation, or invention of an

---

[55]Schaeffer, *How Should We Then Live? The Rise and Decline of Western Thought and Culture*, p. 48.
[56]Because it was a Christian era, the unbelieving historians of modernity wrongly ascribe to it scientific, mathematical, and intellectual sterility.
[57]Jean Gimpel, *The Medieval Machine: The Industrial Revolution of the Middle Ages* (New York: Penguin Books, 1976), p. 237.
[58]James Burke, *Connections* (New York: Little, Brown and Company, 1978), p. 89.
[59]Lynn T. White, "Technology and Invention in the Middle Ages," *Speculum*, 15 (1940), 151.

array of devices, from the wheelbarrow to double-entry bookkeeping, climaxed by those two avatars of modern Western civilization, firearms and printing.[60]

Professor Lynn T. White documents between the 6th and 10th centuries that inventions rapidly altered life, provided the base for urbanization, and increased food supplies.[61] Some key inventions during this period were the plough, the horse collar and the horseshoe, the windmill and watermill, and the crank.[62]

A southward drift of the glacial front that had commenced in the 5th century (just after the fall of Rome) reversed itself in the middle of the 8th century.[63] As the frost retreated, northern Europe became more hospitable to agriculture. The introduction of the three-field farming system in the 8th century greatly enhanced agricultural production. Combined with the invention of the heavy plough in the 6th century, the invention of the padded horse collar in the 7th century, the organization of the self-contained tenant-farmed estate in the 8th century,[64] and the invention of the iron horse shoe in the early 9th century, the three-field system's advantages were, quite naturally, three-fold:

1. Only one-third of the land lay fallow in any year, as against one-half with the old Roman two-field system. This meant that a higher proportion of land was under cultivation each year.

2. Two crops could now be harvested at different times of the year. This was insurance against crop failures and spread the ploughing more evenly throughout the year.

3. It allowed farmers who wanted to plough with the more efficient horse (made viable by the invention of the padded horse collar and iron horseshoes) to have a spring crop of oats (for horse feed).

---

[60]Frances and Joseph Gies, *Cathedral, Forge, and Waterwheel: Technology and Invention in the Middle Ages* (New York: Harper Collins, 1994), p. 15.

[61]Lynn T. White, *Medieval Technology and Social Change* (Oxford: Clarendon Press, 1962). See also Gimpel and Gies.

[62]For the plough, see White, *Medieval Technology and Social Change*, pp. 41-56. For the horseshoe, see pp. 57-69. For the windmill and watermill, see pp. 79-103. For the crank, see pp. 103-134.

[63]Since the Bible reveals God to be in control of all aspects of the weather (see Job 37:9-13), we cannot help but to discern both His providential judgments and blessings in its apparent (to us, that is) vagaries. It is also important to note that in the 6th century, Europe was still reeling from the Barbarian invasions and the 7th century saw repeated attacks from Muslim hordes. Little, if any, effective progress can be made in either science or mathematics (or learning in general) during times of political chaos, military invasion, and/or food production/distribution snarls. The late 8th and early 9th century saw a providential respite in the Carolingian Renaissance under the tutelage of Charlemagne the Great (768-814).

[64]This was a market-based economy that took the place of a marginal and declining slave economy inherited from the Romans. Tenants performed farm labor by dividing their time between their lord's land and their own small property. Both free tenants and unfree (called serfs or villeins) had a recognized right to the use of their land for their own welfare (which was also inheritable).

3.10.1 THREE-FIELD FARMING

Field	Year One	Year Two	Year Three
Field 1	Winter planting	Spring planting	Fallow
Field 2	Spring planting	Fallow	Winter planting
Field 3	Fallow	Winter planting	Spring planting

Concomitant with the clearing of much of Northern Europe's forests, the increased food production generated from such agricultural innovation resulted in a significant rise in population.[65]

The northern region of Europe possessed, in addition to abundant forests, fast-growing vegetation, obtainable metal ores, and numerous rivers and streams, many swift flowing and ice free, with potential beyond mere transportation and communication. A multitude of waterwheels began to dot the European landscape beginning in the 5th century. By the 11th century, according to the *Domesday Book,* England alone had six thousand grain mills in three thousand locations.[66] During the Viking invasions in the 9th century the camshaft (initially a 3rd century BC Greek invention) appeared on the scene. This mechanism could transform circular motion into linear motion and vice versa operating on the simple binary (up/down) principle. A cam on the driveshaft can operate many machines (e.g., a trip hammer for crushing metal ores or a bellows for use in a blast furnace where the metal would be smelted) and generate a multiplicity of products and services. It could raise water from wells, operate knives and saws, crush malt for beer, convert hemp to linen, forge hammers, mill oil and silk, crush sugar cane, tan animal hides, grind stones, crush ore, operate lathes, string wire, mint coins, split metal, manufacture paper from linen, and increase the weight and bulk of cloth by shrinking and beating (called fulling).

The hydropowered sawmill, along with its multitasking camshaft, touched the medieval man in every area of his life. The house he lived in, the flour he ate, the oil he put on his bread, the leather of the shoes he put on his feet and the textiles he wore on his back, the iron of his tools, and the paper he wrote on all were produced in part with the aid of waterpower.

Science historian Alistair C. Crombie (1915-1996) delineates the impor-tance of the invention of the crank:

> The combined crank and connecting-rod was a medieval invention ... With the crank it became possible for the first time to convert reciprocat-ing into rotary motion and *vice versa,* a technique without which modern machinery is inconceivable.[67]

---

[65]Seventy million by 1300. See Gimpel, p. 57.
[66]The *Domesday Book* is the written record of a census and survey of English landowners and their property made by order of William the Conqueror in 1085-1086.
[67]Alistair C. Crombie, *The History of Science from Augustine to Galileo* (New York: Dover Publications, [1959, 1970, 1979] 1995), 1:203.

As Stanley L. Jaki concludes, "It can indeed be said without exaggeration that the Western world lived until the advent of the steam engine on technological innovations made during the medieval centuries."[68]

Jean Gimpel remarks on the impact of the camshaft on Western Civilization:

> The engineers of the classical world – men like Hero of Alexandria – knew the use that could be made of the cam, but applied it only to animate toys or gadgets. Although the Chinese operated trip-hammers for hulling rice as early as A.D. 290, the use of the cam evidently failed to spread to other industries in the following centuries. In fact, it is a feature of Chinese technology that its great inventions – printing, gunpowder, the compass – never played a major evolutionary role in Chinese history. The introduction of the cam into medieval industry, on the other hand, was to make an important contribution to the industrialization of the Western Hemisphere.[69]

Alistair C. Crombie compares the technical devices of the Greeks with medieval Christendom:

> ... it is characteristic of medieval Christendom that it put to industrial use technical devices which in classical society had been known but left almost unused or regarded simply as toys.[70]

Crombie continues his analysis:

> The mechanical devices and instruments invented in classical times, pumps, presses and catapults, driving wheels, geared wheels and trip hammers, and the five kinematic 'chains' (screw, wheel, cam, ratchet and pulley) were applied in the later Middle Ages on a scale unknown in earlier societies.[71]

Frances and Joseph Gies make the following astute observation:

> Armed with innovative technology, both borrowed and homegrown, the European civilization that Edward Gibbon believed had been brought to a long standstill by "the triumph of barbarism and religion" had in reality taken an immense stride forward. The Romans so congenial to Gibbon would have marveled at what the millennium following their own era had wrought. More perceptive than Gibbon was the English scientist Joseph Glanville, who wrote in 1661: "The last Ages have shewn us what Antiquity never saw; no, not in a dream."[72]

---

[68]Stanley L. Jaki, *Christ and Science* (Royal Oak, MI: Real View Books, 2000), p. 22.
[69]Gimpel, pp. 13-14.
[70]Crombie, 1:196.
[71]*Ibid.*, 1:203.
[72]Gies, p. 16.

As Carol Cipolla observes, "What the Europeans sowed from the sixth to the eleventh centuries was not so much inventive ingenuity as a remarkable capacity for assimilation. They knew how to take good ideas where they found them and how to apply them on a large scale to productive activity."[73] Jean Gimpel describes the medieval soul, "The period was characterized by a sense of optimism, a rationalist attitude, and a firm belief in progress."[74] *Why this aptitude for progress and from whence comes this capacity to assimilate and innovate?* A sense of progress suggests a sense of history, something markedly missing among the Egyptians, Greeks, and Romans. "Lacking any objective understanding of the past – that is, lacking history," says D. S. L. Cardwell, "the hierarchical and slave-owning societies of classical antiquity failed to appreciate the great progress that had been achieved by and through technics."[75] Instead of looking forward, the ancients indulged in looking back to what they conjectured as a vanished "Golden Age." This basic worldview of history denies the very concept of forward progress. Frances and Joseph Gies credit the idea of progress to the biblical worldview of history, "The Christian Church, whose pioneering monastic orders made many practical and material contributions to medieval technology, also supplied a noncyclical, straight-line view of history that allowed scope for the idea of progress."[76]

Kenneth Scott Latourette (1884-1968), eminent historian of Christianity, comments on the significance of the monastic orders, particularly the Benedictine order that began in the 6th century:

> It is clear that through monasticism Christianity did something to give dignity to labour and added greatly to agriculture and so to the increase in the supply of food. Under the Benedictine rule work was obligatory. Although in many of the Benedictine houses food and clothing came from estates cultivated by serfs and while in several monastic orders manual work in the fields was assigned to lay brothers and the choir monks gave themselves to prayer and study, in others all the monks, even those of aristocratic birth, toiled in their gardens or on the lands of the monastery. Whether by all members of the community or only by the lay brothers, monasteries did much to clear land, bring it under cultivation, and develop improved crops and methods of tillage. The first use of marl to enrich the soil is attributed to them and they were noted for their vineyards and their wines.[77]

The dignity that the Benedictine monastic order (including the many derived from it) gave to labor, particularly manual labor in the fields, stood in striking contrast to the aristocratic conviction of the servile status of manual

---

[73]Carlo Cipolla, *Before the Industrial Revolution* (New York, 1980), p. 113. Cited in Gies, p. 41.
[74]Gimpel, p. ix.
[75]D. S. L. Cardwell, *Turning Points in Western Technology* (New York, 1972), p. 1. Cited in Gies, p. 288.
[76]Gies, p. 288.
[77]Kenneth Scott Latourette, *A History of Christianity* (New York: Harper & Row, [1953] 1975), 1:556-557.

work (illiberal) which prevailed in much of ancient society. This same menial attitude toward manual work was also prevalent among the warriors (knights) and the non-monastic ecclesiastics who constituted the upper middle classes of the Middle Ages. As Latourette notes, the monasteries were responsible for the clearing of land (and a concomitant reduction in the number of wild animals) and improvement in methods of agriculture. In the midst of the chaos of barbarism, the monasteries were centers of orderly and settled life. The monks were responsible for road building and road repair. Until the rise of the towns in the 11th century, they were pioneers in industry and commerce. The shops of the monasteries preserved the industries of Roman times. The French Cistercian monastic orders of the 11th through the 13th centuries led the way in the agricultural colonization of Western Europe. The Cistercian monasteries made their houses centers of agriculture and contributed to improvements in that occupation. With their lay brothers and their hired laborers, they became great landed proprietors. German monasteries also set advanced standards in agriculture and produced many artisans and craftsmen. Some serious doctrinal flaws did exist in monastic theology (e.g., advocacy of celibacy, the calling of God to the ecclesiastical life is the highest and holiest of callings). In spite of this shortcoming, Western civilization should thank the monasteries for leaving a preponderant inheritance. They were the precursors of hospitals, hotels, publishing houses, libraries, law courts, art academies, music conservatories, places of refuge, markets for barter and exchange, centers of culture, newspaper offices, orphan asylums, general stores, and, ultimately, through their educational endeavors, the university.

Alistair C. Crombie emphasizes the indispensable nature of the monasteries during the early stages of the Frontier Age:

> That so much was preserved in spite of the gradual collapse of Roman political organization and social structure under the impact, first, of Goths, Vandals and Franks, and then, in the 9th century, of Norsemen, was due to the appearance of monasteries with their attendant schools which began in eastern Europe after the foundation of Monte Cassion by St Benedict in 529 (here St Benedict had also established an infirmary. The care of the sick was regarded as a Christian duty for all such foundations). The existence of such centres made possible the temporary revivals of learning in Ireland in the 6th and 7th centuries, in Northumbria in the time of Bede, and in Charlemagne's empire in the 9th century.[78]

Because of an enduring philosophical prejudice and academic high-mindedness, most histories picture this era as savage, cruel, and primitive – a dead end in every respect. Jean Gimpel rebuts this caricature, one of the myths of history that has an unusually high endurance value:

---

[78]Crombie, 1:32. For a discussion of the impact of the Irish monasteries on progress of Western civilization, see Thomas Cahill, *How the Irish Saved Civilization: The Untold Story of Ireland's Heroic Role from the Fall of Rome to the Rise of Medieval Europe* (New York: Doubleday, 1995).

If this picture [of progress and development - J.N.] of the medieval world does not sound like the Dark Ages ... it is because the history of technology has been so universally neglected, thanks largely to the age-old attitude of academics and intellectuals toward manual work and engineering.[79]

Like any other era of history, savage acts of cruelty did exist. There were wars, famines, acts of injustice, confusion (doctrinal and societal), and disorder. But these things have been, and are common to *every* age. An analysis of the history of the 20th century should affirm this fact to any reader.[80] Note William Carroll Bark's riveting summary:

We know now that the Dark Age was not that dark. Ignorance, lethargy, and disorder existed then as now, but they were far from blighting an age eager for learning, vigorous in living and in expressing itself, and idealistically constructive. Perhaps it is not too much to say that medieval society was functional in ways not even dreamed of by antiquity and leading to ends beyond the imagination of earlier times. By "functional" I mean that it was a working, striving society, impelled to pioneer, forced to experiment, often making mistakes but also drawing upon the energies of its people much more fully than its predecessors, and eventually allowing them much fuller and freer scope for development. That conditions, events, and peoples came together as they did in the early Middle Ages was extremely fortunate for the present heirs of the Western tradition.[81]

In regards to the Middle Ages, we must re-evaluate it from a new perspective, especially when we look at what happened to mathematics and science.

3.11 THE STATUS OF MATHEMATICS AND SCIENCE

We have already made note of the fact that the mathematical heritage bequeathed to Europe came from Greece via the auspices of the Roman Empire. Reijer Hooykaas (1906-1994), professor of History and Science at the University of Utrecht, the Netherlands, comments about this heritage:

For the building materials of Science (logic, mathematics, the beginning of a rational interpretation of the world) we have to look to the Greeks; but the vitamins indispensable for a healthy growth came from the biblical

---

[79]Gimpel, p. x.
[80]See Gil Elliot, *Twentieth Century Book of the Dead* (New York: Charles Scribner's Sons, 1972) and Paul Johnson, *Modern Times: The World from the Twenties to the Nineties* (New York: Harper & Row, [1983] 1991). According to one-time Russian exile Aleksandr Solzhenitsyn (1918-), at the height of the Spanish Inquisition (late Middle Ages) about ten persons per month were executed. During the eighty years before the Russian revolution seventeen persons per year were executed. In the first two years of Lenin's revolution more than one thousand persons per month were executed without due process of law. At the height of Stalin's terror an estimated forty thousand persons per month were executed. See Aleksandr Solzhenitsyn, "America: You Must Think About the World," *Solzhenitsyn: The Voice of Freedom* (June 30, 1975), p. 9.
[81]Bark, pp. 99-100.

concept of creation. The fact that the victory of Christianity did not bring an immediate liberation from the bonds of Greek metaphysics in no way disproves this statement. The compromise of Christian religion, first with Platonism, then with Aristotelianism, strongly influenced not only secular learning but also theology.[82]

Rousas J. Rushdoony remarks, "NeoPlatonic ... contempt of the world and of material things ... infected every area of the church in the West, but it was fought bitterly in every area also."[83]

Lynn Thorndike (1882-1965), medieval science historian, observes that the scientists of this era:

> ... had to struggle against a huge burden of error and superstition which Greece and Rome and the Arabs handed down to them; yet they must try to assimilate what was of value in Aristotle, Galen, Pliny, Ptolemy, and the rest. Crude naïve beginners they were in many respects. Yet they show an interest in nature and its problems; they are drawing the line between science and religion; they make some progress in mathematics, geography, physics and chemistry; they not only talk about experimental method, they actually make some inventions and discoveries of use in the future advance of science. Moreover, they themselves feel that they are making progress. They do not hesitate to disagree with their ancient authorities, when they know something better.[84]

Concerning the Middle Ages, William M. Ivins, Jr. believes that it "acted as a catharsis that was necessary to bring into existence another set of basic postulates and to make possible new vision and new thought."[85] He views this period as the "long struggle of Western Europe to free itself from the inhibiting burden of the Greek tradition."[86]

German historian Oswald Spengler (1880-1936) remarks that the "history of Western knowledge is thus one of *progressive emancipation* from Classical thought."[87] Alistair C. Crombie links this emancipation not only to the technological innovations of the time, but to the scientific instruments of the future:

> In Western Christendom during the early Middle Ages men were concerned more to preserve the facts which had been collected in classical

---

[82]Hooykaas, p. 85.
[83]Rushdoony, *World History Notes*, p. 131. As we have seen, some fruits of neo-platonism in the medieval church would be its occasional asceticism and the sacerdotal practice of celibacy.
[84]Lynn Thorndike, *A History of Magic and Experimental Science: During the First Thirteen Centuries of Our Era* (New York: Columbia University Press, 1923), 2:979. The "line in the sand" dividing science from religion was a *functional* separation; this does not mean that these seminal medieval thinkers separated science from the Creator God.
[85]Ivins, p. 56.
[86]*Ibid.*, p. 113.
[87]Oswald Spengler, "The Meaning of Numbers," *The World of Mathematics*, ed. James R. Newman (New York: Simon and Schuster, 1956), 4:2335.

times than to attempt original interpretations themselves. Yet, during this period, a new element was added from the social situation, an activist attitude which initiated a period of technical invention and was to have an important effect on the development of scientific apparatus.[88]

Alfred Crosby, a sagacious and perceptive historian, amplifies Crombie's observations:

> Westerner's advantage, I believe, lay at first not in their science and technology, but in their utilization of habits of thought that would *in time* enable them to advance swiftly in science and technology and, in the meantime, gave them decisively important administrative, commercial, navigational, industrial, and military skills. The initial European advantage lay in what French historians have called *mentalité* ... these people were thinking of reality in quantitative terms with greater consistency than any other members of their species.[89]

It was the slow but certain maturation of a new corporate mental frame of reference, engendered by the progressive prospects acquired from the biblical view of linear time and an appreciation of a rational creation reflective of "measure, number and weight" (Wisdom of Solomon 11:20-21),[90] that laid the foundation for an explosion in quantitative analysis. Aristotle, "the Philosopher," as medieval Europe called him, had deemed scientific analysis more useful in qualitative terms than in quantitative ones. Although Augustine emphasized qualitative analysis (the sacramental nature of creation and numbers as symbolic of spiritual truths or moral lessons), he did forge a bridge between the quantitative attributes of the created order and God, as the Supreme Craftsman of that order. He said, "In everything where you find measures, numbers and order, look for the craftsman. You find none other than the One in whom there is supreme measure, supreme numericity, and supreme order, that is God, of whom it is most truly said that He arranged everything according to measure, and number, and weight."[91] During the one thousand-year interval after this statement by Augustine, the reality of the Christian Gospel broke the bonds of Greek metaphysics and eventually set man free to develop mathematics on an immense scale. Although this emancipation process was slow (with multiple pe-

---

[88]Crombie, 1:25.

[89]Alfred W. Crosby, *The Measure of Reality: Quantification and Western Society, 1250-1600* (Cambridge: Cambridge University Press, 1997), pp. x-xi.

[90]In Latin, "you have ordered all things in measure, number, and weight" is *"omnia in mensura, numero et pondere disposuisti."* This phrase was the most often quoted and alluded to phrase in Medieval Latin texts. See E. R. Curtius, *European Literature and the Latin Middle Ages*, trans. W. R. Trask (London: Routledge and Kegal Paul, 1953), p. 504. See also Ivor Grattan-Guiness, *The Rainbow of Mathematics: A History of the Mathematical Sciences* (New York: W. W. Norton, [1997] 2000), p. 127.

[91]Cited in Jaki, *Science and Creation: From Eternal Cycles to an Oscillating Universe*, p. 181. Jaki cites his source as *De Genesi contra Manicheos*, Book I, chap. 16, Latin text and French translation in *Œuvres complètes de Saint Augustin*, ed. Péronne, et al. (Paris: de Louis Vives, 1873), 3:440-441.

riods of advances and retreats), Alfred Crosby documents the distinctive intellectual achievement of the West:

> The record indicates that cycles of advance and retreat, in this case of combining abstract mathematics and practical measurement, and then of nodding and napping and forgetting, is the norm of human history. The West's distinctive intellectual accomplishment was to bring mathematics and measurement together and to hold them to the task of making sense of a sensorially perceivable reality, which Westerners, in a flying leap of faith, assumed was temporally and spatially uniform and therefore susceptible to such examination.[92]

From one perspective, one might comment, like Howard Eves, that "very little in mathematics ... was accomplished in the West."[93] From another perspective, one might comment, like reformed theologian Gary North, that "however ignorant of theology they may have been, however erroneous in their perception of things spiritual, not to mention things scientific, they nevertheless succeeded in reshaping the history of mankind."[94]

In the early part of this era of emancipation, the usage of mathematics was primarily limited to the needs of trade, keeping accounts, keeping time by the hour (for appointed seasons of prayer as per King David's example in Psalm 119:164), and reckoning calendar dates.[95] The monks of the ecclesiastical monasteries, who diligently copied both Greek and Scripture manuscripts, were the "mathematics professors" of their day.

The first recorded examples of the weight-driven mechanical clock are found in the 14th century. This clock probably appeared much earlier on the European landscape. Whether the motivation for its development was governed by economic or ecclesiastical (e.g., monastic) concerns is up for debate. We do know that these clocks were immediately "referred to as small replicas of the Creator's great clockwork, the universe."[96] We also know that medieval engineers perfected its foundational principle, the mechanical escapement, as a result of the transformation of accelerated motion, the falling of weights, to circular motion at a constant velocity. These unknown tinkers had no idea of the decisive role that their invention would play in the development of mathematics in Western civilization. These clocks (which divided the hour into sixty minutes, the minute into sixty seconds) "completed the first stages in the scientific measurement of time, without which the later refinements of both physics and machinery would scarcely have been possible."[97]

---

[92]Crosby, p. 17.
[93]Eves, pp. 207-208.
[94]Gary North, *Backward Christian Soldiers* (Tyler, TX: Institute for Christian Economics, 1984), pp. 131-132.
[95]See David E. Duncan, *Calendar: Humanity's Epic Struggle to Determine a True and Accurate Year* (New York: Avon Books, 1998).
[96]Jaki, *Science and Creation: From Eternal Cycles to an Oscillating Universe*, p. 243.
[97]Crombie, 1:219.

3.12 MUSLIM AND BYZANTINE INPUT

In the monasteries of Europe, usually only the clergy had access to Greek manuscripts. Two other civilizations, the Byzantine and the Saracen, also preserved these writings.

In 640, Muslim armies ransacked Egypt forcing any remaining Greek scholars in Alexandria to migrate to Constantinople, the capital of the Byzantine Empire. In 641, an Arab commander gave an order that the Alexandrian library be burnt to the ground. The local populace made such energetic protests that the commander referred the matter to the local caliph, named Omar. The caliph sent a reply, "As to the books you have mentioned, if they contain what is agreeable with the book of God, the book of God is sufficient without them; and, if they contain what is contrary to the book of God, there is no need for them; so give orders for their destruction."[98] The books were placed in the public baths of the city and it took six months to consume all of them. In spite of this unfortunate incident, the Muslims generally respected and absorbed Greek learning. They were also responsible for introducing to Europe several valuable concepts that they borrowed from the Hindus of India: the base 10 decimal system, positional notation,[99] the number zero, and negative numbers.

The decimal system of writing numbers (e.g., 1, 2, 3, 4, etc.) proved to be much more efficient than the old, cumbersome Greek letters or Roman numerals. This method of number writing acted as an indispensable asset in later developments of European mathematics and science.

Beginning in the 7th century, the Hindus began working with the number zero, negative numbers, and the idea of positional notation. All three of these concepts were foreign to Greek mathematics.

Arabic mathematicians not only enhanced the rudiments of algebra, but gave it the name. In 830, the astronomer Mohammed ibn Musa al-Khowarizmi (ca. 780-ca. 850) wrote a book titled *Hisab al-jabr w'al-muqabalah*, which literally means "science of reunion and opposition" or more freely "science of transposition and cancellation."[100] The title refers to the two principle operations that al-Khowarizmi used in solving an equation:
1. "al-jabr" – the transposition of terms from one side of an equation to the other.
2. "al-musqabalah" – the cancellation of equal

Figure 33: Mohammed ibn Musa al-Khowarizmi

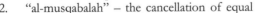

[98]Cited in Ball, p. 115.

[99]Positional notation is a system of writing numbers in which the position of a digit affects its value.

[100]Eves, p. 193. An English translation was compiled long ago under the title, *The Algebra of Mohammed ben Musa*, ed. and trans. F. Rosen (London: Printed for the Oriental Translation Fund, 1831). al-Khowarizmi's work was very popular among medieval scholars and they slightly changed his name to "algorithm" in their representation of the subject matter.

terms appearing on opposite sides of the equation.

Given an equation (using modern symbolic algebra):

$$2x + 3 = 4x^2 + 2x - 9$$

Applying "al-jabr" transforms the equation to:

$$2x + 12 = 4x^2 + 2x$$

Applying "al-musqabalah" transforms the equation to:

$$12 = 4x^2$$

This text became known in Europe through Latin translations and was shortened to the one word, "algebra," which became synonymous with the science of equations.[101] Both Hindu and Arab algebraists were still tied to the syncopated algebra of Diophantus. They did not use simple symbols (e.g., x, y, etc.) but words, or at best, abbreviations of words.

Omar Khayyam (ca. 1050-1122), Persian mathematician, astronomer, and author of one of the world's best-known works of poetry, *Rubáiyát*, wrote an original work on algebraic equations and made a systematic effort at correlating algebra with geometry.[102] In the area of trigonometry, Al-Battani (877-918) displayed much original thinking and derived the formulae $\sin x = \dfrac{\tan x}{\sqrt{1 + \tan^2 x}}$ and $\cos x = \dfrac{1}{\sqrt{1 + \tan^2 x}}$. It was through his influence that many of the Hindu contributions to trigonometry became established. Abu-al-Wafa's (940-ca. 997) creative trigonometric work unearthed a very important relationship: sin (A + B) = sinAcosB + sinBcosA.[103]

Despite their contributions in many areas, the Hindu and Muslim methods of multiplying and dividing were still in the very crude stage. The modern methods that we now employ had to wait for the European innovations of the later Middle Ages. Reflecting on this innovation, Alistair C. Crombie remarks, "This made division into an ordinary matter for the counting house, whereas it had formerly been a formidably difficult operation even for skilled mathematicians."[104]

Arabic translations, which included many Greek works and especially those of Aristotle and the innovations of the Hindus, became known to European merchants and traders through their mercantile transactions and also through the unfortunate auspices of the Crusades (1095-1272). According to Alistair C. Crombie:

> In the field of mathematics the Arabs transmitted to Western Christendom a body of most valuable knowledge which had never been available to the Greeks, though here the Arabs were not making an original contribution

[101]Eves, p. 19.

[102]See *The Algebra of Omar Khayyam*, trans. David S. Kasir (New York: Columbia University, 1931).

[103]Carra de Vaux, "Astronomy and Mathematics," *The Legacy of Islam*, ed. Sir Thomas Arnold and Alfred Guillaume (London: Oxford University Press, 1931), pp. 389-390.

[104]Crombie, 2:22.

but simply making more widely known the developments in mathematical thought which had taken place among the Hindus.[105]

Byzantine preservations were introduced later into Europe as a result of the conquest of that empire by the Turks in 1453. Alistair C. Crombie notes the impact that these Greek and Arabic translations had on European scholars of the 13th century:

> The Western scholars were trying to make the natural world intelligible and they seized upon the new knowledge [the Greco-Arabic scientific system] as a wonderful, but not final, illumination of mind and as a starting-point for further investigation.[106]

3.13 THE UNIVERSITY

In Europe, the entrance of these Arabic manuscripts created fresh interest in science and mathematics. Late in the 12th century, a phenomenon unique to Europe appeared on the scene, the university. Known then as *universitas magistrorum et scholarium*, these schools find their roots in the earlier monastic schools and developed as a sort of guild or trades union. Originally, these schools owned no real estate. According to David C. Lindberg, "It is important to note ... that a university was not a piece of land or a collection of buildings or even a charter, but an association or corporation of teachers (called "master") or students."[107] The Seven Liberal Arts formed the foundational core of the curriculum. It was rounded out by three philosophy courses: moral philosophy, natural philosophy, and metaphysics. Courses in medicine, law, and theology constituted advanced studies.

Although not always theologically or scholarly accurate, what undergirded the university was the idea that all the subjects were united by an all-encompassing worldview and it was Christianity that provided this unity for all departments of study.[108] The university could arise and flourish in the late Middle Ages only because this civilization believed in a cultural unity or dogma; a worldview in which every aspect of life had its place within a unifying theological framework.

Alistair C. Crombie notes, "... it was within a general framework of philosophy closely bearing on theology, and specifically within the system of uni-

---

[105]*Ibid.*, 1:65.

[106]*Ibid.*, 1:80.

[107]Lindberg, p. 208.

[108]Theology (the study of God) was believed to be foundational – the queen of the sciences (knowledge). Today, pluralism and fragmentation are demolishing the underlying philosophical foundation of the university. These schools have become intellectual smorgasbords where a student picks up bits of knowledge from the Physics department, bits from the History department, bits from the English department, etc. There is little or no attempt to relate the parts to the whole. The universities of modernity teach masses of detail without focus. They have, in reality, become multiversities because there is no longer any "uni" (unity; i.e., the biblical God and His revelation) to the "versity" (diversity; i.e., academic disciplines).

versity studies run by clerics, that the central development of medieval science took place."[109] Friedrich Heer confirms Crombie's observation:

> In the universities were laid the foundations of the scientific culture of our modern world, in them grew up the habit of disciplined thinking, followed by systematic investigation, which made possible the rise of the natural science and of the technical civilization necessary to large industrial societies.[110]

3.14 THE SCHOLASTICS AT OXFORD

Scholars in these universities were known as scholastics. In the 13th century, the best mathematicians of Europe came out of Oxford University in England.[111]

One was Robert Grosseteste (1170-1253). He claimed that nothing could be known in science without good knowledge of mathematics, by which he meant geometry.[112] He made the appeal to observational experience, not just book learning, as the true basis for science. Here, he made a revolutionary break with Greek views as Friedrich Heer reveals:

> Western thinkers were now submitting Greek views to the test of experiment, which meant making use of those manual arts the Greeks themselves had disdained as servile.[113]

Due to the influence of Platonic realism and Aristotelian dogma, the Greeks disdained "messing with Mother Nature." Reijer Hooykaas amplifies the practical implications of this philosophy:

> The learned cultivators of mathematics and theoretical mechanics considered it below their dignity to occupy themselves with the practical applications of their inventions.... Moreover, to philosophers of the Platonic school the investigation of *material* things was inferior to the pursuit of *spiritual* things. Manual work, even for a scientific end, was considered beneath the dignity of the philosopher.[114]

The Greeks thought it to be an utmost act of impiety to trespass and violate the bounds of the eternal world order. Their "rationalism, the deification of nature ... disregard for manual work were all factors which militated against the

---

[109]Crombie, 2:126. It must be noted that much of the *application* of the academic sciences, such as astronomy (the determination of the calendar and its reform), arithmetic (the work of the exchequer and of local business), anatomy, physiology and chemistry (surgery and medicine), were put into practice *outside* the universities. The development of technology, art, and architecture (all of profound importance for science) also originated *outside* the university system.
[110]Heer, p. 235.
[111]*Ibid.*, p. 252.
[112]*Ibid.*, p. 295.
[113]*Ibid.*
[114]Hooykaas, p. 78.

use of experiments."[115] In the 13th century, the Oxford scholars were making a clear break with this tradition.

In the late 13th century, a decision made by Etienne Tempier (Bishop of Paris) encouraged many others to break with Aristotle. Scholastic followers of the Muslim scholar Averroës (1126-1198) had taken a stand on the "irrefutable rational truth of Aristotelian philosophy and accepted the consequence that Christian theology was irrational or even untrue."[116] Tempier condemned this "determinist" interpretation in 1277.[117] David C. Lindberg comments, "The position of Tempier ... was that Aristotle and the philosophers must not be allowed to place a lid on God's freedom or power to act...."[118] With the rejection "that Aristotle had said the last word on metaphysics and natural science, the bishops in 1277 left the way open for criticism which would, in turn, undermine his system."[119] Natural philosophers "because of the attitude of Christian theologians ... were made free to form hypotheses regardless of Aristotle's authority, to develop the empirical habit of mind working within a rational framework, and to extend scientific discovery."[120]

Roger Bacon (1214-1291), another key Oxford scholar, continued to encourage this freedom of inquiry, observation, and experimentation. He was convinced that *the* door to verification in science could be found *only* through the key of experiments. Hence, he challenged the accepted dogmas of Aristotelian physics at every level. Insatiably curious, he studied the rainbow, diagrammed the anatomy of the human eye, and even devised a secret formula for gunpowder. Two hundred years before Leonardo da Vinci (1452-1519), he predicted the invention of the telescope, eyeglasses, airplanes, high-speed engines, self-propelled ships, and power motors. He also noted flaws in the way the current calendar kept time; flaws which were finally corrected by Pope Gregory XIII in 1582. He maintained that the study of God was the goal of all learning and did not disregard the first chapter of Genesis where man was directed to take dominion over and subdue all of God's creation (see Genesis

Figure 34: Roger Bacon

---

[115]*Ibid.*, p. 82.

[116]Crombie, 1:76.

[117]Other Aristotelian propositions condemned were the eternity of the world, denial of personal immortality, denial of divine providence, denial of free will, that secondary causes of natural events are autonomous (i.e., they could continue to act even if the first cause, God, ceased to participate), the astrological ideas that the heavens influence the soul as well as the body, that celestial bodies are moved by intelligences (or souls), and that time is cyclical.

[118]Lindberg, p. 238.

[119]Crombie, 1:78-79.

[120]*Ibid.*, 1:79.

1:26 and Psalm 8:6). To him, mathematics was a handmaid to the natural sciences and theology.[121]

In his *Opus Majus* (published in 1267), Part 4, he proclaimed:

> Mathematics is the gate and key of the sciences.... Neglect of mathematics works injury to all knowledge, since he who is ignorant of it cannot know the other sciences or things of this world.[122]

As Alistair C. Crombie reflects, "With Roger Bacon the programme for mathematicizing physics and a shift in the object of scientific inquiry from the Aristotelian 'nature' or 'form', to laws of nature in a recognizably modern sense, becomes explicit."[123]

3.15 A CHAMPION OF ARISTOTLE

Thomas Aquinas (1226-1274), Dominican theologian, did most of his work out of the University of Paris. He disagreed with Roger Bacon and sought to reconcile Aristotelian thought with the Christian faith; i.e., to reconcile reason with revelation.[124] In *Summa Theologica* (1266-1273), he used deductive reasoning to its fullest to construct a massive theological system that sought to account for and give answer to every possible question that could ever be raised about any aspect of life. In other words, Aquinas embarked on the noble task of worldview analysis and application.[125] The axioms that Aquinas used were a synthesis of biblical principles, church tradition, and Aristotelian philosophy. He and others engendered much debate and controversy over the merits of Aristotelian thought, revelation from Scripture, and the function and power of human reason. Alistair C. Crombie comments on the basis of the "double-revelation" theory of Aquinas:

Figure 35: Thomas Aquinas

> St Thomas realized, as Adelard of Bath had done a century earlier, that theology and natural science often spoke of the same thing from a different point of view, that something could be both the work of Divine Providence and the result of a natural cause. In this way they established a dis-

---

[121]Robert Steele, "Roger Bacon and the State of Science in the Thirteenth Century," *Studies in the History and Method of Science*, ed. Charles Singer (New York: Oxford University Press, 1921), p. 145.

[122]Cited in Robert Edouard Moritz, *On Mathematics and Mathematicians* (New York: Dover Publications, 1958), p. 41.

[123]Crombie, 2:39.

[124]Aquinas disagreed with Aristotle in three key areas: the existence of a transcendent God, the creation of the world out of nothing, and the freedom of man rooted in the immortality of his soul.

[125]He did this at the behest of Pope Urban IV (died 1264) who reminded scholars like Aquinas of the decree of Pope Gregory IX (ca. 1147-1241) in 1231 which, while forbidding the Arabic deterministic embodiment of Aristotelianism, called for the scholastics to interpret Aristotle for the Christian faith.

tinction between theology and philosophy which assigned to each its appropriate methods and guaranteed to each its own sphere of action. There could be no real contradiction between truth as revealed by religion and truth as revealed by reason.[126]

For the historian of science, the emphasis of Aquinas on the contingency of the universe and man's cognitive unity with it is of supreme importance. The concept of contingency should be understood to mean "an ultimate dependence of something upon someone else." In context, this means that the operation of the universe is dependent upon the Creator (a rejection of the *a priori* determinism of Aristotle). The concept of cognitive unity means that the rational order of creation can be understood by the reason of man. Stanley L. Jaki reflects on the consequences of these two emphases of Aquinas:

> The contingency of the universe obviates an a priori discourse about it, while its rationality makes it accessible to the mind though only in an a posteriori manner. Hence the need for empirical investigations. The contingency of the universe as a whole serves in turn as a pointer to an ultimate in intelligibility which though outside the universe in a metaphysical sense, is within the inferential power of man's intellect.[127]

Hence, even though Aquinas disagreed with Roger Bacon on the place of Aristotelian thought in the Christian understanding of reality, they both agreed on the necessity of empiricism as a tool of scientific analysis.

3.16 FORERUNNERS OF INNOVATIVE THOUGHTS

Leonardo of Pisa (ca. 1170-1250), also called Fibonacci, was the most talented mathematician of the Middle Ages.[128] His father was a mercantile businessman and as a boy Leonardo spent time in various places like the north coast of Africa, Egypt, Sicily, Greece, and Syria. Due to his father's occupation and contacts, arithmetic roused his interests. His most famous work was titled *Liber Abaci* (1202). This book exerted great influence in Europe, especially toward the acceptance of the Hindu-Arabic notation for numbers.[129] A celebrated numerical sequence, called the Fibonacci sequence, is named in honor of him (see section 7.4.2). He also made a small step toward symbolic algebra when in one

Figure 36: Leonardo of Pisa

---

[126]Crombie, 1:78.

[127]Jaki, *The Road of Science and the Ways to God*, p. 38. The emphasis of Aquinas on the power of man's reason (he believed that it was unaffected by the fall) runs counter, as many Protestant theologians believe, to biblical revelation (we will return to this in section 5.1).

[128]Eves, p. 210.

[129]During the 13th and 14th centuries, the popularity of almanacs and calendars helped spread the knowledge of Hindu-Arabic numerals throughout Western Europe.

instance he used a letter in place of a number in his algebra, but he did not vigorously pursue this innovative idea.

Thomas Bradwardine (1290-1349), a theologian known as Doctor Profundus, was one of the most prominent Oxford mathematicians of the 14th century. He wrote four works on mathematics dealing with a wide variety of topics: the theory of numbers, stellar (star shaped) polygons, isoperimetric figures,[130] ratio and proportion, irrational numbers, and loci in space. He also made halting, but important steps in the development of symbolic algebra. According to Alistair C. Crombie:

> Two main methods of expressing functional relationships were developed. The first was the 'word-algebra' used in mechanics by Bradwardine at Oxford, in which generality was achieved by the use of letters of the alphabet instead of numbers for the variable quantities, while the operations of addition, division, multiplication, etc. performed on these quantities were described in words instead of being represented by symbols as in modern algebra.[131]

Oxford scholars Richard of Wallingford (1292-1336) and John Manduith (ca. 1320), both lecturers in trigonometry, headed up an important school of astronomy at Merton College. It was these two men that were the progenitors of Western trigonometry. They initiated the Hindu-Arabic practice, already found in the *Toledan tables* of Arab astronomer al-Zarqali and other astronomical tables in wide circulation, of basing plane trigonometry on sines instead of chords, as had been done in the old Greco-Roman tradition dating from Hipparchus.[132]

Jean Buridan (ca. 1295-1358) of Sorbonne, France, anticipated Newton's first law of motion (a body will continue to be in a state of rest or of uniform velocity unless acted upon by an external force) by three centuries by striking at the root of Aristotle's theory of the eternity of motion. He articulated his rudimentary theory of inertia, also an echo of the 6th century work of Philoponus, in his commentary on Aristotle's *Physics*:

> ... God, when He created the world, moved each of the celestial orbs as He pleased, and in moving them He impressed in them impetuses which move them without His having to move them any more except by the method of general influence whereby He concurs as a co-agent in all things which take place.... And these impetuses which He impressed in the celestial bodies were not decreased nor corrupted afterwards, because there was no inclination of the celestial bodies for other movements. Nor was there resistance which would be corruptive or repressive of that impetus.[133]

---

[130]Isoperimetric figures are figures with equal perimeters.
[131]Crombie, 2:101.
[132]See *Ibid.*, 1:110.
[133]Cited in Marshall Clagett, *The Science of Mechanics in the Middle Ages* (Madison: University of Wisconsin Press, 1959), p. 536.

Although there is much that is good and much that is incorrect in this analysis (you cannot expect perfection in initial formulations), what strikes Aristotle at his roots is Buridan's assertions that motion, like the heavens, had a beginning. For Aristotle and Ptolemy the concept of an absolute beginning of time and creation was inconceivable given their *a priori* commitment to the eternity of the heavens. For Buridan, to believe in an absolute beginning was a natural expression of the truths of the first chapter of Genesis.

Nicole Oresme (ca. 1323-1382), Bishop of Lisieux and a teacher in the Parisian College of Navarre, developed many germinal concepts. A disciple of Buridan and his initial successor at Sorbonne, he denied the heavens were eternal, asserting instead that the heavens had a beginning due to God's creative fiat.[134] Responding to Aristotle's insistence on the perfection of the heavens, Oresme emphasized that the perfection of the laws of nature was merely a modest reflection of the infinitely perfect attributes of the Creator.[135] He especially noted that the clockwork precision of the heavenly motions evidenced God's power and wisdom.[136] Note the innovating, mechanistic ring in his challenge to the celestial intelligences of Aristotle:

> If we assume the heavens to be moved by intelligences, it is unnecessary that each one should be everywhere within or in every part of the particular heaven it moves; for, when God created the heavens, He put into them motive qualities and powers just as He put weight and resistance against these motive powers in earthly things. These powers and resistances are different in nature and in substance from any sensible thing or quality here below. The powers against the resistances are moderated in such a way, so tempered, and so harmonized that the movements are made without violence; thus, violence excepted, the situation is much like that of a man making a clock and letting it run and continue its own motion by itself. In this manner did God allow the heavens to be moved continually according to the proportions of the motive powers to the resistances and according to the established order.[137]

---

[134]Nicole Oresme, *Le Livre du ciel et du monde*, ed. Albert D. Menut and Alexander J. Denomy, trans. Albert D. Menut (Madison: University of Wisconsin Press, 1968), p. 85.

[135]*Ibid.*, p. 57.

[136]*Ibid.*, p. 283.

[137]*Ibid.*, p. 289. Oresme did not dispense with the concept of celestial intelligences. He differed with Aristotle in his rejection of Aristotle's "world as an organism" view. To Oresme, the heavenly bodies are neither animate nor intelligent. Neither were his intelligences neo-platonic emanations of Deity. To him, the intelligences (possibly angels?) were created servants who imparted the exact amount of motion to each celestial body.

Note that Oresme and his medieval brethren did not absolutize the "God as a watchmaker" concept contrary to the Deism that flourished in the 17th and 18th centuries.[138] In Oresme's time, a clock was never viewed as autonomous. Both the water clock and the sundial needed continual maintenance and recalibration to keep them functioning properly. Although mechanical clocks were first developed during Oresme's lifetime, it was not until the 18th century that they achieved the regularity that enabled philosophers to use them to picture a fully autonomous and mechanistic universe. Oresme understood this phrase as an expression illustrating the autonomous mechanical laws given by the transcendent Creator at the creation of all things. Concerning this autonomous mechanical law concept, Oresme

Figure 37: Mechanical clock

posited that God was powerful enough to dispense these laws of nature without impairing His power over nature. Oresme's clockwork cosmology reflected a relative and secondary autonomy (laws dependent upon God), not an absolute and primary autonomy.

At this point in our discussion of Oresme, we need a parenthetical respite for some commentary and analysis. The mechanistic picture presented by Oresme, although powerful in its refutation of Aristotelian panorganism, does open the door to the absolutization of the laws of nature. As we shall see in sections 5.3 through 5.7, it was very easy for the Deists of the 17th and 18th centuries to drop God from the picture and embrace the absolute

Figure 38: Sundial

autonomy of natural law. For Oresme, this "dropping of God from the picture" was inconceivable. As Jean Gimpel recognizes, "Their society [medieval – J.N.] professed a faith, and they would not have been able to imagine that one day men would be living in Europe without such a faith."[139] Remember that the mechanistic picture is a model reflecting how God operates in this universe and

---

[138]Deism is the *a priori* belief in a God who created the universe and then abandoned it, assuming no control over life, exerting no influence on natural phenomena, and giving no supernatural revelation.

[139]Gimpel, p. 171.

models can never offer an absolute picture of the essence of this operation. What the mechanist picture does positively is to de-deify the ancient organismic view of the world.[140] The universe is not God; God created it and it is separate from Him. What the machine model infers negatively is that a machine can exist independently of its maker and that a maker can abandon his work. The God of the biblical authors never abandons His work; He continuously and actively sustains it by the decreed word of His power (Hebrews 1:3) and He may employ His angels in the maintenance and governance of His creation (a possible inference from Psalm 104:1-4). In spite of its inadequacies, the mechanistic world picture proved to be a potent model for the scientific enterprise. Reijer Hooykaas reflects:

> This adaptation ... led to a positive and empiricist conception of science ... It formed the basis of that rational empiricism which has become the legitimate method of modern science. The scientist of today, when using mechanical or other pictures or models, considers them as means of rational description and not as explications of the essence of the world. The world of the physicist is a translation of the world of phenomena into symbols that are more liable to mathematical manipulation and whose consequences may be easily translated back into external phenomena ... Most scientists of the nineteenth and twentieth centuries, when taking this view, may have been unconscious of the fact that the metaphysical foundations of their discipline stemmed, in spite of all secularization, in great part from the biblical concept of God and creation.[141]

Returning from our parenthetical interlude, in mathematics Oresme introduced a notation for fractional exponents giving rules for their operation.[142] He introduced into geometry the idea of motion that Greek geometry had lacked. By clarifying functional relationships through reference to geometrical figures, he contributed to the formation of the critical concept of mathematical functions and their use in the description of physical laws.[143] He developed the seed thought of coordinate geometry, a graphical method of describing such functions using "latitude of forms." A "form" was any variable quantity (e.g., motion, heat, or light) and its "latitude" was the measure of this quantity in association to a given value of "longitude" (e.g., time). Here we find the rudiments of the modern mathematical concepts of independent and dependent variables and their functional contingency.[144] Finally, he pioneered mathematical meth-

---

[140]For a survey of the "world as a mechanism" picture, see Jaki, *The Relevance of Physics*, pp. 52-94.

[141]Hooykaas, pp. 25-26.

[142]Simon Stevin (1548-1620), Flemish mathematician, expanded Oresme's work in fractional exponents (e.g., $x^{1/2}$ or $\sqrt{x}$).

[143]For example, Oresme graphed uniformly accelerated motion as a straight line.

[144]Oresme's graphs were primarily linear. The deficiencies in geometrical knowledge and algebraic technique of the time prevented him from extending this seminal idea to curvilinear figures. This annexation had to wait for the 16th and 17th century work of François Viète (1540-1603), René Descartes (1596-1650), and Pierre de Fermat (1601-1665).

ods that dealt quantitatively with change and rate of change. His work in this area foreshadowed many of the methods of the calculus.

Concerning the meditations of the scholastic philosophers, Howard Eves observes:

> ... [they] led to subtle theorizing on motion, infinity, and the continuum, all of which are fundamental concepts in modern mathematics. The centuries of scholastic disputes and quibblings may, to some extent, account for the remarkable transformation from ancient to modern mathematical thinking.[145]

In this context, Stanley L. Jaki gives comment:

> Inertia, momentum, conservation of matter and motion, the indestructibility of work and energy – conceptions which completely dominate modern physics – all arose under the influence of theological ideas.[146]

Mathematics historian Carl Boyer remarks that, in this theorizing, "there was perhaps as much originality in medieval times as there is now."[147] The Greeks, although they faced the concept of infinity, could never develop any workable theory of infinity. In essence, they "shrank before its silence." They contented themselves with the study of static forms, not dynamic change. The application of the concept of infinity to the study of change in motion is foundational to the calculus. Carl Boyer comments about the impact of the input of the scholastics in this area:

> The blending of theological, philosophical, mathematical, and scientific considerations which has so far been evident in Scholastic thought is seen to even better advantage in a study of what was perhaps the most significant contribution of the fourteenth century to the development of mathematical physics ... a theoretical advance was made which was destined to be remarkably fruitful in both science and mathematics, and to lead in the end to the concept of the derivative.[148]

Given modernity's abhorrence of anything that smacks of the supernatural and its arrogant premise that those who believe in Scripture believe in fairy tales, how, then, could the theology of Scripture (a fairy tale at best), be the source of such remarkable fruit in science and mathematics? The answer of modernity is an answer of silence.

3.17 THE ROLE OF SCRIPTURE

Stanley L. Jaki details the historical conditions that acted as a necessary prerequisite for the viable birth of modern science:

---

[145]Eves, p. 213.
[146]Jaki, *The Road of Science and the Ways to God*, p. 157.
[147]Boyer, *The History of the Calculus and Its Conceptual Development*, p. 65.
[148]*Ibid.*, pp. 70-71.

The rise of science needed the broad and persistent sharing by the whole population, that is, an entire culture, of a very specific body of doctrines relating the universe to a universal and absolute intelligibility embodied in the tenet about a personal God, the Creator of all.[149]

He continues:

While biblical monotheism owed nothing to Greek science, that science could develop into a true science only within a monotheistic matrix, which happened to be biblical through the mediation of Christianity.[150]

During the latter period of the Middle Ages, as Stanley L. Jaki documents, Christianity became, for the first time, a broadly shared cultural matrix.[151] Before the 14th century, the Scriptures were not generally in the hands of the common man. The Bible, in its Latin translation, was honored by an elite few, church officials, who were literate and therefore leaders in society.[152]

In the second half of the 14th century, two important events occurred that were to change the whole face of European society.

First, a sudden change in weather (it got much colder) buffeted Europe at the beginning of the 14th century. It caused a failure in the grain harvests, resulting in an economic downturn. As James Burke notes, "In this weakened state, Europe was ill-prepared to fight off an invasion – especially as the invader was almost invisible."[153] In 1347, at the Crimean trading port of Caffa (Black Sea), marauding Tatars, infected by the Bubonic Plague that struck China ten years previously, besieged some Genoese merchants returning from Cathay with silks and furs. Before withdrawing, the Tatars catapulted their corpses over the walls into Caffa, infecting the merchants, some who would die on the trip home while others carried the plague to Constantinople, Genoa, Venice, and other ports. The plague, called bubonic because of its characteristic bubo (enlarged lymph glands), is transmitted by fleas, carried by rats, and was harbored perhaps in the baggage of the Genoese merchants. The Bubonic Plague, or Black Death, shook Europe to its knees and brought a lull to the development of mathematics and science. Death toll estimates reveal that one-third to one-half of the total population lost their lives. In some towns there were not enough people to

---

[149]Jaki, *The Road of Science and the Ways to God*, p. 33.
[150]*Ibid.*, p. 153.
[151]Stanley L. Jaki, *The Origin of Science and the Science of Its Origins* (Edinburgh: Scottish Academic Press, 1978), p. 11.
[152]This does not imply that the common man of medieval times was biblically illiterate. The Medieval Church made sure that the common man was instructed in the basic Gospel story through stone (art and architecture) and story (Gospel dramas and plays). The common man was biblically illiterate in the sense that the Medieval Church mediated the Scriptures and their proper interpretation to him; he generally did not have direct access to the written Scriptures.
[153]Burke, p. 98.

bury the dead. The ensuing massive poverty, total confusion, and physical degeneration reflected upon the fact that the end of an age had come.[154]

The stage was now set for the second event, a full-scale reformation that would augment and refine what Stanley L. Jaki terms "a broadly shared cultural matrix of Christianity." Soon, the heart and mind of the common man of Europe would be united with the Bible, not in Latin only, but in his own language. Without a clear understanding of the Scriptures, man cannot take complete, effective, and constructive dominion over himself, his society, or the physical world.

3.18 THE MORNING STAR

The Englishman John Wycliffe (ca. 1320-1384) was a leader of a group of poor priests nicknamed the Lollards. After trying unsuccessfully to externally reform a corrupt English church, he soon realized the dire need for the common man to have the Word of God in his own language. To him, this was the precondition for biblical reformation in the Church and in society. He acted as the catalyst in the massive job of translating the whole Bible from the Latin Vulgate into English, the first portions hitting the presses in 1381.

As he was translating, he began to come to grips with a serious problem. How could he get the Bible into the hands of the people knowing the opposition he would face from the established ecclesiastical hierarchy? God led him to form a religious order of poor preachers who took portions of the Bible, as it was translated, and distributed them to the people throughout England. But a new problem presented itself immediately. Many people

Figure 39: John Wycliffe

could not read! So, Wycliffe began a massive literacy campaign. On the university campuses, he and others like Jan Hus (1369-1415) introduced a dynamic of spiritual instruction that was to mark the dawn of a new age.[155]

Little did Wycliffe know that his efforts would prove to herald the greatest reformation yet seen in the Christian movement. Indeed, he was a morning star signaling the dawn of a new day, a day that would shed its light on every aspect of human endeavor, including mathematics.

---

[154]Medical historians also note that, by the late 15th century (around the time of Columbus), one third to one half of Europe was infected by venereal diseases, far more virulent then than now. See Rushdoony, *The One and the Many: Studies in the Philosophy of Order and Ultimacy*, p. 26.
[155]Heer, p. 258.

QUESTIONS FOR REVIEW, FURTHER RESEARCH, AND DISCUSSION

1. Explain the impact of stoicism on Roman culture.
2. How did the Roman view of mathematics affect the development of mathematics?
3. Develop the components of the biblical worldview from the Gospel of John.
4. Detail the worldview antithesis between Greek philosophy and biblical Christianity.
5. Critique the differing perspectives that the Church fathers had with respect to Greek philosophy.
6. In terms of their perspectives on the interpretation of reality, compare and contrast the approach of Augustine with that of Plato.
7. Explain this statement, "Augustine warned everyone to beware of mathematicians."
8. How did Christianity compromise its doctrines with Greek philosophy? Give some examples of modern accommodations of Christian doctrines with non-Christian philosophies.
9. Provide a biblical and historical rationale establishing or criticizing the following statement, "Christian education should plunder and use of the tools of learning derived from the Trivium."
10. Critique the following statement, "The characteristic of the Dark Ages was sterility in every area of life."
11. Explain how to understand and apply the seeming vagaries of weather in terms of Scripture and God's providence.
12. Explain why Christianity did not bring an "immediate liberation" from the bonds of Greek metaphysics.
13. Detail the cultural impact of the camshaft.
14. Detail the cultural impact of the three-field farming system.
15. Detail the contributions of the monasteries to Western civilization.
16. What part did the Arabic culture play in the development of mathematics?
17. What part did the Hindu culture play in the development of mathematics?
18. Explain how the Hindu-Arabic number system filtered into European civilization.
19. Explain the impact that European universities had on the development of modern science.
20. Explain the link between the technological innovations of the Middle Ages, biblical theology, and the development of quantitative analysis.
21. What significance did the mechanical clock play in the development of mathematics and science?
22. Explain the significance of Etienne Tempier's condemnation of Aristotelianism.
23. Explain the place of Aristotelianism in the thought of Thomas Aquinas.

24. Explain the impact that Robert Grosseteste, Roger Bacon, and Thomas Aquinas had on scientific empiricism.
25. Explain the part that the "meditations of the scholastics" had on the development of mathematics.
26. Explain the scientific and mathematical heritage bequeathed by Jean Buridan and Nicole Oresme.
27. In reference to the "clock work" mechanistic model of the world and scientific law, explain the difference between relative autonomy and absolute autonomy.
28. Explain the cultural ingredients necessary in order to give a viable birth to the scientific enterprise.

4: FROM WYCLIFFE TO EULER

The heavens declare the glory of God; and the firmament shows His handiwork. Day unto day utters speech, and night unto night reveals knowledge (Psalm 19:1-2).

King David

In review, the scientific heritage of Medieval Christendom can be summarized in the following five points. These principal and original contributions greatly enhanced the development of natural science and mathematics.[1]

4.1 MEDIEVAL RECAP

1. The recovery of the Greek idea of theoretical explanation, especially the "Euclidean" form of such explanations, and its use in mathematical physics raised the issue of how to construct and to verify or falsify theories. The result was the development of the use of induction and experimentation (both in the imagination and with equipment). These procedures formed the rudiments of the scientific method.

2. The extension of mathematics to the whole of physical science (in principle if not in fact) using quantitative measurement.

3. A radically new approach to the question of space and motion. Greek mathematics, especially Greek geometry, was the mathematics of static forms. The ideas that space might be infinite and void, and the universe without a center, undermined Aristotle's cosmology with its qualitatively different directions and led to the idea of relative motion. The theory of impetus (inertial motion), founded upon the theological distinctive of God creating the heavens and the earth in the beginning (Genesis 1:1), was used to explain many different phenomena – the motion of projectiles and falling bodies, bouncing balls, pendulums, and the rotation of the heavens or

---

[1] From Crombie, 2:117-120.

the earth. Discussions of the nature of the continuum and of maxima and minima problems also began in the 14th century.

4. This period exhibited a remarkable progress in technology. New methods of exploiting animal-, water- and wind-power appeared and new machines were developed for a variety of purposes, often requiring considerable precision. Some of the inventions – the mechanical clock and magnifying lenses – were to be later used as scientific instruments. The astrolabe and quadrant were greatly improved as a result of the demand for accurate measurement. In chemistry, the balance came into general use.

5. A fresh understanding regarding the nature and purpose of science. Its purpose was understood in the context of the dominion mandate of Genesis 1:26-28. Science was seen as a tool that enabled man to gain power over nature and scientific results were to be used to meet the practical needs of society. Regarding the nature of scientific theories, neither God's action nor man's speculation could be constrained within any particular system of scientific or philosophical thought. This brought out the relativity of all scientific theories and the fact that others more successful in fulfilling the requirements of the rational and experimental methods might replace them.

The reality and inherent power of the Gospel, no matter how diluted it may have been doctrinally in the Middle Ages, acted like a leavening influence in European culture and thereby generated the contributions listed above. Concerning this time period, William M. Ivins, Jr. states:

Figure 40: Equilateral Gothic arch

Slowly and gradually all sorts of new things, new ideas, and interests were being discovered. Among these was the knowledge of stresses and strains that finally resulted alike in the architecture of the Gothic cathedrals....[2]

Oswald Spengler remarks about the influence of the architecture of the Gothic cathedrals on later mathematical thought:

The idea of the Euclidean geometry is actualized in the earliest forms of Classical ornament, and that of the Infinitesimal Calculus in the earliest

---

[2]Ivins, p. 61. The flying buttress, one of the great inventions of the medieval architect-engineer, was a mainstay of Gothic architecture. These engineers conceived this ingenious and revolutionary building technique to help solve the technical problems created by the desire to give maximum light to the churches while raising the vaults higher and higher. These technical solutions must have amazed men of the time as they still amaze visitors today. For an idea as to how these cathedrals were constructed, see David Macaulay, *Cathedral: The Story of Its Construction* (Boston: Houghton Mifflin, 1973) and J. James, *The Contractors of Chartres*, 2 vol. (London: Croom Helm, [1979] 1981).

forms of Gothic architecture, centuries before the first learned mathematicians of the respective Cultures were born.[3]

These majestic cathedrals, taking at times over one hundred years to construct, were purposely designed to use light to reflect the awe and grandeur of the infinite God. Sitting in such cathedrals, a simple upward gaze exhibits a modicum of the principle of infinity. Cardinal Nicholas Cusa (1401-1464) wrote a few tracts on the rudimentary ideas of the calculus. According to Stanley L. Jaki, his "real contributions to science are in some of his opuscula, in which he time and again ties the need for quantitative accuracy to the words of the Book of Wisdom about the Creator having arranged everything according to weight, measure, and number."[4] In Spengler's opinion, it was "instinct" that guided Cusa "from the idea of the unendingness of God in nature to the elements of the Infinitesimal Calculus."[5]

4.2 MODERN HERITAGE

These elementary thoughts concerning the concepts of the calculus would eventually prove to be extremely fruitful, not only in further mathematical work, but also in the technology of the 20th century. In his autobiography, historian Arnold Toynbee (1889-1975) notes the significance of the calculus:

> Looking back, I feel sure that I ought not to have been offered the choice [whether to study Greek or calculus – J.N.] ... calculus ought to have been compulsory for me. One ought, after all, to be initiated into the life of the world in which one is going to live. I was going to live in the Western World ... and the calculus, like the full-rigged sailing ship, is ... one of the characteristic expressions of the modern Western genius.[6]

Concerning the effects of the calculus, Sir Oliver Graham Sutton, applied mathematician, remarks that "the rapid advances towards the conquest of nature made in the nineteenth century and in our own time are the direct consequences of this great upheaval in mathematics."[7] Russian-born mathematician Tobias Dantzig (1884-1956) reflects on the importance of infinite processes to modern technology:

> The importance of infinite processes for the practical exigencies of technical life can hardly be overemphasized. Practically all applications of arithmetic to geometry, mechanics, physics, and even statistics involve these processes directly or indirectly. Indirectly because of the generous use these sciences make of irrationals and transcendentals; directly because the most fundamen-

---

[3]Spengler, 4:2318.
[4]Jaki, *The Road of Science and the Ways to God*, p. 44.
[5]Spengler, 4:2330.
[6]Arnold Toynbee, *Experiences* (New York: Oxford University Press, 1969), pp. 12-13.
[7]Oliver Graham Sutton, *Mathematics in Action* (New York: Dover Publications, [1954, 1957] 1984), p. 37.

tal concepts used in these sciences could not be defined with any conciseness without these processes. Banish the infinite process, and mathematics pure and applied is reduced to the state in which it was known to the pre-Pythagoreans.[8]

In the late 19th century, people were still riding horses to work and reading at night by candlelight. Within the scope of fifty or so years, man built giant skyscrapers and sturdy bridges that spanned many a harbor. He harnessed the invisible power of electricity and deciphered the mystery of radio waves. He discovered atomic particles too small for even the most powerful

Figure 41: Jet travel

microscope to see. He built supersonic jets able to cross the Atlantic in the time it takes to drive two hundred miles. He set foot on the moon and penetrated the vast outreaches of the universe. With the invention of the silicon chip, the computer became a household item.

4.3 INGREDIENTS IN THE MIX OF THE SCIENTIFIC REVOLUTION

This advancement did not happen by some sort of mysterious osmosis. There were many ingredients in the mix that produced what is today known as the Scientific Revolution.

Although the Black Plague drastically depopulated Europe, those who survived inherited the remaining wealth. By 1450, poor people could now own chairs, fireplaces, etc. Production of goods rose 30% per capita. The demand for luxuries by the upper classes increased, especially for silk. Peasants were now able to afford linen (in plentiful supply thanks to the invention of the horizontal loom and the spinning wheel). The local "rag and bone" man collected any discarded linen. Bone was used for fertilizer and rag (the discarded linen) became the raw material for high quality, durable *paper*.

Figure 42: Johann Gutenberg

Enter a goldsmith from Mainz, Germany, Johann Gutenberg (ca. 1400-ca. 1468), who invented (or innovated) the moveable type printing press seventy

---

[8]Tobias Dantzig, *Number: The Language of Science* (Garden City: Doubleday Anchor Books, [1930] 1954), p. 139.

years after Wycliffe's death.[9] Although paper mills were functioning in Italy as early as 1280, most printing was still done by the hand of monks using expensive parchment (it took 200-300 sheepskins to produce one Bible). Gutenberg, an experienced metalworker, knew how to manipulate soft metals, the screw press, and moveable type. With the readily available and durable linen paper, the stage was set for mass production of information. His primary desire was to print Bibles for distribution to the masses. The earliest dated example (1457) of his work is the Mainz Psalter, the introduction of which states:

> This volume of the Psalms, adorned with a magnificence of capital letters and clearly divided by rubrics, has been fashioned by a mechanical process of printing and producing characters, without use of a pen, and it was laboriously completed, for God's Holiness, by Joachim Fust, citizen of Mainz, and Peter Schoeffer of Gersheim, on Assumption Eve in the year of Our Lord, 1457.[10]

Gutenberg's first major production run was three hundred copies of a 1,282 page Bible. Listen to his reflection on the role of the printing press:

> Religious truth is captive in a small number of little manuscripts, which guard the common treasures instead of expanding them. Let us break the seal which binds these holy things; let us give wings to truth that it may fly with the Word, no longer prepared at vast expense, but multitudes everlastingly by a machine which never wearies – to every soul which enters life.[11]

The printing press engendered an immense spread of knowledge facilitating the interaction of ideas. In addition to the book most in demand, the Bible, off the presses came many "how to" books explaining craft technique.[12] Printing launched the Reformation. The best selling author in 16th century Europe was Martin Luther (1483-1546). Out from the presses of Europe came translation after translation until even the farmer ploughing the soil knew more about the Scripture than church leaders. In the 1470s, mathematical and scientific works were released from those same presses. By 1515, most

Figure 43: Martin Luther

---

[9]The place of Gutenberg's seniority in the invention of the moveable type press has been questioned. See Pierce Butler, *The Origin of Printing in Europe* (Chicago: University of Chicago Press, [1940] 1966).

[10]Cited in Burke, p. 104.

[11]Cited in William J. Federer, *America's God and Country: Encyclopedia of Quotations* (Coppell, TX: FAME Publishing, 1994), p. 758.

[12]Some knowledge was lost. Not all of the old hand-copied manuscripts were reprinted. Since people generally tend to gravitate to new things, some valid and useful knowledge contained in these old manuscripts was lost to the general public. Also, the printing press augmented a flourishing underground trade in pornography.

Greek works were translated and published.

More and more it was the common man of Europe, generally free from the current ecclesiastical and scientific dogmas, who observed the physical world and obtained new and productive mathematical insights. With Greek scientific works finding new homes on European soil, the common man, in the form of merchants and traders, began to take action. Having direct access to these writings and being of economic persuasion, they had a keen interest in the development of new materials and skills. They hired men, called free artisans, whose sole task was to produce such materials and in so doing a multitude of new scientific problems appeared. The applications of mathematics provided solutions to these problems, so much so, that it became "the soul, or ... the 'main spring', of modern science."[13]

Other key inventions, besides printing, were the magnetic compass and gunpowder. The compass aided men like Christopher Columbus (1451-1506), Vasco da Gama (ca. 1460-1524), and Ferdinand Magellan (ca. 1480-1521) in their worldwide ocean travels. These explorations not only opened up new trade routes, but also called for accurate mapping of the globe.[14] Cartography, ever dependent upon mathematics, ably met these needs.[15] Gunpowder introduced a whole new range of problems dealing with principles of motion. In a short time, Aristotelian mechanics was shot down.

4.4 THE CATALYST: NATURE'S NEW PERSPECTIVE

All of the above ingredients contributed directly to the popularization of the old Greek idea that nature reflects mathematical principles. It is significant to note that, at this time in history, the common man of Europe was becoming literate in both Scripture and the physical world. It is the Scriptural perspective of the physical world that proved to be the catalyst that produced the Scientific Revolution.

In review, classical Greece not only viewed history in terms of ever-recurring cycles, but also perceived nature and mathematical reasoning in deifying overtones. As we have seen, when the Greeks absolutized what God had created, their culture, including mathematics and science, became fragmented and eventually stagnated. The worldview of the Greeks allowed them to go only so far in the development of mathematics and science. Their understanding of

---

[13]Gordon H. Clark, *The Philosophy of Science and Belief in God* (Nutley, NJ: The Craig Press, 1977), p. 46.

[14]Some scholars mock the inaccuracy of the maps of the medieval times. According to David C. Lindberg, "Such maps as medieval people produced were not necessarily intended to portray in exact geometrical terms the spatial relationships of the topographical features indicated on them, and the notion of scale was almost nonexistent. Their function may have been symbolic, metaphorical, historical, decorative, or didactic" (Linberg, p. 254). Portolan charts (embodying the practical knowledge of sailors), invented perhaps in the second half of the 13th century, had both the mathematical and topographical rudiments of modern cartography.

[15]For a clear explanation of the relationship of mathematics to cartography, see H. L. Resnikoff and R. O. Wells, Jr., *Mathematics in Civilization* (New York: Dover Publications, [1973] 1984), pp. 153-178.

nature and history could not give them sustaining perspective, motivation, and purpose. When faced with concepts that they could not handle, like infinity, they "shrank before its silence."

At this time in Europe, when modern science arose, "religion was one of the most powerful factors in cultural life."[16] When the people of this culture looked at nature, they did not see it as absolute, for every scholar had been trained in the scholastic universities where theology had been hailed as the Queen of all knowledge. In other words, the men of learning who contributed to the advancement of both mathematics and science were men whose thoughts were infused with the belief that nature was not only the creation of the Biblical God, but also distinct from Him. To them, pantheism was out. So too was the idea that nature is eternal or uncreated. An infinite, yet very personal God, had, in the beginning, created the heavens and the earth. The finite universe, including the reason of man, now had an infinite reference point.

Concerning the reason of man, the biblical perspective exposed the fallacious Greek belief that said, "Man is the measure of all things." The ultimate in intelligibility no longer belonged to man or nature, but to the only true and wise God of Scripture. In order for the mind of man to work effectively, it must be submitted to the liberating truths of a biblical worldview.

The Bible commands man to take dominion over and subdue the physical creation (Genesis 1:26-28; Psalm 8:3-8). As Reijer Hooykaas narrates, "The biblical conception of nature liberated man from the naturalistic bonds of Greek religiosity and philosophy and gave a religious sanction ... to the dominion of nature...."[17] Rousas J. Rushdoony notes the liberating impact of the Christian doctrine of dominion, "Under the influence of Christianity, science escaped from magic. The purpose of science ceased gradually to be an attempt to play god and became rather the exercise of dominion over the earth under God. The redeemed Christian is God's vice regent over the earth, and science is one of man's tools in establishing and furthering that dominion."[18] Stanley L. Jaki explains that the originality of the scientists of Europe resulted from "the *conviction* that the universe was the rational product of the Creator and that as Christians they had to become masters and possessors of nature."[19]

Alfred North Whitehead identifies the "greatest contribution of medievalism to the formation of the scientific movement" as "the *inexpugnable belief* that every detailed occurrence can be correlated with its antecedents in a perfectly definite manner, exemplifying general principles. Without this belief, the incredible labours of these men would have been without hope."[20] To him, this conviction "must come from the medieval insistence on the rationality of

---

[16]Hooykaas, p. xiii.
[17]*Ibid.*, p. 67.
[18]Rousas J. Rushdoony, *The Mythology of Science* (Nutley, NJ: The Craig Press, [1967] 1979), pp. 1-2.
[19]Jaki, *The Origin of Science and the Science of Its Origin*, p. 21 (emphasis added).
[20]Whitehead, *Science and the Modern World*, p. 15 (emphasis added).

God."[21] This insistence implied that "Every detail was supervised and ordered: the search into nature could only result in the vindication of the *faith* in rationality."[22] This was not the faith of a few isolated scientists, but of the whole of medieval European culture; a faith ingrained in a "broadly shared cultural matrix" using the words of Stanley L. Jaki. Whitehead confirms this analysis, "Remember that I am not talking of the explicit beliefs of a few individuals. What I mean is the impress on the European mind arising from the unquestioned faith of centuries. By this I mean the instinctive tone of thought and not a mere creed of words."[23]

Francis Schaeffer gives affirmation:

> Living within the concept that the world was created by a reasonable God, scientists could move with *confidence*, expecting to find out about the world by observation and experimentation. This was their epistemological base – the philosophical foundation with which they were sure they could know.... Since the world had been created by a reasonable God, they were not surprised to find a correlation between themselves as observers and the thing observed.... Without this foundation, Western modern science would not have been born.[24]

The biblical view of nature, time, and history gave the scientists of this era conviction, inexpugnable belief, faith, and confidence to pursue the course of taking dominion over the physical world. The result? A revolution in mathematics and science. According to English philosopher Robin G. Collingwood (1889-1943), "The possibility of an applied mathematics is an expression, in terms of natural science, of the Christian belief that nature is the creation of an omnipotent God."[25] Whitehead remarks that "the mathematics, which now emerged into prominence, was a very different science from the mathematics of the earlier epoch."[26] As Morris Kline observes, "The consequences for mathematics was a burst of activity and original creation that was the most prolific in its history."[27]

---

[21]*Ibid.* Stanley L. Jaki's remarks are insightful, "Half a century has passed since these words startled a distinguished audience at Harvard University and indeed the whole intellectual world. The magnitude of the shock merely corresponded to the impenetrable density of a climate of opinion for which the alleged darkness of the Dark Ages represented one of the forever established pivotal truths of the 'truly scientific' interpretation of Western intellectual tradition." (*Science and Creation: From Eternal Cycles to an Oscillating Universe*, p. 146).

[22]*Ibid.*, pp. 15-16 (emphasis added). Whitehead, however, maintains that the source of medieval belief in rationality is Greek philosophy. To him, a belief in God provided *only one thing* for the medievals; i.e., personal motivation. See *Ibid.*, pp. 19-20.

[23]*Ibid.*, p. 16.

[24]Schaeffer, *How Should We Then Live? The Rise and Decline of Western Thought and Culture*, p. 134 (emphasis added).

[25]Robin G. Collingwood, *An Essay on Metaphysics* (London: Oxford University Press, 1940), p. 253.

[26]Whitehead, *Science and the Modern World*, p. 38.

[27]Kline, *Mathematical Thought from Ancient to Modern Times*, p. 230.

4.5 TWO INDISPENSABLE TOOLS OF APPLIED MATHEMATICS

4.5.1 SYMBOLIC ALGEBRA

Since the very soul of science consists of theoretical generalizations leading to the formulation of quantitative laws and systems of laws, applied mathematics needed a refinement in algebra before it could proliferate into a plenitude of scientific uses. As Tobias Dantzig notes:

Algebra ... enables one to transform literal expressions and thus to paraphrase any statement into a number of equivalent forms ... it is this power of transformation *that lifts algebra above the level of a convenient shorthand* ... the literal notation made it possible to pass from the individual to the collective, from the 'some' to the 'any' and the 'all'.... It is this that made possible the general theory of functions, which is the basis of all applied mathematics.[28]

Figure 44: François Viète

It was in 16th and 17th century Europe that symbolic algebra came into its own. The decimal system needed the work of the Dutch mathematician Simon Stevin (1548-1620) and the Frenchman François Viète (1540-1603) in the late 16th century to acquire the conceptual precision demanded by systematic scientific work.[29] According to Alfred Crosby:

Algebraic notation remained a mishmash of words, their abbreviations, and numbers until French algebraists, particularly Francis Vieta [Viète – J.N.] late in the sixteenth century, took the step of systematically using single letters to denote quantities. Vieta used vowels for unknowns and consonants for knowns.[30]

Crosby continues:

Because the algebraist could concentrate on the symbols and put aside for the moment what they represented, he or she could perform unprecedented intellectual feats.[31]

---

[28]Dantzig, p. 89.
[29]These two men built upon the interest generated by the Italian mathematician, physician, gambler, and astrologer Girolamo Cardano (1501-1576) who once cast a horoscope of Christ. In 1545, he published *Ars magna sive de regulis algebraicis*, generally known as *Ars magna*. Algebra in this era was often referred to as *ars magna* (the great art) as opposed to the lesser art of arithmetic. The book contains the first reference to negative numbers and complex numbers (i.e., square roots of negative numbers). It also included solutions to cubic (third degree) and quartic (fourth degree) equations.
[30]Crosby, pp. 119-120.
[31]*Ibid.,* p. 120.

Alfred North Whitehead confirms the indispensable power of symbolic algebra for the future developments of 17th century science:

> The point which I now want to make is that this dominance of the idea of functionality in the abstract sphere of mathematics found itself reflected in the order of nature under the guise of mathematically expressed laws of nature. Apart from this progress of mathematics, the seventeenth century developments of science would have been impossible.[32]

4.5.2 LOGARITHMS

The invention of logarithms in the early 17th century by the Scottish mathematician John Napier (1550-1617) gave future scientists, and especially astronomers, an indispensable calculating tool. Symbolic algebra and logarithms, along with decimal notation, decimal fractions, and trigonometry, all combined to produce a powerful mathematical and scientific synergy. Note Napier's own analysis of the situation in 1614:

Figure 45: John Napier

> Seeing there is nothing that is so troublesome to mathematical practice, nor that doth more molest and hinder calculators, than the multiplications, divisions, square and cubical extractions of great numbers.... I began therefore to consider in my mind by what certain and ready art I might remove those hindrances.[33]

Napier was a fervent and zealous Protestant who, in 1593, wrote a commentary on last book of the Bible, Revelation.[34] In it, he associated the Pope as the Antichrist and he predicted that the end of the world would come between 1688 and 1700.[35]

Like many men of his time, Napier was a man of varied interests, a literal "jack of all trades and master of every one of them." Some of his "trade interests" were improvement of crops, development of new fertilizers, invention of a mechanical device called the hydraulic screw, and prototyping (at least in

---

[32]Whitehead, *Science and the Modern World*, p. 40.

[33]Cited in George A. Gibson, "Napier and the Invention of Logarithms," *Handbook of the Napier Tercentenary Celebration, or Modern Instruments and Methods of Calculation,* ed. E. M. Horsburgh (Los Angeles: Tomash Publishers, [1914] 1982), p. 9.

[34]Napier's *A Plaine Discovery of the whole Revelation of Saint John* was translated into several languages and went through ten editions in his lifetime (twenty-one total).

[35]Most Protestants of Napier's time identified the Pope as the Antichrist (note carefully the dedication in the 1611 English translation of the Bible to King James). For an historical analysis of this type of interpretation, called by some as "pin the tail on the Antichrist," see Richard Kyle, *The Last Days Are Here Again: A History of the End Times* (Grand Rapids: Baker, 1998). For an exegetical analysis of modern-day end times hysteria, see Gary DeMar, *Last Days Madness: The Folly of Trying to Predict When Christ Will Return* (Brentwood, TN: Wolgemuth and Hyatt, 1991).

thought if not in reality) an artillery piece that prefigured the modern machine gun. He developed a mechanical calculating device, called "Napier's bones." You could solve multiplication and division problems using this apparatus. He also simplified the writing of decimal fractions by advocating the use of a decimal point to separate the whole part of a number from its fractional part.

A subject of many stories that reflect his astuteness, if not in biblical exegesis at least in human nature, Napier once suspected one of his servants of theft. He brought all of his servants together and announced that he had a black rooster that would identify the guilty party. He ordered his servants into a dark room and asked each to pat the rooster on its back. Unknown to them, Napier had coated the bird's back with a layer of chimney soot. Upon leaving the room, each servant had to give a "show of hands." The guilty one, fearing to touch the rooster, came out "clean handed" thus sealing his culpability. Yes, as Napier predicted, his rooster did identify the thief.

Neither Napier's biblical exegesis, nor mechanical acumen, nor insight into human nature guaranteed him a place in history. It was an abstract mathematical idea, an idea and the calculation tables that took him *twenty years* to laboriously compute, which secured him an honored spot. That idea was logarithms.[36]

First, Napier must have noted the use of prosthaphaeretic[37] rules in some of the commonly known trigonometric relations. For example, $\sin A \sin B = \frac{1}{2}[\cos (A - B) - \cos (A + B)]$.[38] Here we see that the multiplication of two terms ($\sin A \sin B$) can be computed by adding or subtracting other trigonometric expressions, $\cos (A - B)$ and $\cos (A + B)$.

Second, Napier was also aware of the work of the German mathematician Michael Stifel (1487-1567).[39] A follower of Luther, Stifel, like Napier, toyed with biblical exegesis and dabbled in cabbalism.[40] For example, he calculated the number of the beast (Revelation 13:18) and "pinned the tail" on Pope Leo X (1475-1521). He also succumbed to the tenets of some of the radical reformers and predicted the end of the world to come on October 3, 1533. The date "fortunately" passed without the apocalypse appearing and in 1544 he published *Arithmetica integra* (the complete science of arithmetic). In this book, Stifel

---

[36]Logarithm literally means the "word concerning number" (*logos + arithmos)* and is generally translated to mean "ratio number" or "number of the speech or expression."

[37]Prosthaphaeretic is Greek for "addition and subtraction."

[38]The German mathematician and astronomer Johann Müller (1436-1476), also known by the Latin Regiomontanus (because he was born in Königsberg, which in German means "royal mountain"), was the first to write a treatise devoted wholly to trigonometry. It first appeared in manuscript about 1464 under the title *De triangulis omnimodis libri quinque* (Of triangles of every kind in five books). After being printed in 1533, it served as a medium for spreading knowledge of trigonometry throughout Europe. Bartholemäus Pitiscus (1561-1613), a German clergyman, wrote the first satisfactory textbook on trigonometry entitled *Trigonometriae sive de dimensione triangulorum libri quinque* (On trigonometry, or, concerning the properties of triangles, in five books) in 1595.

[39]Stiffel called negative numbers "absurd."

[40]Cabbalism deals with esoteric rabbinical teachings of the Scriptures and includes numerology in its analysis.

compares a sequence of consecutive integers with a sequence of corresponding powers of 2 having these integers as their exponents:

n	0	1	2	3	4	5	6	7	8	9	10
2^n	1	2	4	8	16	32	64	128	256	512	1024

Stifel noted that the sums and differences of the power indices correspond to the products and quotients of the powers themselves. For example, to find the product of $2^2 \cdot 2^5$, all you need to do is add the exponents ($2 + 5 = 7$) and read the value of 2^7 to be 128. To find the quotient of $2^9/2^4$, subtract the exponents ($9 - 4 = 5$) and read the value of 2^5 to be 32.

Any student of arithmetic knows that addition and subtraction are much easier than multiplication and division, so the prosthaphaeretic trigonometric formula and Stifel's law of exponents provided Napier with a rudimentary and innovative idea. Napier used this inspiration, and there is no other word to effectively describe it, to eventually compute a table of logarithms that could be used to translate multiplication and division problems into addition and subtraction problems.[41] In 1914 at Edinburgh, Lord Moulton, in his inaugural address commemorating the 300th anniversary of this strategic invention, paid Napier this tribute:

> The invention of logarithms came on the world as a bolt from the blue. No previous work had led up to it, foreshadowed it or heralded its arrival. It stands isolated, breaking in upon human thought abruptly without borrowing from other work of intellects or following known lines of mathematical thought.[42]

After Napier died, Henry Briggs (1561-1631) took up the torch and developed a table of logarithms that we now call "common" or base 10 logarithms in 1624.[43] He also invented the modern method of long division. In 1619, Briggs became the first Savilian Professor of Geometry at Oxford University. After him, a distinguished line of British scientists also held this chair at Oxford: John Wallis (1616-1703) who introduced the mathematical symbol for infinity (∞), Edmund Halley (1656-1742) of comet fame, and Christopher Wren (1632-1723) who designed St. Paul's Cathedral in London.

---

[41]The three basic laws of logarithms, where the letter "a" represents the base, are: (1) $\log_a(xy) = \log_a x + \log_a y$, (2) $\log_a(x/y) = \log_a x - \log_a y$, and (3) $\log_a(x^n) = n \log_a x$. The logarithmic tables compiled by Napier and others turned out to be an immediate boon to astronomers in that they transformed their somewhat complex and "astronomical" (they dealt with very large numbers) calculations into a simple matter of addition or subtraction.

[42]Inaugural address, "The Invention of Logarithms," *Napier Tercentenary Memorial Volume*, ed. Cargill Gilston Knott (London: Longmans, Green, and Company, 1915), p. 3. Napier published his invention in 1614 in a work entitled *Mirifici logarithmorum canonis descriptio* (Description of the wonderful canon of logarithms).

[43]In modern terminology, if $N = 10^L$, then L is the Briggsian or "common" logarithm of N. It is written as $\log_{10} N$ or simply log N.

Whole flurries of mechanical devices were developed in order to perform logarithmic calculations. The basic idea was to engrave numbers on a ruler that were spaced in proportion to their logarithms. The first such primitive device appeared in 1620 and was invented by Edmund Gunter (1581-1626), an English minister and later to become a professor of Astronomy at Gresham College (the school where Briggs held his first professorial post). Gunter's device consisted of a single logarithmic scale. With it distances could be measured and then added or subtracted with a pair of dividers. Gunter also introduced the trigonometric terms cosine and cotangent *(sinus complementi* and *tangens complementi).* As early as 1622, another clergyman and mathematician, William Oughtred (1574-1660), developed a device that had two logarithmic scales that could be moved along each other.[44]

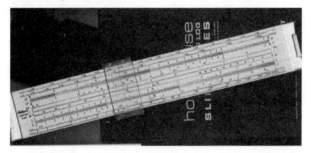

Figure 46: Slide rule

Anyone born before 1960 should recognize these descriptions – they are descriptions of slide rules. From the 1620s to the early 1970s, the slide rule was the faithful companion, the "Tonto" of every mathematician, scientist, and engineer. If the slide rule could talk, it would refer to these companions as "kemo sabe" (trusting brave).[45] By 1980, however, with the invention of hand-held electronic calculators, the slide rule became a museum piece (the author still proudly possesses his © 1960 Pickett Model N 3-ES power log exponential slide rule). Also, logarithmic tables have virtually disappeared (they may still be found in the back of some mathematics textbooks). In the words of mathematics professor Eli Maor, "But if logarithms have lost their role as the centerpiece of computational mathematics, the logarithmic *function* remains central to almost every branch of mathematics, pure or applied. It shows up in a host of applications, ranging from physics and chemistry to biology, psychology, art, and music."[46]

---

[44]Oughtred also introduced the symbol "x" to stand for multiplication in 1631. The German mathematician Johann Heinrich Rahn (1622-1676) introduced the symbol "÷" for division in 1659.
[45]Anyone born before 1960 will readily recognize where the metaphors Tonto and kemo sabe come from (i.e., the Lone Ranger radio and television series).
[46]Eli Maor, *e: The Story of a Number* (Princeton: Princeton University Press, 1994), p. 16.

✳ Napierian logarithms, more accurately called natural logarithms, also uncovered e, a very intriguing and famous transcendental number.[47] This number finds its home in an extensive variety of mathematical relationships – the calculation of interest earned (in fact, any functional relationship of growth or decay), the arrangement of seeds in a sunflower, the chambered nautilus, and the shape of the Gateway Arch in St. Louis.[48]

Figure 47: Gateway Arch

4.6 TESTIMONIALS

It is a revealing exercise to read the writings of the mathematicians and scientists of this era. From the writings of Greek scientists, it is hard at times to determine what really motivated them, but one is not left in doubt when it comes to the motivations and the epistemology undergirding the work of these European scientists.

4.6.1 NICHOLAUS COPERNICUS

The Polish cathedral canon, lawyer, and physician, Koppernigk (1473-1543), better known by its Latinized form, Copernicus (first name Nicholaus), marveled when he observed the hand of God in the correlation between mathematical thought and the actions of nature. To him, the universe was "built for us by the Best and Most Orderly Workman of all."[49] He lauded the Creator in praise, "How exceedingly fine is the godlike work of the Best and Greatest Artist!"[50] He articulated his debt to his Christian heritage that motivated him in the cultivation of his worldview outlook:

---

[47]Transcendental numbers, the existence of which was proven by the French mathematician Joseph Liouville (1809-1882) in 1844, are technically defined as numbers that are not roots of any algebraic equation. pi (π), the ratio of the circumference to the diameter of a circle, is also a transcendental number. e is defined as follows:

$$e = \lim_{n \to \infty} \left(1 + \frac{1}{n}\right)^n \approx 2.718281828459045\dots$$

[48]The shape of the Gateway Arch is a catenary curve. In architecture, this shape makes an extremely stable arch because the force of the arch's weight acts along its legs *directly into the ground*. The general catenary curve is a form of the hyperbolic cosine:

$$y = \frac{x}{a}\left(e^{x/a} + e^{-x/a}\right) = a \cosh\frac{x}{a}, \text{ where } a \neq 0. \text{ The abbreviation "cosh" means hyperbolic cosine.}$$

[49]Nicholaus Copernicus, *The Revolution of the Heavenly Spheres*, trans. Charles Glenn Wallis (Amherst: Prometheus Books, [1543, 1939] 1995), p. 6.

[50]*Ibid.*, p. 27.

For the divine Psalmist surely did not say gratuitously that he took pleasure in the workings of God and rejoiced in the works of His hands, unless by means of these things as by some sort of vehicle we are transported to the contemplation of the highest Good.[51]

Figure 48: Nicholaus Copernicus

Although he faced opposition from the ecclesiastical establishment, Copernicus rejected the geocentric theory of Ptolemy and opted for the much simpler heliocentric view.[52] Note that many in the established church had a difficult time adjusting to these new views, since it had for centuries established the priority of the authority of Aristotelian dogma over any other challenger.[53]

Scholars of modernity have put a presuppositional spin on his heliocentric views. This spin is that when Copernicus disengaged himself from the Ptolemaic view by hypothesizing that a rotating earth circles the sun he dethroned man from a place of significance in the universe. Copernicus would never have inferred this and would have rejected the "dethroning of man" through heliocentrism. His theory did not concern itself with the place of man in the universe. He saw his theory, in contrast to geocentrism, as better evidencing the wisdom of the Creator.

4.6.2 FRANCIS BACON

Francis Bacon (1561-1626), Lord Chancellor of England and a noted lawyer and writer, fought hard to dethrone Aristotelian absolutism by proclaiming that the only way to unlock the secrets of nature was by diligent observation and experimentation.[54] He said:

---

[51]*Ibid.*, p. 8. The biblical source for this quote is Psalm 111:2, "The works of the Lord are great, studied by all who have pleasure in them." Although the "highest Good" has Platonic overtones, Copernicus mentioned Plato after this statement, not before it. Every concept mentioned before this phrase has the solid ring of biblical heritage.

[52]The simplicity was conceptual, not computational. He still made use of epicycles to handle anomalies. His system had thirty-four circles compared to Ptolemy's eighty.

[53]The ecclesiastical opposition was a cautious one. Copernicus did dedicate *The Revolution of the Heavenly Spheres* to the Pope. Many Aristotelian academics mocked him, however.

[54]There was a tendency in Francis Bacon's writings to absolutize the empirical method. For a thorough analysis, see Jaki, *The Road of Science and the Ways to God*, pp. 50-64. This imbalance laid the groundwork for Auguste Comte's (1798-1857) positivism, a doctrine contending that sense perceptions are the only admissible basis of human knowledge and precise thought.

For as all works do shew forth the power and skill of the workman ... so it is of the works of God; which do shew the omnipotency and wisdom of the maker....[55]

Bacon echoed the utilitarian emphasis of the medieval scientists by insisting the true end of scientific activity to be the "glory of the Creator and the relief of man's estate."[56]

4.6.3 Johannes Kepler

Figure 49: Francis Bacon

Johannes Kepler (1571-1630), a German Lutheran, developed three famous laws of planetary motion after many years of arduous experimentation and observation in which he compared facts with theory.[57] He was one of the first astronomers to avail himself of Napier's logarithms to aid him in calculating the orbits of the planets. Stanley L. Jaki notes that in Kepler we find the "heroic groping of a great man of science with facts, with ideas, with perspectives, and not least with the need to arrive at a law which enabled the prediction of planetary positions with the greatest possible accuracy."[58] Albert Einstein (1879-1955), regarding Kepler's approach as a basic principle of scientific work, comments, "Knowledge cannot spring from experience alone but only from the comparison of the inventions of the intellect with observed fact."[59]

Kepler did not hide the motivations for his work under a bushel. He understood his discoveries and writings as hymns to his Creator and Redeemer and he often punctuated his writings with prayers and psalms of praise to God as illustrated from *The Harmonice mundi* (The Harmonies of the World), published in 1619:

Accordingly let this do for our *envoi* concerning the work of God the Creator. It now remains that at last, with my eyes and hands removed from the tablet of demonstrations and lifted up towards the heavens, I should pray, devout and supplicating, to the Father of lights: O Thou Who dost by the light of nature promote in us the desire for the light of grace, that by its

---

[55]Francis Bacon, *The Works of Francis Bacon*, ed. J. Spedding, R. L. Ellis, D. D. Heath (Boston: Taggard and Thompson, 1857-1874), 6:212.

[56]*Ibid.*, 6:134. Bacon, however, tended to equate truth with utility reflecting the grievous error of embracing a pragmatic standard for truth.

[57]Kepler relied on the meticulous astronomical observations made by the Danish astronomer Tycho Brahe (1546-1601). One of the amazing "coincidences" of history is that, in a sword duel, Tycho lost only his nose. A difference of a mere inch would have ruined his eyesight and deprived Kepler (and the world) of his uniquely accurate planetary measurements. Brahe clung to a revised version of geocentricity in which the five known planets were supposed to revolve around the sun which, with the planets, circled the earth each year.

[58]Jaki, *The Road of Science and the Ways to God*, p. 152.

[59]Albert Einstein, *Essays in Science* (New York: Philosophical Library, 1934), p. 27.

means Thou mayest transport us into the light of glory, I give thanks to Thee, O Lord Creator, Who hast delighted me with Thy makings and in the works of Thy hands have I exulted. Behold! now, I have completed that work of my profession, having employed as much power of mind as Thou didst give to me; to the men who are going to read those demonstrations I have made manifest the glory of Thy works, as much of its infinity as the narrows of my intellect could apprehend.[60]

Great is our Lord and great His virtue and of His wisdom there is no number: praise Him, ye heavens, praise Him, ye sun, moon, and planets, use every sense for perceiving, every tongue for declaring your Creator. Praise Him, ye celestial harmonies, praise Him, ye judges of the harmonies uncovered ... and thou my soul, praise the Lord thy Creator, as long as I shall be: for out of Him and through Him and in Him are all things ... [both the sensible and the intelligible]; for both whose whereof we are utterly ignorant and those which we know are the least part of them; because there is still more beyond. To Him be praise, honor, and glory, world without end. Amen.[61]

The ethos of Kepler's work is expressed in the *Epitome astronomiae Copernicanae* (Epitome of Copernican astronomy) published between 1618 and 1621, where he suddenly interrupts his scientific research with this paean:

Figure 50: Johannes Kepler

With a pure mind I pray that we may be able to speak about the secrets of His plans according to the gracious will of the omniscient Creator, with the consent and according to the bidding of His intellect. I consider it a right, yes a duty, to search in cautious manner for the numbers, sizes and weights, the norms for everything He has created. For He Himself has let man take part in the knowledge of these things and thus not in a small measure has set up His image in man. Since He recognized as very good this image which He made, He will so much more readily recognize our efforts with the light of this image also to push into the light of knowledge the utilization of the numbers, weights and sizes which He marked out at creation. For these secrets are not of the kind whose research should be forbidden; rather they are set before our eyes like a mirror

[60]Johannes Kepler, *Epitome of Copernican Astronomy & Harmonies of the World*, trans. Charles Glenn Wallis (Amherst: Prometheus Books, [1618-1621, 1939] 1995), p. 240.
[61]*Ibid.*, p. 245.

so that by examining them we observe to some extent the goodness and wisdom of the Creator.[62]

4.6.3.1 KEPLER'S THREE LAWS OF PLANETARY MOTION

1. The orbit of a planet traces out the path of an ellipse with the sun at one of its two foci.

2. A planet changes its speed according to its distance from the sun and the line joining the planet to the sun sweeps out equal areas in equal times. In Figure 51, the time that it takes for a planet to travel from one point to the next is the same and the two shaded areas ABS and GSH are equal. This means that a planet moves faster as its orbit brings it closer to the sun (the nearest point is called the perihelion, point A in Figure 51) and slower as it moves farther away from the sun (the farthest point is called the aphelion, point F in Figure 51).

3. The square of the time, T, taken by a planet to complete one orbit varies in direct proportion to the cube of its mean distance, D, from the sun. Mathematically expressed, $T^2 = kD^3$, where k is a constant that is the same for all the planets.

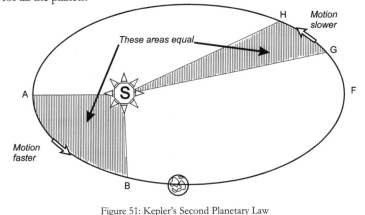

Figure 51: Kepler's Second Planetary Law

To Kepler, the works of God reflect the "wonderful example of His wisdom."[63] He was motivated to make "as many discoveries as possible for the glorification of the name of God and sing unanimous praise and glory to the All-Wise Creator."[64] In his studies, he often exclaimed, "I feel carried away and

---

[62]Cited in Max Caspar, *Kepler*, trans. C. Doris Hellman (New York: Dover Publications, [1959] 1993), p. 381.

[63]*Ibid.*, p. 63.

[64]*Ibid.*, p. 64. Although Kepler was attracted to astrology all of his life (in a pragmatic sense, he used it to earn much needed money), his commitment to the works of God saved him from falling into the pit of false speculations.

possessed by an unutterable rapture over the divine spectacle of the harmony."[65] After Kepler reduced Copernicus' heliocentric model to seve lipses with each planet traveling at variable speeds, he "contemplated its beau with incredible and ravishing delight."[66]

Biographer Max Caspar (1880-1956) summarizes Kepler's views of God and scientific research:

> God is the beginning and end of scientific research and striving. Therein lies the keynote of Kepler's thought, the basic motive of his purpose, the life-giving soil of his feeling. His deep religiousness expresses itself not only in occasional bents and passions of a pious soul; it feeds not only on reminiscences from the time of his theological studies. It penetrates his entire creativity and spreads out over all the works he left behind. It is this feeling for religion which above all lends them the special warmth which we experience with such pleasure when reading them. All of its own accord at every opportunity the name of God crosses his lips; to Him he turns now with a request, now with praise and thanks; before Him he examines his deeds and omissions, his thoughts and words, to discover whether they can pass the test and are directed toward the proper goal.[67]

Concerning Kepler's view of man's chief purpose in life, Caspar states, "As the bird is created to sing, so, according to his convictions, is man created for his pleasure both in contemplating the magnificence of nature and in inquiring into her secrets, not for the purpose of extracting practical uses but to arrive at a deeper knowledge of the Creator."[68]

Kepler spoke much of "celestial harmonies" revealing his appreciation for the Pythagorean concept of the "harmony of the spheres." Caspar reflects on Kepler's cosmology and its link to music and geometry:

> They [the orbits of planets – J.N.] have the form of ellipses with the sun in one focus. But the eccentricities of these ellipses are no more arbitrary and without rule than any other measures. No, in this fine construction the highly artistic formative hand of the Creator is shown in a very special way. Since the eccentricities determine the rates of the planets at aphelion and perihelion, they have been so measured by the Creator that between them appear the harmonic proportions which are to be presented by geometry and which are the foundation of music. So a divine sound fills the whole world. To be sure, sensual hearing is unable to perceive the wonderful

---

[65]*Ibid.*, p. 267.

[66]Cited in Edwin A. Burtt, *Metaphysical Foundations of Modern Science* (London and New York: Routleluge & Kegal Paul, 1925), p. 47.

[67]Caspar, p. 374.

[68]*Ibid.* By this statement Kepler did not negate the practicality of science. To him, the study of God's works served a transcendent purpose – to reveal to the observer a tiny glimpse of the grandeur and character of the Creator.

ual ear perceives it, just as it is also the spiritual eye oveliness of the sizes.[69]

matics to Kepler provided the key that unlocked the y of God's creation. Mathematics, particularly ge- onies of God's creation and this harmony was also mind along with God's image. Therefore, man, with ...., ... p.......... and discover the harmonies of the world.

4.6.4 GALILEO GALILEI

Galileo Galilei (1564-1642), an Italian Catholic, established many new physical principles on the basis of carefully thought out experiments.[70] In the words of Alistair C. Crombie:

> In his use of 'thought experiments' – but not of impossible imaginary experiments – Galileo also carried on established practices. But he made one advance of the greatest importance. He insisted, at least in principle, on making systematic, accurate measurements, so that the regularities in phenomena could be discovered quantitatively and expressed in mathematics.[71]

Figure 52: Galileo

Galileo was responsible for the discovery of the laws of falling bodies and of the parabolic path of projectiles. He studied the motions of pendulums, and he investigated mechanics and the strength of materials. According to Stanley L. Jaki:

> Unlike Archimedes, they [the forerunners of Galileo – J.N.] applied, from almost the very start, geometry to problems of motion. Such a contrast will not be understood unless one keeps in mind the contrast between Plato's god; who merely cultivated geometry, and the God of Christian theism, who created everything according to weight, measure, and number even in a world of motion.[72]

---

[69]*Ibid.*, p. 386.
[70]A great many of his experiments were "thought" experiments. In the 1930s it was pointed out that Galileo never dropped weights from any tower and that he had derived the time-squared law of free fall long before he experimented with balls and inclined planes. See Jaki, *The Road of Science and the Ways to God*, p. 230.
[71]Crombie, 2:148.
[72]Jaki, *The Road of Science and the Ways to God*, p. 48.

With the refracting tele-scope, Galileo provided evidence for confirming the theories of Copernicus and Kepler, again much to the chagrin of church hierarchy.[73] It is commonly thought that the Roman Catholic Church silenced him, and it did, but *only upon the instigation of Academics blindly committed to Aristotelianism.* Charles Dykes comments:

Figure 53: Refracting telescope

> Aquinas had Christianized Aristotle so well that when the authority of Aristotle in the area of astronomy and physics was called into question, many Christian theologians thought biblical truth was being denied. So completely had Aristotle and medieval Christian theology been harmonized in the Thomistic synthesis that the threatened overthrow of Aristotelian cosmology seemed to many theologians to be a rejection of biblical revelation as well. Hence the move against Galileo: the Aristotelian theologians realized that if the Copernican doctrines were sanctioned, this would seriously damage their own authority as guardians of orthodoxy by proving false what they had taught was true. The real issue in the trial of Galileo was not the truth of Holy Scripture, but rather the truth of Aristotle and the authority of the Aristotelian theologians.[74]

Galileo had no intention of questioning Scriptural doctrine but only the scientific framework inherited from Aristotelian philosophy. Galileo brought this condemnation upon himself by his own belligerence. He did not realize that it takes time for academics to shift paradigmical gears. There is evidence showing that he may have been framed by jealous academics possibly because Galileo was taking his views to the public via publication and hence bypassing

---

[73]A Dutch spectacle maker, Jan Lippershey (ca. 1570-ca. 1619), patented the first telescope in 1608 initially intending it for use in military applications. It did not take long for Galileo to make one and lift it toward the heavens (unfortunately, he presented it to the Venetian senate as his own invention). This instrument brought distant objects within the realm of measurement making it possible for scientists to analyze in a quantitative manner, using existing trigonometric formulae, the far outreaches of space. With the telescope Galileo observed the mountains and craters on the moon. He also saw that the Milky Way was composed of stars, and he discovered the four largest satellites of Jupiter. By 1610 he had observed the phases of Venus, which confirmed his preference for the heliocentric system. In essence, the *a posteriori* evidence obtained from the telescope shattered the *a priori* cosmology of Aristotle and Ptolemy.

[74]Charles Dykes, "Medieval Speculation, Puritanism, and Modern Science," *The Journal of Christian Reconstruction: Symposium on Puritanism and Progress*, ed. Gary North (Vallecito, CA: Chalcedon, 1979), 6:1, p. 35.

their "stamp of approval."[75] His excommunication consisted only of his being placed under "house arrest" (instead of "burning at the stake") and had no basis in Canon law.[76]

Writing to a friend at the Tuscan Court on January 30, 1610, Galileo said, "I give thanks to God, who has been pleased to make me the first observer of marvelous things unrevealed to bygone ages...."[77] Writing to the Grand Duchess Christina in 1615, he said, "God is known ... by Nature in His works, and by doctrine in His revealed word."[78] He understood "the phenomena of nature ... as the observant executrix of God's commands."[79] He possessed a high view of science and "experimental philosophy."[80] Continuing his comments to the Grand Duchess:

> And to prohibit the whole science would be but to censure a hundred passages of holy Scripture which teach us that the glory and greatness of Almighty God are marvelously discerned in all his works and divinely read in the open book of heaven. For let no one believe that reading the lofty concepts written in that book leads to nothing further than the mere seeing of the splendor of the sun and the stars and their rising and setting, which is as far as the eyes of brutes and of the vulgar can penetrate. Within its pages are couched mysteries so profound and concepts so sublime that the vigils, labors, and studies of hundreds upon hundreds of the most acute minds have still not pierced them, even after continual investigations for thousands of years.[81]

He took the human mind as the greatest and finest product of the Creator and is therefore designed and equipped to be able to fathom, through mathe-

---

[75]See George Sim Johnston, *The Galileo Affair* (Princeton: Scepter Press, 1995). Pietro Redondi's *Galileo: Heretic* (Princeton: Princeton University Press, 1987) provides intriguing evidence for the conjecture that Galileo's trial for heliocentrism was "stage-managed" in order to divert attention from the challenges that a revival of atomism presented to the Roman Catholic doctrine of the Eucharist; i.e., transubstantiation. See also J. L. Heilbron, *The Sun in the Church: Cathedrals as Solar Observatories* (Cambridge, MA: Harvard University Press, 1999). In this fascinating work, Heilbron shows how the Roman Catholic Church actually fostered the burgeoning scientific enterprise in a wonderful, yet discreet way – by incorporating solar observing instruments into the structure of several European cathedrals. By doing so, Italian astronomers like Giovanni Domenico Cassini (1625-1712) were able to garner some confirming evidence of the heliocentric hypothesis and the elliptical nature of planetary orbits. The official agents of the Roman See allowed such research and publication only if the authors included a disclaimer that the conclusions were entirely hypothetical.
[76]In 1992, a Papal commission acknowledged error in Galileo's case.
[77]Cited in J. J. Fahie, "The Scientific Works of Galileo," *Studies in the History and Method of Science*, ed. Charles Singer (New York: Oxford University Press, 1921), p. 232.
[78]Galileo Galilei, *Discoveries and Opinions of Galileo*, trans. Stillman Drake (Garden City: Doubleday, 1957), p. 183.
[79]*Ibid.*, p. 182.
[80]The term, scientific method, did not become popular until the 19th century. Until then, it was known as "experimental philosophy." This was an excellent phrase in that it tied together the *a priori* and the *a posteriori* elements of science in perfect balance.
[81]Galileo Galilei, *Discoveries and Opinions of Galileo*, p. 196.

matics, all His other products.[82] In the *Assayer* (1623), he wrote with convincing authority linking the study of God's works to the principles of mathematics:

> Philosophy is written in this grand book, which stands continually open to our gaze. But the book cannot be understood unless one first learns to comprehend the language and read the letters in which it is composed. It is written in the language of mathematics, and its characters are triangles, circles, and other geometrical figures without which it is humanly impossible to understand a single word of it; without these, one wanders about in a dark labyrinth.[83]

Stanley L. Jaki summarizes Galileo's foundations for scientific work:

> The creative science of Galileo was anchored in his belief in the full rationality of the universe as the product of the fully rational Creator, whose finest product was the human mind, which shared in the rationality of its Creator.[84]

4.6.4.1 PARALLAX: PROOF OF HELIOCENTRISM

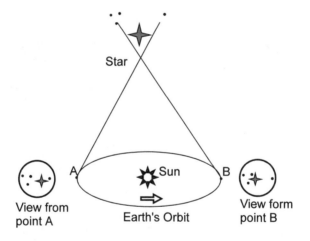

Figure 54: Parallax illustrated

If earth were in motion around the sun (heliocentricity), then a nearby star would exhibit a parallax (apparent angular shift) with respect to the more distant background stars if observed from two different positions of the earth's orbit. The instrumentation of Galileo's time could not measure a star's parallax. With accurate instruments and the principles of trigo-

---

[82]Galileo Galilei, *Dialogues Concerning the Two Chief World Systems: Ptolemaic and Copernican*, trans. Stillman Drake (Berkeley: University of California Press, 1962), p. 104.
[83]Galileo Galilei, *Discoveries and Opinions of Galileo*, pp. 237-238.
[84]Jaki, *The Road of Science and the Ways to God*, p. 106.

nometry, a star's parallax can be determined as follows. Given 1 degree (1°) = 60 minutes (60') and 1 minute (1') = 60 seconds (60"), then 1° = 3600" or 1" = (1/3600)°. A parsec (pc) is the distance to a celestial object that exhibits a *parallax* of one *second* (1") when viewed from the earth at a right angle to the sun. One astronomical unit (AU) is the average distance from the earth to the sun (1 AU = 93,000,000 miles). Using trigonometry, we can calculate the distance, in miles, of one parsec (pc). Let r = 1 pc. Then sin 1" = (1 AU/r). Or, r sin 1"= 1 AU and r = (1 AU/sin 1") = 1.914 x 10^{13} miles = 206,265 AU. Since light travels at approximately 186,000 miles per second, it travels 5.87 x 10^{12} miles per year. Therefore, 1 pc = (1.914 x 10^{13})/(5.87 x 10^{12}) = 3.26 light-years.

In 1838, the Prussian astronomer Friedrich Wilhelm Bessel (1784-1846), who also developed a class of mathematical functions (called Bessel functions) based upon the study of planetary perturbations, measured the parallax of the star 61 Cygni to be 0.29". Remember, 1 pc = 206,265 AU = 1" measured parallax. If a star's parallax is less than 1", then the star is greater than 1 pc away. If a star's parallax is greater than 1", then the star is less than 1 pc away. 61 Cygni's distance, d, is calculated as follows: d = (1 AU)/sin (0.29") = 711,258 AU = 3.45 pc = 11.25 light-years.

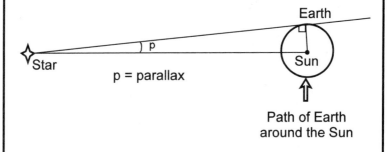

Figure 55: Calculation of Stellar Parallax

The actual determination of parallax, as given in this historical example, is considered by many astronomers to be the principal proof that heliocentricity is the cosmology of our solar system.

4.6.5 ISAAC NEWTON

Born in the year that Galileo died, Sir Isaac Newton (1642-1727) was going to be a farmer but turned out to be universally ranked as the greatest mathematician of all time. According to Howard Eves, "His insight into physical problems and his ability to treat them mathematically has probably never been excelled."[85] For the modern heirs of his genius, it is fortunate and providential

---

[85]Eves, p. 326.

that young Newton was not lost to the plague that ravaged London and the university town of Cambridge from 1665-1667.

Tributes paid to him border the sublime. A famous contemporary and sometimes opponent, Gottfried Wilhelm Leibniz (1646-1716), said, "Taking mathematics from the beginning of the world to the time when Newton lived, what he did was much the better half."[86] The poet Alexander Pope gave tribute with these famous lines,

> Nature and Nature's laws lay hid in night;
> God said, "Let Newton be," and all was light.

Albert Einstein remarks that "his great and lucid ideas will retain their unique significance for all time as the foundation of our whole modern conceptual structure in the sphere of natural philosophy."[87] In the preface of Newton's book, *Opticks*, Einstein continues his praise, "Nature to him was an open book, whose letters he could read without effort."[88]

Figure 56: Sir Isaac Newton

Newton himself took a humbler approach to his accomplishments. Just before his death, he said:

I do not know what I may appear to the world; but to myself I seem to have been only like a boy playing on the seashore, and diverting myself in now and then finding a smoother pebble or a prettier shell than ordinary, whilst the great ocean of truth lay all undiscovered before me.[89]

He recognized his indebtedness to those who preceded him, explaining that the only reason for seeing farther than other men was "because I have stood on the shoulders of giants."[90]

Possessing remarkable powers of concentration, he was able to often spend eighteen or nineteen hours a day in writing.[91] This diligence in thought was the secret of his ability to read nature like an "open book." As English physicist E. N. da Costa Andrade (1887-1971) observes, "I would rather say that Newton was capable of greater sustained mental effort than any man, be-

[86]Cited in Moritz, p. 167.
[87]Einstein, p. 59.
[88]Isaac Newton, *Opticks* (New York: Dover Publications, [1704, 1931, 1952] 1979), p. lix.
[89]Cited in E. N. Da Costa Andrade, "Isaac Newton," *The World of Mathematics*, ed. James R. Newman (New York: Simon and Schuster, 1956), 1:271.
[90]This phrase was not original with Newton. It first appeared in *Metalogicon* (Book 3, chapter 4) by John of Salisbury (ca. 1115-1180). See Stanley L. Jaki, *The Savior of Science* (Grand Rapids: Eerdmans, [1988] 2000), p. 51.
[91]Moritz, p. 170.

fore or since."[92] As Stanley L. Jaki reflects, "Newtonian science was the product of a truly inventive intellect pondering the witness of the senses."[93]

What did Newton accomplish with all of this concentrated effort? With his famous three laws of motion, he mathematically explained forces and movement (remember, the first law, the law of inertial motion, was anticipated by Philoponus in the 6th century and then again by Buridan in the 14th century).[94] He developed theories that explained not only the motion of planets and comets, but how the movement of the moon affects ocean tides on the earth. He worked out the color spectrum of light. He invented the first modern reflecting telescope. He fully developed the method of the calculus, a mathematical method that uses the concept of infinity to understand and calculate instantaneous rates of change and areas of curved surfaces. Hailed as one of the greatest achievements of the human mind, Newton could use it to predict the velocity necessary for a projectile to escape the gravitational pull of the earth. With space age rocketry, his predictions have been marvelously confirmed. Above all, he united terrestrial and celestial motion in one, all encompassing mathematical equation, the law of universal gravitation. Making deductions from this law, Newton brilliantly verified Kepler's three laws of planetary motion.[95]

4.6.5.1 NEWTON'S THREE LAWS OF MOTION

1. Law of inertial motion: a body will continue to be in a state of rest or of uniform velocity (speed and straight-line direction) unless acted upon by an external force. Newton's rendering: "Every body perseveres in its state of rest, or of uniform motion in a right line, unless it is compelled to change that state by forces impressed thereon."[96]
2. The acceleration (rate of change of velocity), a, produced in a body of mass, m, is in direct proportion to the force, F, applied. Mathematically expressed: $F=ma$. Newton's rendering: "The alteration of motion is ever proportional to the motive force impressed; and is made in the direction of the right line in which that force is impressed."[97]

---

[92]Andrade, 1:275.

[93]Jaki, *The Road of Science and the Ways to God*, p. 119.

[94]The link connecting Buridan to Newton was René Descartes (1596-1650). Although Galileo's experiments confirmed this law, Newton took its conceptualization from Descartes without giving him credit. In 1644, Descartes published a book, significantly entitled *Principia*, in which he postulated ten laws of motion. The first two are almost identical to Newton's first two laws while the remaining eight are inaccurate. Descartes learned of Buridan's ideas, if not of Buridan by name, by way of the traditions taught to him during his student years at the Jesuit College in La Flèche (see section 5.2).

[95]For a detailed description of how Newton did this, see Morris Kline, *Mathematics and the Physical World* (New York: Dover Publications, [1959] 1980), pp. 238-255. The bridge between Kepler and Newton was erected by the writings of the Englishman Jeremiah Horrocks (1618-1641), the single competent devotee of Kepler in the 17th century.

[96]Isaac Newton, *The Principia*, trans. Andrew Motte (Amherst: Prometheus Books, [1687, 1848] 1995), p. 19.

[97]*Ibid.*

3. Whenever two bodies experience an interaction, the force of the first
on the second is equal and opposite to the force of the second body on the
first. Newton's rendering: "To every action there is always opposed an equal
reaction: or the mutual actions of two bodies upon each other are always equal,
and directed to contrary parts."[98]

In 1696, Newton became a member of English parliament and acted as
warden and master of the British mint. In 1705, Queen Anne (1665-1714)
knighted him. From 1703 up until his death, Newton was president of the
Royal Society of London, a scientific club founded in 1661 by English Puritans.[99]

Newton not only did work in mathematics and science; he also wrote volumes of material expositing Bible passages. According to E. N. da Costa Andrade:

> The works of the Church Fathers
> [were] prominent in his library. His
> two books the *Chronology of Ancient
> Kingdoms Amended* and *Observations
> upon the Prophecies of Daniel and the
> Apocalypse of St. John* probably cost
> him as much effort as *The Principia.*
> There were over 1,300,000 words in
> manuscript on theology in the
> Portsmouth papers, according to my
> estimate from the catalogue.[100]

Today, scientists view Newton's
work in theology as a waste of time.[101]
This was not so to Newton. To him, the
message of the Bible provided the basic
foundation and motivation for all his scientific works.

Figure 57: Newtonian telescope

His most celebrated work is *Mathe-
matical Principles of Natural Philosophy* or more briefly, *The Principia.* Stanley L. Jaki
comments about the influence of this work:

> The vision of the world it embodied was ultimately a creation of the mind,
> a leap from sensory data far beyond the range of the sense. But because
> that vision was rooted in data provided by nature, the vision could become

---

[98]*Ibid.*

[99]Hooykaas, pp. 95-99.

[100]Andrade, 1:274.

[101]As an example, Robert K. Merton cites Cesare Lombroso's reaction to Newton's exposition on
the book of Revelation, "Newton himself can scarcely be said to have been sane when he de-
meaned his intellect to the interpretation of the Apocalypse." In *Science, Technology & Society in Seventeenth Century England* (New York: Howard Fertig, [1970, 1990] 2001), p. 106, n. 62.

as so vigorous and fruitful that the physical science
es became an ordinary science busy with unfolding
creative science of *The Principia*.[102]

N'S LAW OF UNIVERSAL GRAVITATION

lose to the surface of the earth is attracted as if
M, of the earth (assumed to be spherical) is con-
centrated at the center of the earth. This force, *F*, of attraction is defined
mathematically as follows: $F=GmM/D^2$ where *G* is the constant of gravitation and *D* is the distance between the two bodies of mass, *m* and *M*. In
Figure 58, this force turns out to be the weight of the man.

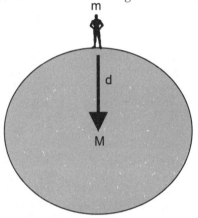

Figure 58: Newton's Law of Universal Gravitation

In the third edition of *The Principia*, Newton explained the true source of
the beautiful system of the universe:

> This most beautiful system of sun, planets, and comets could only proceed
> from the counsel and dominion of an intelligent and powerful Being. And
> if the fixed stars are the centres of other like systems, these, being formed
> by the like wise counsel, must be all subject to the dominion of One; ...
> This Being governs all things, not as the Soul of the world, but as Lord
> over all; and on account of his dominion he is wont to be called *Lord
> God...*, or *Universal Ruler* ...."[103]

He then begins to extol the attributes of God in a manner that would em-
barrass many shallow theologians of today:[104]

---

[102]Jaki, *The Road of Science and the Ways to God*, p. 87.

[103]Newton, *The Principia*, p. 440.

[104]Isaac Newton's doctrinal positions were unorthodox at times. For example, his Arian views on
the nature of Christ (which he kept secret) certainly were at odds with the theology of his fellow

The Supreme God is a Being eternal, infinite, absolutel
ing, however perfect, without dominion, cannot be said
And from his true dominion it follows that the true G
gent, and powerful Being; and, from his other perfections, una.
preme, or most perfect. He is eternal and infinite, omnipotent and omniscient; that is, his duration reaches from eternity to eternity; his presence
from infinity to infinity; he governs all things, and knows all things that are
or can be done.... We know him only by his most wise and excellent con-
trivances of things, and final causes; we admire him for his perfections; but
we reverence and adore him on account of his dominion: for we adore him
as his servants; and a god without dominion, providence, and final causes,
is nothing else but Fate and Nature.... And thus much concerning God; to
discourse of whom from the appearances of things does certainly belong
to Natural Philosophy.[105]

In *Opticks*, Book 3, part 1, he continues to extol the Creator God as the
original cause of the workings of the universe. By doing so, he emphasized the
need of a nonmechanical ultimate cause of mechanical patterns reflected in his
model of the universe:

... The main business of natural Philosophy is to argue from phaenomena
without feigning Hypotheses, and to deduce Causes from Effects, till we
come to the very first Cause, which certainly is not mechanical.... What is
there in places almost empty of Matter, and whence is it that the Sun and
Planets gravitate towards one another, without dense Matter between
them? Whence is it that Nature doth nothing in vain; and whence arises all
that Order and Beauty which we see in the World? To what end are Comets, and whence is it that Planets move all one and the same way in Orbs
concentrick, while Comets move all manner of ways in Orbs very excen-
trick; and what hinders the fix'd Stars from falling upon one another? How
came the Bodies of Animals to be contrived with so much Art, and for
what ends were their several Parts? Was the Eye contrived without Skill in
Opticks, and the Ear without Knowledge of Sounds? .... And these things
being rightly dispatch'd, does it not appear from Phaenomena that there is
a Being incorporeal, living, intelligent, omnipresent, who in infinite Space,
as it were in his Sensory, sees the things themselves intimately, and thoroughly perceives them, and comprehends them wholly by their immediate
presence to himself ... And though every true Step made in this Philosophy brings us not immediately to the Knowledge of the first Cause, yet it
brings us nearer to it, and on that account is to be highly valued.[106]

Englishmen, the Puritans. In spite of his defective Christological views, Newton's biblical and Puritan upbringing retained within him a feeling of awe for the Creator and Ruler of all.
[105]Newton, *The Principia*, pp. 440-442.
[106]Newton, *Opticks*, pp. 369-370.

Newton's scientific method consisted of "the mathematization of material processes and the insistence on the agreement of conclusions with observational data."[107] He perfectly balanced Francis Bacon's radical *a posteriori* empiricism and René Descartes' (1596-1650) *a priori* rationalism (see section 5.3). In him, "we see the use of the mind which transcended empiricism without being trapped in a priorism."[108]

4.6.6 GOTTFRIED WILHELM LEIBNIZ

As already mentioned, the German philosopher and mathematician, Gottfried Wilhelm Leibniz, was a contemporary of Newton. Both he and Newton fine-tuned the methods of the calculus independent of each other.[109] The great controversy of the times was over who did this first. While Newton developed the theory first, Leibniz was first to publish. Unfortunately, both parties discharged some rather vicious and vitriolic volleys across the North Sea in this dispute over priority.

Figure 59: Gottfried Wilhelm Leibniz

Leibniz also constructed the first calculating machine that multiplied and divided (1671) and was a pioneer in the development of mathematical logic. In spite of his work in mathematics, he produced very little in the realm of physics. He also delved into history, politics, jurisprudence, economics, theology, and philology.

Leibniz believed in a "preestablished harmony between thought and reality."[110] He testified that "it is especially in sciences ... that we see the wonders of God, his power, wisdom and goodness; ... that is why, since my youth, I have given myself to the sciences that I loved."[111]

Great pains have been made by some mathematics historians to point out that the above men did not concern themselves with the why of their discoveries, but only with the how. Although Newton, like the others before him, focused on mathematical descriptions, he did not concern himself too much with the why because *he had no problem with it*. In the words of Francis Schaeffer,

---

[107]Jaki, *The Road of Science and the Ways to God*, p. 84.

[108]*Ibid.*, p. 90.

[109]The well-devised symbolism that Leibniz employed, $\dfrac{d^n f(x)}{dx^n}$ for the n-th derivative of f(x), carried the day over Newton's fluxional notation.

[110]Gottfried Wilhelm Leibniz, *New Essays Concerning Human Understanding*, trans. Alfred Gideon Langley (New York: Macmillan, 1896), p. 507. He was a metaphysical atomist believing that the universe is composed of countless conscious centers of spiritual force or energy, known as monads. The universe that these monads constitute is the harmonious result of a divine plan.

[111]Gottfried Wilhelm Leibniz, *Leibniz Selections*, ed. Philip P. Wiener. (New York: Charles Scribner's Sons, 1951), p. 596.

"Newton and these other scientists would have been obsessed with *how* the universe functions, but profess' question 'Why'?"[112] The biblical God, the first cause a provides the final explanations of the workings of the ...

4.7 BIBLICAL THEOLOGY: THE MATCH THAT LIT THE FIRE

In summary, the match that lit the fires of the Scientific Revolution was the conviction that the biblical God had designed the universe in a rational and orderly fashion; in fact, so orderly that it could be described mathematically.[113] This fact of history at times mystifies and bewilders modern historians.

For example, the late Morris Kline, a prolific writer and Professor of Mathematics at New York University's Courant Institute of Mathematical Sciences, writes:

> The work of the 16th-, 17th-, and most 18th-century mathematicians was ... a religious quest. The search for the mathematical laws of nature was an act of devotion which would reveal the glory and grandeur of His handiwork.[114]

> Each of the great intellects possessed a combination of mathematical or scientific genius and religious orthodoxy which today are regarded as incompatible and possible only in a period of transition.[115]

> Scientists today have kept apart from their professional work the religious convictions that formerly motivated some of the finest research in the sixteenth to eighteenth centuries. From the modern viewpoint, the inspiration that these mathematicians and scientists drew from religious beliefs seems strange.[116]

It "seems strange" to Kline because of his adamant prejudicial bias against orthodox Christianity. For example, he makes the following conclusion about the initial rise of Christianity:

---

[112]Schaeffer, *How Should We Then Live? The Rise and Decline of Western Thought and Culture*, p. 136.

[113]Recall that the Greeks also believed in the rationality of the universe. Note also that many unbelieving scientists of modernity posit the same (see Chapter 6). Both perspectives have absolutized this rationality either in nature or in man's reason while the biblical perspective rests on the autonomy of God, not man or nature. What is the practical difference? While recognizing God's common grace to all men and cultures allowing them to discover valid laws and produce beneficial results thereby (in spite of their false worldviews), we must look to God's word for the practical difference. Note the very words of God in I Samuel 2:30, "Far be it from Me; for those who honor Me I will honor, and those who despise Me shall be lightly esteemed."

[114]Morris Kline, *Mathematics: The Loss of Certainty* (New York: Oxford University Press, 1980), pp. 3 35.

[115]Kline, *Mathematics in Western Culture* (New York: Oxford University Press, 1953), p. 259.

[116]Kline, *Mathematics and the Physical World*, p. 441.

From the standpoint of the history of mathematics, the rise of Christianity had unfortunate consequences.[117]

On one hand Kline says that Christianity retarded the growth of mathematics and on the other, he reluctantly admits that Christianity expedited the growth of mathematics. The former assertion fails to take into account why, as Reijer Hooykaas states, "the victory of Christianity did not bring an immediate liberation from the bonds of Greek metaphysics."[118] The latter acknowledgment is granted only because of Kline's honest reflection on the writings of Copernicus, Kepler, Galileo, Newton, etc., which is to be applauded.

Stanley L. Jaki comments that, when historians are baffled by what Kline identifies as "the religious convictions that formerly motivated some of the finest research," it is because they "have never experienced what it means to look at the world as the product of a personal, rational Creator."[119] Historians cannot and should not separate the motivations of a scientist from the work of a scientist. To Copernicus, Kepler, Galileo, Newton, and a host of others, belief in a Creator God gave them the momentum for their work in science and mathematics.

Not all scientists during this era were thoroughgoing biblical Christians, but all of them breathed the refreshing and invigorating air that was produced by a culture steeped in biblical perspective. Francis Schaeffer observes that "the majority of those who founded modern science, from Copernicus to Maxwell, were functioning on a Christian base."[120]

4.8 THE REFORMATION AND MODERN SCIENCE

Both Reijer Hooykaas and Rousas J. Rushdoony pinpoint the indispensable role of Protestantism, especially Calvinist Puritanism, in the rise of modern science.[121] It is not by accident that the Reformation triumphed in areas (e.g., England, Germany, the Netherlands) that modern science later attained its greatest development. According to Robert K. Merton (1910-), "… among the original list of members of the Royal Society in 1663, 42 of the 68 for whom information pertaining to religious leanings is available, were clearly Puritan."[122] According to English physicist, inventor, and mathematician Robert Hooke (1635-1703), in a memorandum of 1663, the business and design of the Royal Society is:

> To improve the knowledge of naturall things, and all useful Arts, Manufactures, Mechanick practises, Engynes and Inventions by Experiments – (not

Ancient to Modern Times, p. 180.
presented in sections 3.9 through 3.11.
ys to God, p. 47.
.49 and Rushdoony, The Philosophy of the Christian Curriculum, p. 72.
s and Moderns: A Study of the Rise of the Scientific Movement in Seven-
Dover Publications, [1936, 1961] 1982), pp. 87-182.

meddling with Divinity, Metaphysics, Moralls, Politicks, Grammar, rick, or Logick). To attempt the recovering of such allowable arts and ventions as are lost. To examine all systems, theories, principles, hypoth ses, elements, histories, and experiments of things naturall, mathematicall, and mechanicall, invented, recorded, or practised, by any considerable au- thor ancient or modern. In order to the compiling of a complete system of solid philosophy for explicating all phenomena produced by nature or art, and recording a rationall account of the causes of things. All to advance the glory of God, the honour of the King, the Royall founder of the Society, the benefit of His Kingdom, and the generall good of mankind.[123]

The high proportion of Puritans among English scientists in the 17th cen- tury, also called the "century of genius," is all the more striking when one con- siders the fact that the Puritans were never more than about four percent of the population.[124] It is also important to note the proximity of the date of this memorandum with the completion of the Westminster Confession of Faith in 1646, a powerful and influential Presbyterian document, in which the Shorter Catechism begins, "The chief end of man is to glorify God, and to enjoy Him forever." According to Charles Dykes, "Puritanism created a spiritual climate in England uniquely favorable to the pursuit and development of experimental science, as well as the personal qualities, objectives, and values so essential to progress."[125] Note the observation of Frank Manuel on the English scientists of this era, "So many divines doubled as scientists, the coexistence in one head of expert knowledge in both books [book of God's word and the book of God's works – J.N.] came to be respected, and the capacity of a man to reveal the glory of God in both spheres was taken for granted."[126]

In a paper written in 1962, Thomas F. Torrance (1913-), born in China where his parents were missionaries, asked an important question, "Why was it then that modern empirical science had to wait until the beginning of the 17th century for its real advance?" As we have seen, scientific empiricism certainly existed before this time (beginning in the High Middle Ages) but, like an expo- nential curve, it "shot off the graph" in the 17th century and it did so primarily in Protestant, not Catholic, lands. Torrance notes that, under the influence of Roman Catholicism, there was a tendency to understand God in terms of a "mystic vision" or a pure contemplation of the divine.[127] As a whole, Reformed theology rejected this subjective mysticism in favor of objective obedience to God's written revelation. Also, because of the leavening influence of the teach- ing of Thomas Aquinas, there was a tendency to separate natural revelation in

---

[123]Cited in C. R. Weld, *History of the Royal Society* (London, 1848), 1:146-147 and cited in A. Rupert Hall, *From Galileo to Newton* (New York: Dover Publications, [1963] 1981), p. 132.
[124]See David Little, *Religion, Order and Law* (New York: Harper Torchbooks, 1969), p. 259.
[125]Dykes, p. 45.
[126]Frank Manuel, *The Religion of Isaac Newton* (London: Oxford University Press, 1974), p. 32.
[127]Not all Roman Catholics followed this mystic vision to the detriment of science. See Copernicus and Galileo as examples.

on) from supernatural revelation in Scripture (ac-
…all see in section 5.1, because of this dichotomy
…otomy), it became easy to see the natural order as
…y autonomous natural laws (an unfortunate and un-
…of Nicole Oresme).

…he Reformers "men
… …unk differently of God and of his
relation to creation as something utterly distinct
from him while yet dependent upon his will for
its being and ultimate order, and therefore
learned to think differently of the nature of nature and of the creaturely nature of its order."[128]
The tenets of the Reformation "gave new significance to the world as the object of divine attention, and therefore as the object of human
attention in obedience to the divine."[129] According to Charles Dykes, "The Reformation restored the biblical perspective and thereby
stimulated an interest in God's handiwork."[130]
As Torrance notes, this refined and renewed

Figure 60: John Calvin

perspective "is one of the great contributions of the Reformation to the modern world, for out of it came the spirit and procedure so characteristic of
modern science."[131]

At this point in our analysis, we must take a parenthetical recess. Some dis-
cerning readers may rebut this analysis of the positive impact of Reformation
thought on the progress of science by appealing, "Did not some of the early
Reformers strongly denounce the theories of Copernicus?" This question is
usually answered in the affirmative based upon a host of modern sources. The
most famous popularizer of the "Protestant renunciation of Copernicus" was
Andrew Dickson White (1832-1918). He said that "all branches of the Protestant Church – Lutheran, Calvinist, Anglican – vied with each other in denouncing the Copernican doctrine as contrary to Scripture; and, at a later period, the

---

[128]Thomas F. Torrance, "The Influence of Reformed Theology on the Development of Scientific
Method," *Theology of Reconstruction* (Grand Rapids: Eerdmans, 1965), p. 63.
[129]*Ibid.*, p. 65.
[130]Dykes, p. 38. The quotations in this paragraph should not be misinterpreted to mean that the
worldview perspective of the scientists of the medieval era (especially the High Middle Ages) was
completely inaccurate. As we have seen in sections 3.14 through 3.16, this was not the case. It must
be emphasized that the author believes that the distinctives delineated by the Protestant reformers
and scientists were a biblical *refinement*, *clarification*, and *correction* of the synthesized and compromised
worldview of some of the medieval scholastics (especially the perspectives bequeathed by Thomas
Aquinas).
[131]Torrance, p. 67.

Puritans showed the same tendency."[132] White goes on to make the following remarks about the French-born Swiss Reformer John Calvin (1509-1564):

> Calvin took the lead [against Copernicus – J.N.] in his Commentary on Genesis, by condemning all who asserted that the earth is not at the centre of the universe. He clinched the matter by the usual reference to the first verse of the ninety-third psalm, and asked, "Who will venture to place the authority of Copernicus above that of the Holy Spirit?"[133]

Philosophy historian Bryan Magee mimics White by quoting Calvin as saying, "Who will venture to place the authority of Copernicus above that of the Holy Spirit?"[134] James Burke repeats this perspective:

> For twenty years before Copernicus' theory was published it was discussed all over Europe. Ironically, the earliest attacks came from the Protestants. Luther said: "People give ear to an upstart astronomer who strove to show that the earth revolves, not the heavens or the firmament, the sun and the moon ... the fool wishes to reverse the entire science of astronomy." His fellow revolutionary, Philipp Melanchthon [1497-1560], went further: "Fools seize the lover of novelty." Calvin looked to the Bible, "The Bible says, 'the world is also stablished that it cannot be moved'."[135]

Even Stanley L. Jaki, a meticulous science historian, notes this about Martin Luther (1483-1546):

> On the basis of Ockhamism, Copernicus's action was sheer foolishness. It was logical that an admirer of Ockham, Martin Luther, should be the first to call Copernicus a fool.[136]

---

[132]Andrew Dickson White, *A History of the Warfare of Science with Theology in Christendom* (Albany, OR: Sage Digital Library, [1896] 1996), p. 150. The Puritan that he is likely referring to is John Owen (1616-1683).

[133]Ibid., p. 151. White's *secondary* source for this quote was probably F. W. Farrar, *History of Interpretation* (London, 1886), p. xviii.

[134]Bryan Magee, *The Story of Thought: The Essential Guide to the History of Western Philosophy* (London: DK Publishing, 1998), p. 65. The quote is not documented.

[135]James Burke, *The Day the Universe Changed* (Boston: Little, Brown and Company, 1985), pp. 135-136. The dates for Melanchthon, a friend of Luther and tellingly denoted a revolutionary by Burke, are added. Burke does not document these quotes but they are taken verbatim from White.

[136]Jaki, *The Road of Science and the Ways to God*, p. 46. At least Jaki gives his source, albeit a *secondary* one: Alexandre Koyré, *La Révolution astronomique* (Paris: Hermann, 1961), p. 77. William of Occam (ca. 1285-ca. 1349), an English philosopher and scholastic, was a nominalist philosopher. By this, he believed that all abstractions, known as universals, are without essential or substantive reality, and that only individual objects have real existence (a harbinger of positivism). He also embraced what is called the "double-revelation theory" (following Aquinas). There is a truth that comes from revelation and a truth that comes from reason. Since these truths are radically separate, they can contradict each other and this is to be expected. His famous proposition, known as Occam's razor, states that when we have two hypotheses that both give an adequate account of the facts, then we should accept the simpler of the two hypotheses (it is more likely to be true). Occam's nominalism limited reality to what was conveyed by direct sensory perception. Hence, Copernicus' "rape of the senses"

Note that the *original* source documentation for the Luther and Calvin quotes is conspicuously absent. In discharging their fusillade of insults, most modern anti-Christian reciters of these quotes are drawing their ammunition from Andrew Dickson White. It would also be natural for Stanley L. Jaki, a Benedictine priest, to take Luther to task without adequate substantiation. The analysis of Reijer Hooykaas is critical at this point:

> "There is no lie so good as the precise and well-detailed one", and this one has been repeated again and again, quotation-marks included, by writers on the history of science, who evidently did not make the effort to verify the statement. For fifteen years, I have pointed out in several periodicals concerned with the history of science that the "quotation" from Calvin is imaginary and that Calvin never mentioned Copernicus; but the legend dies hard. It seems strange that Farrar [the probable source of White's quote – J.N.], who in the body of his work did full justice to the scholarly character of Calvin's method of exegesis, could go so far astray in the Introduction. I became suspicious of his statement because it does not fit in with Calvin's exegetical principles and because a parallel quotation allegedly from the Independent divine John Owen could immediately be proven to be spurious.[137]

Concerning the Luther quote, Hooykaas continues:

> Much stress is often laid on Luther's attitude in order to corroborate the statement that the Reformers and the Protestants, because of their biblicism, were in general less favourably inclined towards Copernicus' system than the Roman church before the condemnation of Galileo. Luther indeed in one of his table-talks rejected the opinion of an astronomer according to whom the sun was standing still, as a mistaken effort to be original: "I believe Holy Scripture, for Joshua told the sun to stand still, not the earth". But in his authorized works, Luther never mentioned the problem; it was just a commonsense remark, made when only rumours about Copernicus' work (not even his name is mentioned in the reminiscence of the reporter) were circulating (1539), and it was only printed (from the memory of one of his guests) twenty-seven years afterwards (1566). So this attitude could hardly have exerted much influence, the more so as it does not play a role in Lutheran doctrine.[138]

Finally, the remarks of Hooykaas on Melanchthon:

> Only Melanchthon, who always remained faithful to Aristotelian philosophy, at first condemned the doctrine of the motion of the earth, and said

---

(our sense tells us that the sun revolves around the earth, not vice versa) would have been anathema to Occam.

[137]Hooykaas, p. 121.

[138]*Ibid.,* pp. 121-122.

that the magistrates ought to punish its proclamation. But one year afterwards, in his second edition [entitled *Initia doctrinae physicae* – J.N.], this passage was omitted. Melanchthon was on very friendly terms with Petreius, the printer of Copernicus' work, and in an oration (1549) on his lately dead friend Caspar Cruciger (1504-1548), he mentioned that the latter was an admirer of Copernicus. Moreover, he gave protection to Rheticus, Copernicus' only immediate pupil.[139]

What about Psalm 93:1 and Joshua 10:12-13? Defenders of geocentricism, both Roman Catholic and Protestant, now and during the time of Copernicus, use these two passages to support their thesis. Supporters of heliocentrism use these passages, now and during the time of Copernicus, as confirmation of the "double-revelation" theory of Occam and Aquinas. That is, science has authority in the areas of reason and experience and Scripture has the authority in the realm of faith and we can expect contradictions like this. Is there an answer to this apparent riddle? Let us look at each passage in turn. First, the passage from Psalms reads:

> The Lord reigns, He is clothed with majesty; The Lord is clothed, He has girded Himself with strength. Surely the world is established, so that it cannot be moved [or shaken – J.N.] (Psalm 93:1; see also Psalm 96:10).

The passage commences with a declaration of the sovereignty of God and His "clothing of majesty and strength." Is God *literally* clothed with these immaterial attributes? Obviously, the Psalmist is using clothing as a figure of speech. In the same context, are we to take the next phrase in an absolute, *literal* sense? Most, if not all, supporters of geocentrism would answer with a resounding affirmative. Note first that the Hebrew word translated "moved" actually means "to waver, to slip, fall or shake." The idea of the earth moving (in a cosmological sense) makes no sense if move really means shake. If one still wants to take this passage as a cosmological proof of geocentricity (the earth does not move), then he is faced with a multitude of other Scriptural passages that say that earth "shall be moved" or it "shall be shaken!" Only one counterexample is necessary. Psalm 60:1-2, "O God, You have cast us off; You have broken us down; You have been displeased; Oh, restore us again! You have made the earth tremble; You have broken it."[140] The Hebrew word translated "tremble" is the same word translated "moved" in Psalm 93:1. *A careful study of all of these pertinent Scriptural passages reveals their primary ethical or providential, not metaphysical or cosmological, character.* "Shaking" implies judgment, a sense of the foundations being removed. "Not shaken" suggests blessing or a sense of foundations solid and sure. Those rightly related to the reign of God, His kingdom, will not be shaken in an ethical sense (Hebrews 12:26-27). Also, the reign of God guarantees the

---

[139]*Ibid.*

[140]See also Job 9:6; Psalm 18:7; Isaiah 2:19-21; Isaiah 13:13; Isaiah 24:18 (cf. Psalm 104:5); Joel 3:16; Haggai 2:6-7, 21.

ethical stability of the earth under His righteous and providential governance. That is, the forces of instability (evil, disorder, or chaos) will never have final and absolute sway in the world.

Now to the Joshua passage:

> Then Joshua spoke to the Lord in the day when the Lord delivered up the Amorites before the children of Israel, and he said in the sight of Israel: "Sun, stand still over Gibeon; And Moon, in the Valley of Aijalon." So the sun stood still, and the moon stopped, till the people had revenge upon their enemies. Is this not written in the Book of Jasher? So the sun stood still in the midst of heaven, and did not hasten to go down for about a whole day (Joshua 10:12-13).

If the Bible teaches heliocentricity, then, according to geocentrists, Joshua should have said, "Earth, stop rotating!" instead of "Sun, stand still!"[141] To a geocentrist, commanding the sun to "stand still" implies that it was normally in motion. The Hebrew word for "stood" means "to be dumbfounded, to be astonished, to holds its peace, to be quiet, to be silent." Yes, the sun did not appear to move in the sky (as detailed in verse 13), but was it *literally* dumbfounded, astonished, holding to its peace, quiet, and silent? Also, such a statement would hardly necessitate a geocentric view, since the sun could move without the earth being at the center of the solar system. As with the Psalm 93:1 passage, we need to use sound principles of biblical hermeneutics before we impulsively use Scripture to prove or disprove a cosmological model (as many people did both during the time of Copernicus and today).[142] The Bible speaks to cosmology providing some general detail, *but not exhaustive detail.* More importantly, the Bible shelters cosmology with an overarching metaphysical canopy. As stated in section 1.6.1, in the beginning, the unbeginning God created all things (cosmogony). There is an infinite chasm between transcendent God and the finite creation. All things, including the orderly heavenly movements, "hold together" in Him (cosmology).

Returning from our rather lengthy parenthetical hiatus, why was the Puritan emphasis so fruitful in 17th century scientific activity? There were four specific doctrines (or emphases) embraced by the Puritans that, when combined together, created a potent dynamism for scientific progress.

First, the Puritans believed in the absolute authority of Scripture. They rejected the "nature/grace" or "double-revelation" theory of Aquinas and placed

---

[141]One should note carefully, in this context, even though 99.99% of the scientists of today believe in heliocentrism, they do not "bat an eyelash" when they mention sunrise and sunset (these terms are technically geocentric conceptions).

[142]A proverbial lesson from science history: he who marries the science of the moment better be ready to become a widow tomorrow. Theologians in the past (and present) have been all too quick to either marry a scientific theory with biblical revelation or to bring God into the unexplained "gaps" or "loopholes" in scientific theory. We know this about scientific theories – they are always liable to updates or revisions in the future and eventually those "gaps" are filled in. When that happens, what happens to the "God of the gaps"? He becomes an unnecessary and irrelevant "fill in."

Scripture in a paramount position. By doing so, reason and were not negated. Reason, mathematics, and science were se minion. Resting under the fundamental and liberating worl delineated by Scripture, man was free to explore, ratiocinate, Both science and theology (nature and grace) have their place in God's all encompassing scheme of things. Science deals specifically with man in relation to God's *world* and theology deals specifically with man in relation to God's *word*. Where they differ is only in a matter of degree. Both, in reality, reflect *one revelation* of God and both presuppose and supplement each other.

Second, the Puritans believed that God is the Creator and Lawgiver. His faithfulness in sustaining His created order is the bedrock of, and the very confidence for, the derivation of scientific laws. Man was under divine duty to explore every aspect of God's mighty works in nature. The English naturalist John Ray (1627-1705) said, "because the Almighty created man able to study nature, He intended that man ought to study nature. The pursuit of natural philosophy is a religious duty."[143] Not only was it a religious duty, it was a duty of joy and pleasure. The words of Psalm 111:2 – The works of the Lord are great, studied by all who have pleasure in them – not only identified the spring of Kepler's creative genius; it was also the mainstay confession of the Puritan scientists of the 17th century.

Third, the Puritans were noted for their emphasis on vocation and calling. All men, they believed, are called of God to a specific task in this life and to benefit other men thereby. Vocational callings should not be pursued purely for reputation or for profit, but, as Francis Bacon declared, "for the glory of the Creator and the relief of man's estate."[144] Called by God, the Puritan scientist saw that his vocational activities, done for the glory of God, *were* the good works that God had prepared beforehand for him to walk therein (see Ephesians 2:8-10).[145] The Puritans enhanced and refined the sacred calling to secular work (pertaining to this age) that was introduced first by Martin Luther. About the emphasis of Luther, Karl Holl comments:

> Luther changed not only the content of the word 'calling'; he recoined the word itself. What is new is that in his mature years, he sees the 'call' of God exclusively in secular duties; i.e., he united just those two elements which for Catholic thought were contradictions that could scarcely coexist. Only timidly had the viewpoint dared to present itself in Catholicism that one could also heed the call of God in the world and in secular work. But by decisively including secular activities under this exalted viewpoint as a

---

[143]Cited in Richard S. Westfall, *Science and Religion in Seventeenth-Century England* (Ann Arbor: The University of Michigan Press, 1973), p. 46.
[144]Francis Bacon, 6:134.
[145]Contra to Roman Catholicism with its emphasis on good works that were *extraneous* to one's vocation.

God-given obligation, Luther diverts to it all the religious energy that here-tofore was exhausted in 'good works' alongside work in a vocation.[146]

The practical results of a Puritan "this-worldly" focus founded upon an "another-worldly" hope is best described by the historian William Haller (b.1885):

> ... nevertheless a lively expectation of becoming 'somebody' does not conduce to a willingness to remain nobody. Men who have assurance that they are to inherit heaven have a way of presently taking possession of earth. The practice of the Puritan code, with its insistence upon active use of individual abilities in the pursuit of an honest calling, was, as a matter of fact, already putting a solid share of the fruits of the earth into the coffers of the elect.[147]

Fourth, Puritan eschatology engendered an optimistic outlook for the fu-ture.[148] The biblical doctrine of progress in time and history spurred the Puritan scientist with a confidence that both his explorations and discoveries would be useful to mankind now and in the future – *Soli Deo Gloria* (to the glory of God *alone*).

4.9 WHY EUROPE?

Although the rudiments of science and mathematics can be found in the Egyptian, Babylonian, Greek, Chinese, Hindu, and Saracen civilizations, why was Europe such a powerful force in its full development? Reijer Hooykaas ob-serves:

---

[146]Karl Holl, *The Cultural Significance of the Reformation* (New York, 1959), pp. 34-35.

[147]William Haller, *The Rise of Puritanism* (New York: Harper Torchbooks, [1938] 1957), pp. 162-163. See also Leland Ryken, *Worldly Saints: The Puritans As They Really Were* (Grand Rapids: Zondervan, 1986).

[148]Eschatology refers to the "doctrine of last things." Specifically and primarily, it relates to issues like the Second Advent of Christ, the future day of judgment and resurrection, and the eternal state. Secondarily, it relates to issues in time and history. In this context, the Puritans self-consciously embraced what is known as a postmillennial perspective on time and history. That is, the Gospel will progressively have a positive healing impact, intensively and extensively, on society in time and in history (Matthew 13:31-33). See St. Athanasius, *On the Incarnation* (Crestwood: St Vladimir's Seminary Press, [1944] 1996), Benjamin B. Warfield, *The Savior of the World* (Edinburgh: Banner of Truth Trust, [1916] 1991), James R. Payton, Jr. "The Emergence of Postmillennialism in English Puritanism," *The Journal of Christian Reconstruction* (Vallecito, CA: Chalcedon, 1979), 6:1, pp. 87-106, Iain Murray, *The Puritan Hope: Revival and the Interpretation of Prophecy* (Edinburgh: Banner of Truth Trust, [1971] 1991), James A. DeJong, *As the Waters Cover the Sea: Millennial Expectations in the Rise of Anglo-American Missions 1640-1810* (Kampen: J. H. Kok N.V., 1970), Greg L. Bahnsen, "The Prima Facié Acceptability of Postmillennialism," *The Journal of Christian Reconstruction* (Vallecito, CA: Chalcedon, 1976-1977), 3:2, pp. 48-105, John Jefferson Davis, *Christ's Victorious Kingdom: Postmillennialsim Reconsidered* (Grand Rapids: Baker, 1986), Keith Mathison, *Postmillennialism: An Eschatology of Hope* (Phillipsburg: Presbyterian and Reformed, 1999), and Greg Bahnsen, *Victory in Jesus: The Bright Hope of Postmillennialism*, ed. Robert Booth (Texarkana: Covenant Media Press, 1999).

Without claiming any intellectual superiority for the scientists of the Renaissance and Baroque periods over their ancient and medieval European predecessors or over Oriental philosophers, one has to recognize as a simple fact that "classical modern science" arose only in the western part of Europe in the sixteenth and seventeenth centuries.[149]

Alfred North Whitehead summarizes the developments of science in the various civilizations:

Chinese science is practically negligible. There is no reason to believe that China if left to itself would have ever produced any progress in science. The same may be said of India. Furthermore, if the Persians had enslaved the Greeks, there is no definite ground for belief that science would have flourished in Europe. The Romans showed no particular originality in that line. Even as it was, the Greeks, though they founded the movement, did not sustain it with the concentrated interest modern Europe has shown.[150]

Figure 61: Europe

Alistair C. Crombie provides this technological analysis of ancient civilizations:

It was the Greeks who invented science as we now know it. In ancient Babylonia, Assyria and Egypt, and in ancient India and China, technology had developed on a scale of sometimes astonishing effectiveness, but so far as we know it was unaccompanied by any conception of scientific explanation.[151]

Concluding these observations, the 18th century French historian of astronomy and politician, Jean Sylvain Bailly (1736-1793), who met his death on a French guillotine, said, "When one considers with attention the state of astronomy in Chaldea, in India, and in China, ONE FINDS THERE RATHER THE DEBRIS THAN THE ELEMENTS OF SCIENCE."[152] We will next briefly analyze the scientific debris of three civilizations in particular.

---

[149]Hooykaas, p. 161.

[150]Whitehead, *Science and the Modern World,* p. 7.

[151]Crombie, 1:24.

[152]Jean Sylvain Bailly, *Historie de l'astronome ancienne depuis son origine jusqu'a l'etablissment de l'Ecole d'Alexandrie* (Paris: chez les Freres Debure, 1775), p. 18.

4.9.1 CHINA

Joseph Needham (1900-1995) is perhaps the world's expert on the development of science in China.[153] He compares the Western mind with the Chinese mind:

In order to believe in the rational intelligibility of Nature, the Western mind had to presuppose (or found it very convenient to presuppose) the existence of a Supreme Being who, himself rational, had put it there.... Such a supreme God had inevitably to be personal. This we do not find in Chinese thought.[154]

He continues:

It was not that there was not an order in Nature for the Chinese, but rather that it was not an order ordained by a rational personal being, and hence there was no guarantee that other rational personal beings would be able to spell out in their own earthly language the preexisting divine code of laws which he had previously formulated.[155]

Because of this worldview defect, W. W. Rouse Ball notes that "... the Chinese made no serious attempt to classify or extend the few rules of arithmetic or geometry with which they were acquainted, or to explain the causes of the phenomena which they observed."[156]

Stanley L. Jaki explains why, "behind the unfathomable forces of the Yin and Yang there was nothing to look for, certainly not a Lawgiver, or a Governor of all."[157]

Whitehead adds to the mounting evidence:

In Asia, the conceptions of God were of a being who was either too arbitrary or too impersonal for such ideas to have much effect on instinctive habits of mind. Any definite occurrence might be due to the fiat of an irrational despot, or might issue from some impersonal, inscrutable origin of things. There was not the same confidence as in the intelligible rationality of a personal being.[158]

4.9.2 INDIA

Concerning the Hindu culture, an important question needs to be asked, Why did science fail to flourish in India after the formulation of the decimal

---

[153]For a thirteen-volume exposition, see Joseph Needham, *Science and Civilization in China* (Cambridge: Cambridge University Press, 1954).
[154]Joseph Neeham, *The Grand Titration: Science and Society in East and West* (London: George Allen and Unwin, 1969), pp. 325-326.
[155]*Ibid.*, p. 327.
[156]Ball, p. 9.
[157]Jaki, *Science and Creation: From Eternal Cycles to an Oscillating Universe*, p. 31.
[158]Whitehead, *Science and the Modern World*, p. 16.

system, negative numbers, and zero? Hindu mathematicians also understood fractions, problems of interest, the summation of arithmetical and geometrical series,[159] the solution of determinate and indeterminate equations[160] of the first and second degrees, permutations and combinations (see section 7.7.7), and other operations of simple arithmetic and algebra. They also developed trigonometric techniques for expressing the motions of the heavenly bodies and introduced trigonometric tables of sines. But, according to Howard Eves, "The astronomy [of the Hindus – J.N.] itself is of poor quality and shows an inaptness in observing, collecting and collating facts, and inducing laws."[161]

The Hindus loved making calculations; in fact, they gave mathematics the name "ganita" which means the science of calculation.[162] According to Carl Boyer, "they delighted more in the tricks that could be played with numbers."[163] Their work freed arithmetic from its geometrical tradition, a strategic move for future developments of the concepts of the calculus, but, according to Boyer, the Hindu culture did not appreciate or act upon the significance of this change.[164]

Morris Kline echoes the analysis of Boyer:

> There is much good procedure and technical facility, but no evidence that they considered proof at all. They had rules, but apparently no logical scruples. Moreover, no general methods or new viewpoints were arrived at in any area of mathematics. It is fairly certain that the Hindus did not appreciate the significance of their own contributions. The few good ideas they had ... were introduced casually with no realization that they were valuable innovations.[165]

Both Eves and Kline fail to give the reasons why the Hindu culture did not advance itself to a stage of theoretical generalizations leading to the formulation of quantitative laws and systems of laws. According to Stanley L. Jaki, the Hindu culture, in addition to its pantheistic, animistic, and cabalistic (numerologistic) roots, was obsessed with the perennial recurrence of cosmic cycles.[166] In this context, man is no more than a senseless product of an all-pervading biological rhythm. Jaki cites a classic illustrative statement by king Brihadratha: "In the cycle of existence I am like a frog in a waterless well."[167] The implications for scientific progress are ominous. As Jaki observes, the hold that the

---

[159]An arithmetic series is a sum of an arithmetic sequence; i.e., a + (a + b) + (a + 2b) + (a + 3b) +.... A geometric series is the sum of a geometric sequence; i.e., a + ax + ax^2 + ax^3 +....

[160]Determinate equations are equations that lead to a unique solution. Indeterminate equations are equations do not have a limited number of particular solutions; they may even have an infinite number of solutions.

[161]Eves, p. 187.

[162]Boyer, *The History of the Calculus and Its Conceptual Development*, p. 60.

[163]*Ibid.*

[164]*Ibid.*

[165]Kline, *Mathematical Thought from Ancient to Modern Times*, p. 190.

[166]Jaki, *Science and Creation: From Eternal Cycles to an Oscillating Universe*, p. 3.

[167]*Ibid.*, p. 7.

doctrine of eternal recurrence had on the Hindu mind "was strong enough to prevent the emergence of a positive and confident outlook on nature and on the value of man's activities concerning nature and society."[168] Because of the Hindu commitment to radical monism with its concomitant denial of reality, the spirit of experimental method simply could not assert itself.[169] As Jaki concludes, "Science ... cannot arise, let alone gain sustained momentum, without an articulate longing for truth which in turn presupposes a confident approach to reality."[170]

4.9.3 THE MUSLIM WORLD

The Arabs not only borrowed many important algebraic ideas from the Hindus, but they also inherited most of the learning bequeathed by the Greeks. According to S. Pines, "The medieval Arabs failed to add anything substantial to the Greek scientific tradition."[171] Concerning the Saracen civilization, Carl Boyer remarks that "Arabic thought lacked the interest which was necessary to further such fecund ideas."[172] In other words, no revolution in creative science occurred in their culture in spite of their extensive work in such key areas as algebra, the representation of numbers (they gave a more explicit form to the Hindu numerals and decimal notation), trigonometry (the transmission of the Hindu replacement of chords by sines), the theory of light, mechanics, hydrostatics, medicine, and alchemy (chemistry). Note also that an integral part of their scientific method was the attempt to achieve power over nature through alchemy, magic, and astrology.[173] Rousas J. Rushdoony remarks that science did flourish briefly in Arabic culture, only to stagnate later. This brief rise happened, not because of any motivating force in the religion of Islam, but because of a parasitic borrowing from Hellenic and Byzantine cultures.[174]

The theology of Islam is unitarian, static, and fatalistic. In the theology of the Koran, the "will of Allah" is a blind and unfathomable will in that it is not tied to wisdom or rationality. According to Rousas J. Rushdoony, because of this defective theology, stagnation is the order of the day in Islamic culture.[175]

---

[168]*Ibid.*, p. 9.

[169]Monism is the view that reality is a unified whole and that everything existing can be ascribed to or described by a single concept or system. The monism of Hinduism implies that everything that appears to be real is simply an illusion. Hence, the concept of an objective reality outside the mind of man and reflective of lawful purpose is nothing but a chimera.

[170]Jaki, *Science and Creation: From Eternal Cycles to an Oscillating Universe*, p. 19.

[171]S. Pines, "What was Original in Arabic Science?" *Scientific Change*, ed. Alistair C. Crombie (New York: Basic Books, 1963), pp. 204-205. Arab scientists did make significant advancements in ophthalmology. It must be noted, however, that this area of medical science has little or nothing to do with the laws of the physical world at large. As Stanley L. Jaki states, "Up to a certain level the practice of medicine could flourish without the need for an entirely new outlook on the physical world and its regularity" (*Science and Creation: From Eternal Cycles to an Oscillating Universe*, p. 195).

[172]Boyer, *The History of the Calculus and Its Conceptual Development*, p. 63.

[173]Kline, *Mathematical Thought from Ancient to Modern Times*, p. 196.

[174]Rushdoony, *World History Notes*, p. 120.

[175]*Ibid.*, p. 114.

Physician and philosopher Ibn-Rushd (1126-1198), better k
roës, is best known for his uncritical commentaries on Aristot
3.14). His writings reflect an absolute capitulation to Aristotle'
worldview, including the eternity of the heavens and the cyclical nature of time.

The physician and philosopher Ibn-Sina (980-1037), or Avicenna, wrote a
million-word long *Qanun* (Canon) that served for centuries as the standard textbook on Arab medical practice. According to Averroës, Avicenna, like him,
embraced pantheism in believing in the identity of the heavenly bodies with
God.[176]

The pantheistic and cyclical worldview of Averroës and Avicenna, although each made contributions to the science of medicine, nipped "in the bud
the ultimate prospects for science in the Muslim world."[177] In the realm of
mathematics, Morris Kline reflects:

> The Arabs made no significant advance in mathematics. What they did was
> absorb Greek and Hindu mathematics, preserve it, and, ultimately ...
> transmit it to Europe.[178]

4.9.4 ONE VIABLE BIRTH

According to Stanley L. Jaki, "... the history of science with its several
stillbirths and only one viable birth, clearly shows that the only cosmology, or
view of the cosmos as a whole, that was capable of generating science was a
view of which the principal disseminator was the Gospel itself."[179] Loren Eiseley (1907-1977), an evolutionary anthropologist, corroborates Jaki's thesis, "We
must also observe that in one of those strange permutations of which history
yields occasional rare examples, it is the Christian world which finally gave birth
in a clear articulate fashion to the experimental method of science itself."[180] Jaki
summarizes the basic reasons for this historical permutation:

> All great cultures that witnessed a stillbirth of science within their ambi-
> ence have one major feature in common. They all were dominated by a
> pantheistic concept of the universe going through eternal cycles. By con-
> trast, the only viable birth of science took place in a culture for which the
> world was a created, contingent entity.[181]

By using the word "contingent" Jaki notes that a variety of meanings are
attached to it. In Latin and the early Romance languages, contingent means

---

[176]Jaki, *Science and Creation: From Eternal Cycles to an Oscillating Universe*, p. 211.
[177]*Ibid.*, p. 195.
[178]Kline, *Mathematical Thought from Ancient to Modern Times*, p. 197.
[179]Jaki, *The Origin of Science and the Science of Its Origin*, p. 99. This does *not* invalidate the scientific and
mathematical work of the culture of the Egyptians, Babylonians, Greeks, Hindus, Arabs, Chinese,
or any individual unbeliever. It only means, in faithfulness to the testimony of history, that a culture
steeped in certain key worldview distinctives (i.e., the nature of God and the nature of nature) finally saw the rise of a viable and self-sustaining science.
[180]Loren Eiseley, *Darwin's Century* (Garden City: Doubleday and Company, 1958), p. 62.
[181]Jaki, *Science and Creation: From Eternal Cycles to an Oscillating Universe*, p. 357.

"something accidental, haphazard or random, under the broad umbrella of chance."[182] In the Oxford Dictionary of the English language, "randomness" or "accidental" dominates the shades of the meaning of contingency. Secondarily, the word refers to "the uncertainty of occurrence, fortuitousness, freedom from predetermining necessity, the condition of being at the mercy of accidents, conjecture of events without design, a merely possible future event, and finally a thing or outcome incident or dependent upon an uncertain event."[183] Finally, and this is the context in which Jaki uses the word in the above quote, the Oxford Dictionary notes that the word can mean "not being determined by necessity in regard to existence ... not existing of itself but in dependence on something else."[184] To Jaki, contingency means "the thorough dependence of a thing on another factor [i.e., the dependence of creation upon the Creator – J.N.], without assuring a necessary existence to that thing or, in our case, the universe [God could have made the universe differently and He chose to make it the way that it is – J.N.]."[185] The created order is contingent in that it reflects an ontological dependence on the biblical God, the ultimate metaphysical reality that is "beyond" it and "beneath" it.

The Bible's revelation of this metaphysical reality – an infinite, personal, and reasonable God who created a real universe that can be understood by man made in His image – is the key to the full and complete development of science and mathematics.

4.10 TWO KEY RECOGNITIONS

Common to most researchers during the Scientific Revolution was the understanding that mathematical law was *not* absolute. In writing to his friend, Johann Georg Herwart von Hohenburg, Kepler commented in early April of 1599:

> God wanted to have us recognize these laws when He created us in His image, so that we should share in His own thoughts. For what remains in the minds of humans other than numbers and sizes? These alone do we grasp in the proper manner and, what is more, if piety permits one to say so, in doing so our knowledge is of the same kind as the divine, as far as we at least in this mortal life, are able to comprehend something about these.[186]

Mathematics to Kepler enabled him analogously to "think God's thoughts after Him." Since God's thoughts are higher than man's thoughts (Isaiah 55:8-9), mathematical thought must always be seen as a tool in understanding,

---

[182]Stanley L. Jaki, *God and the Cosmologists* (Fraser, MI: Real View Books, [1989] 1998), p. 93.
[183]*Ibid.*
[184]*Ibid.*, p. 94.
[185]*Ibid.*
[186]Cited in Caspar, p. 380.

developing, and using God's creation; a tool that is always open to further refinement.

Reijer Hooykaas observes that "what strikes one most about the early Protestant scientists is their love for nature, in which they recognize the work of God's hands, and their pleasure in investigating natural phenomena."[187] Exposed to the wonders of God's creation, scientists and mathematicians were deeply impressed with the infinite wisdom of the Creator God. Kepler wrote to von Hohenburg on March 26, 1598, "... astronomers, as priests of God to the book of nature, ought to keep in their minds not the glory of their own intellect, but the glory of God above everything else."[188] Kepler prayed in *Harmonice mundi*:

> My mind has been given over to philosophizing most correctly: if there is anything unworthy of Thy designs brought forth by me – a worm born and nourished in a wallowing place of sins – breathe into me also that which Thou dost wish men to know, that I may make the correction: If I have been allured into rashness by the wonderful beauty of Thy works, or if I have loved my own glory among men, while I am advancing in the work destined for Thy glory, be gentle and merciful and pardon me; and finally deign graciously to effect that these demonstrations give way to Thy glory and the salvation of souls and nowhere be an obstacle to that.[189]

In the first edition (1596) of *Mysterium cosmographicum* (Mystery of the Universe), Kepler, upon contemplating his ability to unravel the mysteries of planetary motion, "breaks into tears and, feeling unworthy of this sign of God's grace, remembers the words which Peter spoke to the Master: 'Withdraw from me, for I am a sinful person.'"[190] Concerning this confession, Max Caspar, reflecting on the state of science in the mid-20th century, pens an insightful query, "Where is there another example of a natural philosopher who would make such a speech?"[191]

These two key recognitions identified the limitations of mathematics and the purpose of mathematics. Mathematics is (1) but a tool, a servant, that (2) aids man in unraveling the wisdom of God found in the harmonies and wonders of His works. Unfortunately, these perceptions, which were really warnings, were not fully heeded by the mathematical heirs of the Scientific Revolution. That story will be detailed in the next chapter.

---

[187]Hooykaas, p. 105.
[188]Cited in Carola Baumgardt, *Johannes Kepler: Life and Letters* (New York: Philosophical Library, 1951), p. 44.
[189]Kepler, p. 240.
[190]Caspar, p. 374.
[191]*Ibid.*, pp. 374-375. Because nature is no longer viewed as God's handiwork, but as Nature (the capital letter means nature is ultimate and understood as a self-contained system consisting of inherent and autonomous laws), the scientists of modernity are incapable of making this kind of confession. Instead of the knowledge of God's handiwork humbling the scientist, the knowledge of Nature inflates his already ostentatious ego.

4.11 SOVEREIGN GOD AT WORK

History is God's story; i.e., how He sovereignly directs the destinies of the nations to accomplish His purposes. From 1400 to 1800, the hand of the sovereign God directed the development of the scientific and technological devices, especially the mariner's compass and the marine chronometer that eventually enabled European civilization to expand its borders to the very ends of the earth.

The mariner's compass, only a tiny instrument, enabled explorers to transverse the seven seas. It did not take long for Christian missionaries to jump on board for a ride. William Carey (1761-1834), the founder of the Protestant missionary movement, wrote in 1792 about the removal of the impediments that had previously been in the way of carrying the Gospel among the heathen. He said:

Figure 62: The World

First, as to their distance from us, whatever objections might have been made on that account before the invention of the mariner's compass, nothing can be alleged for it, with any color of plausibility in the present age. Men can now sail with as much certainty through the Great South Sea as they can through the Mediterranean.... Yea, and providence seems in a manner to invite us to the trial, as there are to our knowledge trading companies, whose commerce lies in many of the places where these barbarians dwell.[192]

Both Catholic and Protestant missionaries used the technological advances of their day as a springboard in fulfilling God's worldwide missionary mandate. Paul Jehle, Christian educator, proclaims:

Though it is a scriptural principle in our spiritual life that "ye shall know the truth, and the truth shall make you free" (John 8:32), it is also an historical principle that when men had their eyes opened to spiritual truth, freedom to search out all other fields of knowledge to spread that truth appeared as well.[193]

Verna Hall (1912-1987), Christian historian and educator, echoes Jehle:

Almost immediately following Wycliffe's translation of the whole Bible (1380-1384), God began to call forth men to develop the many scientific

---

[192]William Carey, "An Enquiry into the Obligation of Christians to Use Means for the Conversion of the Heathens," *Perspectives on the World Christian Movement: A Reader*, ed. Ralph Winter and Steven Hawthorne (Pasadena: William Carey Library, 1981), p. 233.
[193]Paul Jehle, *Go Ye Therefore and Teach* (Plymouth: Plymouth Rock Foundation, 1982), p. 310.

and economic fields which would be necessary to enable man to seas, explore, and finally settle the lands across the vast Atlantic ocean.[194]

The dominion mandate of Genesis 1 and Christ's great commission of Matthew 28 work hand in hand. Each one serves the other and each one complements the other. When both are united, the heart purposes of God are fulfilled on earth.

4.12 THE BERNOULLI FAMILY

First generation	Second generation	Third generation	Fourth generation	Fifth generation
Nicholaus Senior (1623-1708)	Jakob I (1654-1705)			
	Nicholaus I (1662-1716)	Nicholaus II (1687-1759)		
	Johann I (1667-1748)	Nicholaus III (1695-1726)		
		Daniel I (1700-1782)		
		Johann II (1710-1790)	Johann III (1746-1807)	
			Daniel II (1751-1834)	Christoph (1782-1863)
				Johann Gustav (1811-1863)
		Jakob II (1759-1789)		

According to David E. Smith, "Students of heredity have called attention to the extraordinary number of distinguished scholars who descended from the protestant populations expelled from the catholic countries in the 16th and 17th centuries."[195] Of those scholars, the members of the Bernoulli family furnish one of the "most remarkable evidences of the power of heredity or of early home influence in all the history of mathematics."[196] Smith concludes, "No less than nine of its members attained eminence in mathematics and physics, and four of them were honored by election as foreign associates of the Académie des Sciences of Paris."[197] Listen to Eric Temple Bell's incredulous synopsis of the Bernoulli posterity:

---

[194]Verna M. Hall, *The Christian History of the American Revolution* (San Francisco: Foundation for American Christian Education, 1976), p. xxv.
[195]Smith, 1:426.
[196]*Ibid.*
[197]*Ibid.*, 1:426-427.

The Bernoulli family … in three generations produced eight mathematicians, several of them outstanding, who in turn produced a swarm of descendants about half of whom were gifted above the average and nearly all of whom, down to the present day, have been superior human beings. No fewer than 120 of the descendants of the mathematical Bernoullis have been traced genealogically, and of this considerable posterity the majority achieved distinction – sometimes amounting to eminence – in law, scholarship, science, literature, the learned professions, administration, and the arts. None were failures.[198]

Figure 63: Jakob Bernoulli

During John Napier's lifetime, the Reformation came to England. Concomitant with this portentous event, violent persecution of Protestants by Roman Catholics took place in Antwerp, Belgium. In 1583, a stream of refugees fled the city, one being Jacques Bernoulli who moved to Frankfort, Germany. In 1622, the Bernoulli family settled in Basel, Switzerland, and became successful businessmen. The sons of Nicholaus Bernoulli Sr. (1623-1708), great grandsons of Jacques, began the family mathematical dynasty.

Jakob Bernoulli (1654-1705), also known as Jacques or James, studied first to be a minister but his taste for astronomy, mathematics, and physics led him, against his father's wishes, to change vocational gears. He made significant contributions to coordinate geometry (see section 5.2), the theory of probability, and cultivated the calculus that he learned from Leibniz. He made one of the earliest and clearest statements regarding the limit of an infinite series, published posthumously in 1713:

> Even as the finite encloses an infinite series and in the unlimited limits appear, so the soul of immensity dwells in minutia and in narrowest limits no limits inhere. What joy to discern the minute in infinity! The vast to perceive in the small, what divinity![199]

He was also one of the first to make use of polar coordinates and in his 1690 publication entitled *Acta eruditorum* (Records of learning), we encounter for the first time the word "integral" in its calculus sense. Several topics in mathematics bear Jakob Bernoulli's name. Among the many are:

1. Theory of statistics and probability theory: Bernoulli distribution and Bernoulli theorem.

---

[198]Bell, *Men of Mathematics*, p. 131.
[199]Jakob Bernoulli, "On Infinite Series," *A Source Book in Mathematics*, ed. David E. Smith (New York: Dover Publications, [1929] 1959), p. 271.

2. Differential equations:[200] Bernoulli equation.
3. Number theory: Bernoulli numbers and Bernoulli polynomials.[201]
4. Calculus: lemniscate of Bernoulli.[202]

Among his many discoveries, and perhaps the most quintessential of them all, is the equiangular spiral, the curve found in the tracery of the spider's web, the pearly-lined shell of the chambered nautilus, and the convolutions of distant galaxies. Mathematically, it is related in geometry to the circle and in analysis to the logarithm (the curve is sometimes denoted as the logarithmic spiral). Amazed by the way this spiral reproduces itself under a variety of transformations he came to understand it as portraiture of his life and faith. Near the end of his life, he requested that it be etched on his grave marker, and with it the words *eadem mutat resurgo* (I shall arise the same, though changed). His tombstone bears this image and these words to this day.

Johann Bernoulli I (1667-1748), although a jealous and surly man, was a very successful mathematics mentor. He added fresh material almost continually to the store of mathematical analysis, especially differential equations. His sons, Daniel Bernoulli I (1700-1782) and Nicholaus Bernoulli III (1695-1726) were also very astute mathematicians. Nicholaus' promising career in mathematics was cut short at age 31. Daniel has been called the founder of mathematical physics. All who today work in the field of pure and applied fluid motion know the name Daniel Bernoulli.

At the University of Basel (ca. 1720), a young student came to study theology and Hebrew. In mathematics, his prowess garnered the attention of Johann Bernoulli who mentored him by giving him private lessons once a week. This student diligently spent the rest of the week preparing for the next lesson so that he would have to ask his teacher as few questions as possible. Soon, Daniel and Nicholaus Bernoulli III noticed the diligence and erudition of their father's protégé. They became fast friends for life. That student was Leonhard Euler.

4.13 THE INCREDIBLE EULER

Leonhard Euler (1707-1783) was a key figure in mathematics during his lifetime. Born in Switzerland and the son of a Calvinist pastor, he further exemplifies and illustrates some of the remarkable work accomplished by mathematicians of his time.

His mathematical productivity was truly incredible. He wrote a score of textbooks, covering many mathematical topics. Most of them became standard works for over a hundred years and even today, many university textbooks follow Euler's method and order. His *Introductio in analysin infinitorum*, first pub-

---

[200]A differential equation is an equation containing the derivatives of a function with respect to one or more independent variables.

[201]A polynomial is an algebraic expression consisting of one or more summed terms, each term consisting of a constant multiplier and one or more variables raised to positive integral powers; e.g., $x^2 - 5x + 6$.

[202]A lemniscate resembles the shape of a figure 8 on its side (i.e., ∞).

lished in 1748, is considered by mathematics historians to be one of the most influential textbooks in history.[203] In this text and others like it, we see Euler employing a new configuration of mathematical thinking. Before Euler's innovative style appeared on the mathematical scene, one meets with great difficulties in the attempt to understand the writings of his predecessors. Euler was the first to devise the ingenious teaching art of skillfully letting mathematical formulae "speak for themselves."

In elementary mathematics, we owe to Euler the conventionalization of the following notations:
1. Functional notation: $f(x)$
2. The base of the natural logarithms: e
3. The sides of triangle ABC: a, b, c
4. The semiperimeter of triangle ABC: s
5. The summation sign: Σ
6. The imaginary unit, $\sqrt{-1}$: i

Figure 64: Leonhard Euler

In addition to textbook writing, he wrote highly original research papers (at a rate of about eight hundred pages a year) during most of his lifetime.[204] The last seventeen years of his life were spent in blindness due to cataracts, but that did not stop his productivity. Sustained by his faith, he dictated his research and findings. His powers of concentration also enabled him to do this because he was able to perform complex mathematical calculations in his head, calculations that other competent mathematicians had trouble with on paper! It can be said of Euler, "As he thinks, he calculates." His works are presently being collected. When finished, it is estimated that they will contain seventy-four volumes![205]

All of his mathematical work flowed out of the matrix of nature. He analyzed mechanics, planetary motion, paths of projectiles, tide theory, and the design and sailing of ships. He worked out construction and architectural problems, developed the theory of acoustics, and contributed insights in optics. He also introduced fresh perspectives in fluid theory and thermodynamics. Chemistry, geography, cartography, and of course philosophy and religion also gained his interest. The epitome of Euler's mathematical analysis can be revealed in one small, but incredible, formula:

$$e^{i\pi} + 1 = 0.$$

The mathematician Felix Klein (1849-1925) remarked that all mathematical analysis was centered in this simple, but elegant, equation derived by Euler. Every symbol has its history – the principal whole numbers 0 and 1, the basic

---

[203]See Leonhard Euler, *Introduction to the Analysis of the Infinite, Book II*, trans. John D. Blanton (New York: Springer-Verlag, 1990). Among the many accomplishments of this book, Euler transformed trigonometry into its modern understanding and extended Cartesian geometry to the space of three dimensions.
[204]Kline, *Mathematical Thought from Ancient to Modern Times*, p. 402.
[205]*Ibid.*

mathematical operations of addition and equality, π (originating from the time of the Greeks), i (the symbol representing $\sqrt{-1}$), and e the base of the natural logarithms developed by John Napier.

Euler did not spend all of his life isolated in his study. He married and fathered thirteen children. He loved having them around too. They would play around his feet while he did his mathematics. He taught his own children, and grandchildren, making scientific games for them and instructing them in the Scriptures every evening.

Figure 65: Voltaire

Universally respected, by the end of his life he could claim all the mathematicians of Europe as his students. His biblical faith invited sharp criticism from atheists like the Frenchmen Voltaire (1694-1778) and Denis Diderot (1713-1784). A humorous story, the authenticity of which is doubted by some, revolves around a confrontation between Euler and Diderot. Diderot had been trying to convert people to atheism, so Russia's Catherine the Great assigned Euler to muzzle his efforts. The two men locked horns in the midst of monarchy and court officials. Euler, knowing Diderot's ignorance of mathematics, approached him grimly and said,

"Sir, $(a + b^n)/n = x$. Hence, God exists. You reply!"[206]

Diderot sat in a stunned and embarrassed silence that was soon interrupted by hilarious laughter. He quickly retreated to his home in France.

It is unfortunate to realize that the stream of thinking found in Voltaire and Diderot would soon overflow in a flood. Mathematics, used by Euler as God's ally, would soon be used to show that God at best, was irrelevant to mathematics and science, and at worst, did not exist.

QUESTIONS FOR REVIEW, FURTHER RESEARCH, AND DISCUSSION

1. Explain the scientific and mathematical heritage bequeathed to us by Medieval Christendom.
2. Explain the relationship of the Gothic cathedrals to mathematics.
3. Detail the factors involved that produced the Scientific Revolution.

---

[206]The story was first printed by Thiébault in his *Souvenirs de vingt ans de séjour à Berlin* (1804), twenty years after Euler's death. It was printed in English under the title *Original Anecdotes of Frederic the Second, King of Prussia, and of His Family, His Court, His Ministers, His Academies, and His Literary Friends; Collected During a Familiar Intercourse of Twenty Years with that Prince* (London: J. Johnson, 1805), 2:4. From there it was found as a selection under the title "Assorted Paradoxes," by Augustus de Morgan in D. E. Smith's *A Budget of Paradoxes*, second edition (Chicago: Open Court, 1915), 2:339. James R. Newman reprinted it in *The World of Mathematics* (New York: Simon and Schuster, 1956), 4:2377-2378. A rebuttal to its authenticity, based upon Diderot's knowledge of algebra, is documented by J. Mayer, *Diderot: L'homme de science* (Rennes: Imprimerie Bretonne, 1959), pp. 93-96.

4. Both Greek philosophy and biblical revelation affirm, on the surface, the "rationality of nature." Explain the "root" differences.
5. Explain the foundation upon which Western science and mathematics was born.
6. Explain the impact of symbolic algebra upon applied mathematics.
7. What was timely about the invention of logarithms?
8. Prove or disprove the following statement, "Logarithms are a relic from the past."
9. Compare the theological views of Ptolemy with Copernicus.
10. Compare the motivations of Archimedes with Kepler.
11. What basic principle of scientific work did Kepler illustrate?
12. In terms of their views on the "laws of nature," compare and contrast the approach of Galileo with that of Aristotle.
13. Comment on Newton's reasons for claiming that the discourse of God belongs to natural philosophy.
14. Explain why science was "stillborn" in the following cultures:
    a. Greek.
    b. Chinese.
    c. Hindu.
    d. Muslim.
15. Explain the impact of Puritan thought on mathematics and science.
16. Prove or disprove the following statement, "Historically, both Roman Catholicism and Protestantism have been at war with science and mathematics."
17. Explain the historical developments that identified the limitation and purpose of mathematics.
18. Develop the historical relationship between the dominion mandate of Genesis and Christ's Great Commission of Matthew 28.
19. What does the Bernoulli family tree teach us about intergenerational faithfulness? Is it primarily a matter of nature or nurture? Or both?
20. Should commitment to one's vocation override commitment to one's family? Comment using the life of Euler as a model.

5: FROM EULER TO GÖDEL

Beware lest anyone cheat you through philosophy and empty deception, according to the tradition of men, according to the basic principles of the world, and not according to Christ (Colossians 2:8).

Paul, the Apostle

The origin of that stream of thinking found in Voltaire and Diderot finds its ancient roots in Genesis 3. Ever since that dark day in the Garden of Eden, man has cursed himself with the idea that he can be autonomous; i.e., that he does not need God.[1] As you look at the history of mankind, you can trace two faiths: (1) the way of independence or (2) the way of dependence.[2]

In the Scientific Revolution, the early founders did their work in a cultural atmosphere that exhibited dependence upon the biblical God. Unfortunately, the way of independence did not take long to raise its ugly head.

5.1 REASON CONTRA REVELATION

As already mentioned, in the 13th century Thomas Aquinas incorporated Aristotelian thought into the current dogmas of the established church. It should be noted that in his exhaustive work he did call attention to the importance of the investigation of physical phenomena. At the same time, he called attention to something else.

As a theologian, Aquinas held an incomplete view of the fall of man. According to him, man had a corrupted will, but a relatively untainted mind.

---

[1]For man to say that he does not "need" God is a sophism. Man, in his rejection of God, still lives and moves and has his being in the God he foolishly rejects (Acts 17:28).
[2]Man can really only commit himself to one of two faiths: (1) faith in himself – independence or (2) faith in the biblical God – dependence.

Hence, the human mind can be relied upon to prove the existence of God.[3] Because of its cognitive unity with the created order, the human mind could also ratiocinate scientific laws.[4]

We have already seen that the Reformation scholars rejected Aquinas' "double-revelation" theory and posited, in its place, a "one-revelation" theory; i.e., God's revelation of Himself is ultimately one (whether coming from the book of God's word or the book of God's works). They knew that, when the fallen nature of the human mind is denied, then man would naturally gravitate toward rejecting God's revelation as primary and see his own cogitations as primary (or autonomous). In the words of Rousas J. Rushdoony:

> Basic to this double-revelation theory is the Thomistic and Greek concept that the reason of autonomous man is capable of *impartially and objectively* investigating the truths of creation and of establishing them into a valid revelation of nature. The source of "revelation," then, concerning the universe is man's reason and science.[5]

Although the mind of man is tremendously complex and powerful in its abilities, it can end up being his Achilles heel. Aquinas placed human reason on par with biblical revelation. Although he believed that reason must be submitted to revelation, both could be relied upon in order to acquire truth. Contrary to Augustine, who said

Faith and Grace

Reason and Nature

Figure 66: Aquinas' "double-revelation" theory

that we "believe in order to understand,"[6] Aquinas reversed the emphasis to "we understand in order to believe." The sad fact is that eventually reason and revelation confronted each other in war, with reason emerging as the victor. Jeremy Jackson, Church historian, states:

> In harmonizing Aristotle with the Bible, Aquinas ... gave a special place to reason independent of God's intervention, which would help launch the

---

[3]Three stand out in particular: (1) The Teleological proof – the argument for God's existence based upon noted design and purpose in the universe, (2) the Cosmological proof – the argument for God's existence showing that since everything has a cause, there must be a first cause, namely God, of all things, and (3) the Ontological proof – the argument for God's existence based upon the nature of existence.

[4]As noted in sections 3.11 through 3.16, this is a valid and useful concept *as long as man's mind is unfettered by autonomous notions.*

[5]Rushdoony, *The Mythology of Science*, p. 40. Emphasis added. Impartial and objective (i.e., neutral) investigation into the facts of creation is an impossibility. A man's faith will always determine how he views and interprets the facts, even the fact that $2 + 2 = 4$.

[6]Anselm of Cantebury (1033-1109), who wrote the famous work *Cur Deus Homo* (Why God became man) in 1099, also embraced and emphasized Augustine's epistemology that faith in God provides the only valid foundation for human reasoning (cf. John 7:17; Hebrews 11:3; Psalm 36:9).

whole emphasis upon human self-help in the Renaissance period, and which continues right down to modern times.[7]

To the men of the Reformation, the final reference point in all thinking, the ultimate in intelligibility, rested in the biblical revelation of the infinite, personal God. Concomitant with this viewpoint, the men of the Renaissance reintroduced, or rebirthed, the old Greek idea about the autonomy of man's mind. To them, the final reference point in all thinking rested not in God, but in man.

5.2 CARTESIAN GEOMETRY

Figure 67: René Descartes

The Frenchman René Descartes (1596-1650), a philosopher first and mathematician second, was one of those giants that Newton "stood on the shoulders of."[8] Although Descartes carefully avoided any reference to his predecessors (including the work of Nicole Oresme in coordinate geometry), he is commonly hailed as the inventor of that geometry, also known as analytical geometry, a mathematical method that wedded algebra with geometry.[9] This method provided the means by which to write an equation for a geometric curve. The Cartesian coordinate system, named in his honor, proved to be a valuable and almost indispensable tool in future mathematical research and application.

While Descartes owed a small debt to Oresme, the immediate inspiration for his geometrical innovations came from fellow-Frenchman François Viète (1540-1603). Viète provided Descartes with a simplified algebra. The symbols for plus (+) and minus (-) were already in common use by the time of Colum-

---

[7]Jeremy Jackson, *No Other Foundation: The Church Through Twenty Centuries* (Westchester: Cornerstone Books, 1980), p. 110. Not only was reason seen as independent of God's intervention, nature also eventually came to be seen as an independent entity run by the autonomous mechanism of natural law. This again is an unfortunate consequence of the nature/grace dichotomy found in the thought of Aquinas and in the thought of his scholastic heirs, namely Nicole Oresme. The seed planted in medieval times that eventually sprouted into the Scientific Revolution eventually and sadly became a weed resulting in the change from the relative autonomy of scientific law to the absolute autonomy of scientific law. The same can be also said of the Puritan scientific distinctives. See E. L. Hebden Taylor, "The Role of Puritan-Calvinism in the Rise of Modern Science," *The Journal of Christian Reconstruction: Symposium on Puritanism and Progress*, ed. Gary North (Vallecito, CA: Chalcedon, 1979), 6:1, pp. 70-86.

[8]Newton, however, never acknowledged this. In his latter days, Newton kept himself "busy" by erasing references to Descartes from his manuscripts. He would have had to burn his *Opticks* manuscript because Cartesian notions permeate almost every page.

[9]Descartes was most likely familiar with the rudiments of coordinate geometry although he may not have heard of Oresme by name. Descartes, in his student years, was trained at the Jesuit College in La Flèche where his teachers taught from the tradition cultivated especially at Salamanca. Salamanca, in turn, was indebted to the 14th century Sorbonne where Oresme worked before he became bishop of Lisieux.

bus (late 15th century).[10] In the middle of the 16th century, the symbol for equal (=) was introduced in England.[11] It was Viète who introduced convenient notations for variables in algebraic equations. As mentioned in section 4.5.1, he suggested that unknown quantities be represented by vowels and known quantities by consonants. This development enabled mathematicians to completely generalize algebraic functions. What Viète failed to do was to associate his algebraic geometry with a coordinate system.

5.2.1 THE CARTESIAN COORDINATE SYSTEM

$$x + y = 0$$

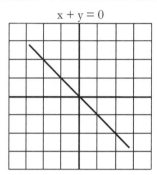

Figure 68: Straight line

$$x^2 + y^2 = 9$$

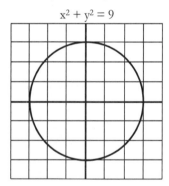

Figure 69: Circle

---

[10]These symbols first appeared in print in 1456 in an unpublished manuscript by the German mathematician and astronomer Johann Müller or Regiomontanus (1436-1476).

[11]English mathematician Robert Recorde (ca. 1510-1558), in his textbook on algebra entitled *The Whetstone of Witte* (1557) introduced two parallel lines (=) as a symbol for equality.

Descartes and a fellow-Frenchman Pierre de Fermat (1601-1665) made this link almost simultaneously. Fermat was a trained lawyer who was fascinated with the geometrical works of antiquity. His short treatise on geometry, *Introduction to Plane and Solid Loci*, introduced the fundamental principle of coordinate geometry:

Figure 70: Pierre de Fermat

Whenever two unknown magnitudes appear in a final equation, we have a locus (set of points – J.N.), the extremity of one of the unknown magnitudes describing a straight line or a curve.[12]

Mathematics historian Carl Boyer notes the significance of Fermat's thoughts:

This brief sentence represents one of the most significant statements in the history of mathematics. It introduces not only analytical geometry, but also the immensely useful idea of an algebraic variable. The vowels in Viète's terminology previously had represented unknown, but nevertheless fixed or determinate, magnitudes. Fermat's point of view gave meaning to indeterminate equations in two unknowns – which had been rejected in geometry – by permitting one of the vowels to take on successive line values (corresponding to Oresme's longitudes), and plotting the values of the other as perpendicular lines (latitudes, Oresme would have called them). Thus Fermat rediscovered the graphical representation of variables, and this time there was an algebra at hand with which to exploit the idea.[13]

Fermat's treatise on coordinate geometry was not printed during his lifetime. It first appeared in print in 1679, some forty years after Descartes' work in the same area was printed and fifty years after Fermat composed it.[14] This is why coordinate geometry is named after Descartes (i.e., Cartesian geometry) and not Fermat.[15]

---

[12]Pierre de Fermat, "An Introduction to Plane and Solid Loci," *A Source Book in Mathematics*, ed. David E. Smith (New York, Dover Publications, [1929] 1959), p. 389.

[13]Carl Boyer, "The Invention of Analytic Geometry," *Mathematics: An Introduction to Its Spirit and Use*, ed. Morris Kline (San Francisco: W. H. Freeman, 1979), pp. 42-43.

[14]In 1637, Descartes also introduced the modern exponential notation; e.g., a^2, a^3, a^4, etc. He also coined the word "imaginary" to represent square roots of negative numbers. In algebra, he conventionalized the use of the last letters of the alphabet (x, y, z) to denote unknowns in an equation and the first letters (a, b, c) of the alphabet to denote known quantities in an equation. He also did work in optics (reflection and refraction of light) and attempted to explain the rainbow in quantifiable terms.

[15]Fermat, who enriched every branch of mathematics known in his time, is acknowledged as the founder of the modern theory of numbers and the one who significantly advanced the study of probability. He is best known for the famous "Fermat's Last Theorem." In the margin of one of his books was found this conclusion (without proof): If x > 2, there are no whole numbers a, b, c, such

5.3 CARTESIAN PHILOSOPHY – MATHEMATICS ABSOLUTIZED

Due to his training by Catholic Jesuits, Descartes firmly believed in the existence of God. His most famous saying was "Cogito, ergo sum" (I think, therefore I am). His most famous work, *Discourse on Method* (1637) in which his treatise on coordinate geometry ("*La Géométrie*") appears as the third appendix, revealed his strong rationalistic faith.[16] In it, he established a foundation for philosophy where the authority of knowledge rests in man, not God. He wanted to base epistemology upon the autonomy of man's "self-evident" ideas, not upon the revelation of God. To begin with oneself, or one's own thinking, is a dangerous epistemological point of departure. Why? Given this starting point, reality becomes questionable. For example, a dream "seems" real to me. If my dreams "seem" real to me, then could the objective world of reality be merely a "dream" also? Beginning only from myself, *I cannot know that anything is real.*

Figure 71: Blaise Pascal

According to Stanley L. Jaki, Cartesian philosophy had only one aim: "to secure absolute certainty for human reasoning and in a measure coextensive with the universe."[17] One of Descartes' contemporary critics, his fellow Frenchman, philosopher, and mathematician Blaise Pascal (1623-1662), noted that the main consequence of Descartes' ideas was the expulsion of God from the world.[18]

According to Cornelius Van Til, Descartes "founded the whole knowledge scheme upon the independent activity of the finite consciousness in its relation to objects that are independent of God."[19] Independent thinking and independent objects provided the foundation for Descartes' philosophical formulations. From thinking, he could determine not only his existence, but also the existence of the world and the existence of God. Descartes, and a legion of like-minded followers, established

---

that $a^x + b^x = c^x$. This theorem stumped mathematicians until Andrew Wiles (1953-) of Princeton University solved it in 1994. See Amir D. Aczel, *Fermat's Last Theorem* (New York: Dell Publishing, 1996).

[16]The full title, in French, is *Discours de la méthode pour bien conduire sa raison et chercher la vérité dans les sciences* (Discourse on the Method of Properly Guiding the Reason in the Search of Truth in the Sciences). For an English translation of the appendix on geometry, see René Descartes, "The Geometry," *The World of Mathematics*, ed. James R. Newman (New York: Simon and Schuster, 1956), 1:239-253.

[17]Jaki, *The Savior of Science*, p. 98.

[18]Blaise Pascal, *Pensées*, trans. W. F. Trotter (New York: E. P. Dutton, 1958), #77, p. 23. At age 19, Pascal invented a mechanical machine that could compute addition. Elementary algebra students know of Pascal's triangle, a numeric scheme that denotes the coefficients of the powers of a binomial. He added considerably to the mathematical corpus in the area of the theory of probability.

[19]Van Til, *A Survey of Christian Epistemology*, p. viii.

man's reason as the neutral (i.e., unprejudiced) and authoritative interpreter of God and the universe. Instead of seeing God as the foundation for ontology and epistemology, Descartes declared man's reason to be the bedrock. He effectively turned epistemology and ontology on their respective heads concentrating both in the autonomy of man instead of the autonomy of God.

As a reductionist, Descartes believed that all realms of knowledge could be submitted to the scheme of geometric rationalism. In his establishment of the supremacy of human reason through the methods of deductive logic, he believed that he could prove the invariability of mathematical law. In reference to the workings of the universe, mathematical law dictated everything. Descartes' view of the universe was mechanistic, not supernaturalistic. Although he applied his mechanistic philosophy first to the workings of the universe, his philosophical heirs eventually applied it to the workings of man. Man became a mere cog in the universal machine, not a person created in the image of God.[20]

The impact of Descartes' absolutization of rationalist thinking throttled scientific endeavor for a time in France. Stanley L. Jaki cites the remarks of French mathematician Jean Baptiste le rond d'Alembert (1717-1783), "d'Alembert put the blame squarely on Descartes for that stagnation in which physical science remained for some time in France while it advanced rapidly in England."[21]

Figure 72: Thomas Hobbes

English political philosopher Thomas Hobbes (1588-1679) took Descartes' mechanistic method and applied it to the body politic (including human motivation and morals). Descartes embraced metaphysical dualism (mind and matter) in a mechanistic context. Hobbes reinterpreted Descartes' mechanistic philosophy in a materialistic context. A reductionist like Descartes, Hobbes advanced metaphysical monism – he reduced everything to matter and equated ratiocination with computation.

Irish prelate and idealist empiricist philosopher George Berkeley (1685-1753) indirectly challenged Descartes' method by questioning Hobbes' materialism. He queried, "Do you think there is hard matter out there?" He answered by proposing that all one has is sense impressions.[22] His epistemology was founded upon sense impressions. To him, our senses "seem" to inform us of the existence of a hard, material world. But, we have no first-hand knowledge of such a

---

[20]For further discussions on the impact of Descartes on modern thought, see Philip J. Davis and Reuben Hersh, *Descartes' Dream: The World According to Mathematics* (Boston: Houghton Mifflin, 1986).

[21]Jaki, *The Road of Science and the Ways to God*, p. 65.

[22]Berkeley's ideas are the basis for the famous query: "If a tree falls in a forest and no one is there to hear it, does it make a sound?" According to Berkeley, the answer is "No." His reasoning? *Sound exists only if an ear hears the sense impression and interprets the sound waves to the brain.*

thing as hard matter. *All we have are sense impressions.* Existence is based upon mere sense perception, *esse est percipi* – "to be is to be perceived." He believed that these sense impressions come from the mind of God to the mind of man, but there is no real world independent of the human mind in between. In essence, Berkeley's philosophical system eliminated any possibility of knowledge of an external material world.

Figure 73: George Berkeley

As a side note, Berkeley wrote *The Analyst* (1734), a brilliant and devastating critique of Newton's infinitesimal method. Its lengthy subtitle might qualify for entry in *The Guinness Book of World Records.* It reads as follows: *A Discourse Addressed to an Infidel Mathematician, Wherein it is Examined Whether the Object, Principles and Inferences of the Modern Analysis are More Distinctly Conceived, or More Evidently Deduced, than Religious Mysteries and Points of Faith. "First cast out the beam of thine own Eye; and then shalt thou see clearly to cast the mote out of thy brother's Eye."* The infidel referred to was astronomer Edmund Halley (1656-1742), Newton's friend and financier of *The Principia.* Halley had supposedly convinced one of Berkeley's friends of the "inconceivability of the doctrines of Christianity."[23] When Halley's "convert" refused Berkeley's spiritual counsel on his deathbed, Berkeley was outraged. Berkeley responded in the text declaring that Newton's infinitesimals were as "obscure, repugnant and precarious" as any tenet in Christian theology.[24] Although not a mathematician, Berkeley made a number of extremely effective points in exposing the weak and confused conceptual foundations of the calculus derisively calling infinitesimals "the ghosts of departed quantities."[25]

Figure 74: Edmund Halley

Note the immediate fruits of Descartes' method. One on hand, we encounter the radical materialism of Hobbes and on the other, we encounter the radical immaterialism of Berkeley. As Stanley L. Jaki reflects, "Berkeley's categorical denial of the existence of an external world is often presented as the necessary end of Cartesian logic."[26] Both forks of the road have one, ultimate

---

[23]There is no evidence of Halley's religious "skepticism" in any of his published writings. His alleged "infidelity" and concomitant atheism rests only upon the established opinion of the time.

[24]George Berkeley, "The Analyst," *The World of Mathematics*, ed. James R. Newman (New York: Simon and Schuster, 1956), 1:288-289.

[25]*Ibid.*, 1:292.

[26]Jaki, *The Road of Science and the Ways to God*, p. 77.

concept in common – an underlying irrationality. We shall encounter more of this irrationality in our discussion of David Hume and Immanuel Kant.

5.4 THE "FRUITS" OF SUCCESS

The application of mathematics to physical phenomena proved to be an extraordinary success. Mathematics enabled painters to paint in three dimensions through the development of principles of perspective. Mathematical method produced laws that both analyzed and explained the intricacies and order of musical sounds. Mathematical theory harnessed the invisible forces of electricity and magnetism revealing the steps necessary to use this power constructively. Confidence was growing in the minds of many mathematicians concerning further applications of the powerful methods of mathematics.

Originally, European mathematicians praised the wisdom of God the Creator in all of their work. So grand and extensive were their successes that their heads began to swim. Many began to see their work as their own creations.[27] Pride made a subtle entrance onto the scene. Unintentional on their part to begin with, mathematicians not only confirmed, but fueled the thrust of the Renaissance; i.e., the belief that the reason of man is autonomous.

Not only was the idea of autonomous human reason on the advance, but the belief in the role of the hand of God in sustaining the workings of the universe was in retreat. This retreat was due, not only to the rising fog of both mathematical absolutism, but also to theological deism.

Deism was a popular mode of thinking among many 18th century philosophers and mathematicians. A Deist believes that God created the universe as a watchmaker would create a watch (mirroring the metaphor originally employed by the medieval scholastic Nicole Oresme). Then, God set the "machine" of the universe in motion, fueled by mathematical law, and only occasionally acted in creation to correct any irregularities. The universe no longer was sustained by the "word of God's power," but by the "watch spring" of mathematical law.

Morris Kline observes:

> Though many a mathematician after Euler continued to believe in God's presence, His design of the world, and mathematics as the science whose main function was to provide the tools to decipher God's design, the further the development of mathematics proceeded in the eighteenth century, and the more numerous its successes, the more the religious inspiration for mathematical work receded and God's presence became dim.[28]

Note also Alistair C. Crombie's remarks:

> The advance of science did in fact give rise to materialist metaphysics, naïve certainly but to become nevertheless influential in the 18th and 19th cen-

---

[27]This line of reasoning does not mean that every mathematician viewed mathematics in this context during the 18th and 19th centuries. This statement reflects the prevailing ambiance of the period.
[28]Kline, *Mathematics: The Loss of Certainty*, pp. 72-73.

turies, and by definition anti-theological. The God of the scientists, of Boyle, the 'intelligent and powerful being' praised by Newton in *The Principia*, when taken over by the 18th century Deists, no longer gave any primacy or uniqueness to Christianity among the religions. Most corrosive of all, the 'fictionalist' or 'conventionalist' policy adopted by Descartes and pressed forward by Berkely [George Berkeley – J.N.], became in the hands of secular philosophers like David Hume (1711-1776), and of Immanuel Kant (1724-1804), the source of a doctrine that was anti-rational and anti-theological alike. Applied universally, as it inevitably was, it ceased to be a defence of theology against science and became a threat to all knowledge, whether rational or revealed. The way was open to the explicitly anti-theological and anti-metaphysical positivism of Auguste Comte (1798-1857) and John Stuart Mill (1806-1873), and to the agnosticism of T. H. Huxley, which became so characteristic a part of the philosophical ambience of science in the 19th century.[29]

And today, modern science and mathematics churn on, driven mostly by the demands of technology, oblivious to the reality of its roots. Stanley L. Jaki gives confirmation:

Science is now in possession of such a vast interconnection of data, laws, and instruments as to continue its progress even if no attention is paid any longer to that faith which played an indispensable role in its rise.[30]

5.5 TWENTIETH CENTURY RAMIFICATIONS

As affirmed by Kline, Crombie, and Jaki, for most early European mathematicians and scientists, mathematics acted as a tool to describe the universe and thereby was revelational of God's power, wisdom, and infinity. In the 18th century, this view of the goal of mathematics slowly began to change. Instead of being a monument to God the Creator, mathematics became the *magnum opus* of the inventiveness of man. The goal of mathematics eventually became the attainment of complex, intellectual theories rather than the inspection of the physical world.

We began to encounter the fruits of this shift of thinking in the middle of the 20th century. Educationally, according to the *Minimum Standards for Ohio Elementary Schools*, mathematics is assumed to be merely human logic and completely divorced from the physical world. To these educators, "the logical structure" is "of mathematics," *not of reality*.[31] In 1961, Marshall Stone (1903-1989), professor at the University of Chicago and a leader in the development of

---

[29]Crombie, 2:320.

[30]Jaki, *Cosmos and Creator*, p. 139. That no attention is paid any longer to that faith which played an indispensable role in its rise is due primarily to the rationalist historians of the 18th century, led by Voltaire, who discounted any possibility of a connection between medieval theology and the rise of modern science.

[31]Virginia M. Lloyd, *Minimum Standards for Ohio Elementary Schools* (Columbus: State Board of Education, 1970), p. 51.

mathematics curricula, said, "Mathematics is now seen to have no necessary connections with the physical world."[32] Needless to say, the development of the new mathematics of the 1960s found its roots in these sentiments: a divorcement of mathematics from the physical world and a placement of mathematics in the context of logic alone.[33]

Professionally, mathematicians can be found in two camps. Most call themselves pure mathematicians while a minority call themselves applied mathematicians or physicists. To the mathematicians of the pure camp, mathematics is studied for its own sake. If the theories formulated happen to apply to the real world, it is okay, but not to be expected. The pure camp looks with disdain upon the applied camp as if they are not fit to be called mathematicians. To the purists, those applied rascals have too much dirt on their shoes and dust on their computers from meddling with the physical world.

In 1944, John L. Synge (1897-1995), world renowned Irish physicist, commented at length about a group of people called mathematicians:

> Most mathematicians work with ideas which, by common consent, belong definitely to mathematics. They form a closed guild. The initiate forswears the things of the world, and generally keeps his oath. Only a few mathematicians roam abroad and seek mathematical sustenance in problems arising directly out of the other fields of science. In 1744 or 1844 this second class included almost the whole body of mathematicians. In 1944 it is such a small fraction of the whole that it becomes necessary to remind the majority of the existence of the minority, and to explain its point of view.
>
> The minority does not wish to be labelled "physicist" or "engineer," for it is following a mathematical tradition extending through more than twenty centuries and including the names Euclid, Archimedes, Newton, Lagrange, Hamilton, Gauss, Poincaré. The minority does not wish to belittle in any way the work of the majority, but it does fear that a mathematics which feeds solely on itself will in time exhaust its interest.
>
> Apart from its effect on the future of mathematics proper, the isolation of mathematics has robbed the rest of science of a support on which it counted in all previous epochs ... Out of the study of nature there have originated (and in all probability will continue to originate) problems far more difficult than those constructed by mathematicians within the circle of their own ideas. Scientists have relied on the mathematician to throw his energies against these problems....

---

[32]Marshall Stone, "The Revolution in Mathematics," *American Mathematical Monthly*, 68 (1961), 716.
[33]Fortunately, in the last two decades of the 20th century we began to see a shift back to establishing mathematics curricula on real life and largely physical problems. We see this shift notably in *some* of the high school curricula of government schools. Many of these newer textbooks and ancillary materials give public acknowledgment to the influence of Morris Kline (who died in 1992) in this regard. Unfortunately, *most* of the mathematics curricula of the modern university (whether public or private) still floats in the ivory tower of pure abstraction.

In all this the mathematician was the directing and disciplining force. He gave science its methods of calculation – logarithms, calculus, differential equations, and so on – but he gave it much more than this. He gave it a blue-print. He insisted that thought be logical. As each new science came up, he gave it – or tried to give it – the firm logical structure that Euclid gave to Egyptian land surveying. A subject came to his hands a rough stone, trailing irrelevant weeds; it left his hands a polished gem.

At present science is humming as it never hummed before. There are no obvious signs of decay. Only the most observant have noticed that the watchman has gone off duty. He has not gone to sleep. He is working as hard as ever, but now he is working solely for himself....

In brief, the party is over – it was exciting while it lasted.... Nature will throw out mighty problems, but they will never reach the mathematician. He may sit in his ivory tower waiting for the enemy with an arsenal of guns, but the enemy will never come to him. Nature does not offer her problems ready formulated. They must be dug up with pick and shovel, and he who will not soil his hands will never see them.

Change and death in the world of ideas are as inevitable as change and death in human affairs. It is certainly not the part of a truth-loving mathematician to pretend that they are not occurring when they are. It is impossible to stimulate artificially the deep sources of intellectual motivation. Something catches the imagination or it does not, and, if it does not, there is no fire. If mathematicians have really lost their old universal touch – if, in fact, they see the hand of God more truly in the refinement of precise logic than in the motion of the stars – then any attempt to lure them back to their old haunts would not only be useless – it would be denial of the right of the individual to intellectual freedom. But each young mathematician who formulates his own philosophy – and all do – should make his decision in full possession of the facts. He should realize that if he follows the pattern of modern mathematics he is heir to a great tradition, but only part heir. The rest of the legacy will have gone to other hands, and he will never get it back....

Our science started with mathematics and will surely end not long after mathematics is withdrawn from it (if it is withdrawn). A century hence there will be bigger and better laboratories for the mass-production of facts. Whether these facts remain mere facts or become science will depend on the extent to which they are brought into contact with the sprit of mathematics.[34]

---

[34]John L. Synge, "Focal Properties of Optical and Electromagnetic Systems," *American Mathematical Monthly*, 51 (1944), 185-187.

The Hungarian-born American mathematician John von Neumann (1903-1957) was a pioneer in the fields of quantum mechanics, game theory, and high-speed electronic computers. In 1947, he gave an articulate, but little-heeded, warning:

Figure 75: John von Neumann

> As a mathematical discipline travels far from its empirical source, or still more, if it is a second and third generation only indirectly inspired by ideas coming from "reality," it is beset with very grave dangers. It becomes more and more purely aestheticizing, more and more purely *l'art pour l'art*. This need not be bad, if the field is surrounded by correlated subjects, which still have closer empirical connections, or if the discipline is under the influence of men with an exceptionally well-developed taste. But there is a grave danger that the subject will develop along the line of least resistance, that the stream, so far from its source, will separate into a multitude of insignificant branches, and that the discipline will become a disorganized mass of details and complexities. In other words, at a great distance from its empirical source, or after much "abstract" inbreeding, a mathematical subject is in danger of degeneration. At the inception the style is usually classical; when it shows signs of becoming baroque, then the danger signal is up.... In any event, whenever this stage is reached, the only remedy seems to me to be the rejuvenating return to the source: the reinjection of more or less directly empirical ideas. I am convinced that this was a necessary condition to conserve the freshness and the vitality of the subject and that this will remain equally true in the future.[35]

The German mathematician Karl Weierstrass (1815-1897), one of the greatest teachers of advanced mathematics that the world has ever known, believed that it was the glory of mathematics to be indispensable to physics. In 1857, he warned that one should think of "the relation between mathematics and physics in a deeper manner than is the case when a physicist sees in mathematics only an indispensable auxiliary discipline, or when a mathematician is willing to see only a rich source of illustrations for his method in the questions posed to him by the physicist."[36]

Morris Kline explains the difference between the so-called pure mathematical developments in the 18th and 19th century with the pure mathematics philosophy of the 20th century:

[35]John von Neuman, "The Mathematician," *The World of Mathematics*, ed. James R. Newman (New York: Simon and Schuster, 1956), 4:2063.
[36]Karl Weierstrass, "Akademische Antrittsrede," *Mathematische Werke* (Berlin, 1894), 1:225.

The motivations [of 18th and 19th century mathematicians – J.N.] were either directly or indirectly physical and the men involved were vitally concerned with the uses of mathematics. In other words, the subjects purportedly created as pure mathematics and found to be applicable later were, as a matter of historical fact, created in the study of real physical problems or those bearing directly on physical problems. What does often happen is that good mathematics, originally motivated by physical problems, finds new applications that were not foreseen ... The unexpected scientific uses of mathematical theories arise because the theories are physically grounded to start with and are by no means due to the prophetic insight of all-wise mathematicians who wrestle solely with their souls.[37]

Stanley L. Jaki concludes this section, "It is there, in an immensely variegated nature and not in his finite intellect where ultimately lies the never-ending challenge for the mathematician."[38]

5.6 "I HAVE NO NEED OF THIS HYPOTHESIS"

Mathematicians of modernity are in this predicament because of the 18th century shift in thinking. Many of the 18th and 19th century mathematicians blatantly denied the need for God in their work.

Joseph-Louis Lagrange (1736-1813), a French Catholic, fine-tuned some of Newton's theories of planetary motion to such an extent that mathematics could account for and explain any irregularities. He also created the calculus notation [i.e., $f'(x)$ and $f''(x)$] that we still use today. During the period of the French Revolution (1789-1799), he directed a commission to establish new weights and measures (i.e., the metric system). An agnostic, he completely rejected any belief in God as the mathematical designer of the universe.[39]

Figure 76: Joseph-Louis Lagrange

Pierre-Simon de Laplace (1749-1827), another Frenchman with the disposition of an "aggressive atheist,"[40] wrote on probability, the calculus, differential equations, and geodesy. As to his style of exposition, the self-made American astronomer Nathaniel Bowditch (1773-1838) remarked, "I never come across one of Laplace's 'Thus it plainly appears' without feeling sure that I have hours of hard work before me to fill up the chasm and find out and show how it plainly appears."[41] Laplace tried to present a complete mechanical explanation of the workings of the solar system in his monumental

---

[37]Kline, *Mathematics: The Loss of Certainty*, p. 295.
[38]Jaki, *The Relevance of Physics*, p. 131.
[39]Bell, *Men of Mathematics*, p. 160.
[40]*Ibid.*, p. 173.
[41]Cited in Smith, 1:487.

work, *Mecanique Celeste*. Unlike Newton's *The Principia*, Laplace was determined not to treat ultimate causes.[42] Arrogant and vain, he often failed to footnote his sources giving the impression that his results were all his own.

The following anecdote is well known in Paris. Laplace presented an edition of his *Systéme du Monde* to the Emperor, Napoleon Bonaparte (1769-1821), an amateur mathematician himself. In it, Laplace mechanically accounted for the planetary perturbations noted by Newton and to which Newton postulated the need for periodic intervention by the Creator. Napoleon had been told that this book contained no mention of the name of God. Being fond of asking embarrassing questions, he received the book with the following comment, "Monsieur Laplace, they tell me you have written this large book on the system of the universe and have never

Figure 77: Pierre-Simon de Laplace

mentioned its Creator." To which Laplace replied, "I have no need of this hypothesis." Napoleon, greatly amused, told this reply to Lagrange, who exclaimed, "Ah, but that is a fine hypothesis. It explains so many things."[43] Laplace also branded the ethos of Kepler as "chimeric speculation."[44] To him it was "distressing for the human spirit" to pursue, in Kepler's ecstatic manner, the idea of world harmony.[45]

It is important to note that both of these mathematicians lived in the atmosphere of the French Revolution and its enthronement of the goddess Reason. There was no philosophical or moral guardian that could forestall either the drive to mathematical absolutism or the eventual blood bath of this revolution. If there was such a guardian in France, it would have more than likely been the Huguenots, the name given to French Protestants in the late 16th century. In the political turmoil of that time, they were expelled from France. In 1572, thousands of Huguenots were killed in the massacre of Saint Bartholomew's day. The protective and directive light that they could have given to the culture of France had been effectively extinguished. Because of this modernistic cultural ambiance, neither Lagrange nor Laplace needed to assume the existence of God in their theories of the workings of the universe.

5.7 PHILOSOPHICAL TOMFOOLERY

Even to consider the question of the existence of God was no longer necessary after the Scottish philosopher David Hume (1711-1776) skeptically de-

[42]Jaki, *The Road of Science and the Ways to God*, p. 263.
[43]Cited in De Morgan, in Newman, *The World of Mathematics*, 4:2376-2377.
[44]Caspar, p. 388.
[45]*Ibid.*

molished basic assumptions like cause and effect. To him, the idea of a first cause, the thrust of Aquinas' cosmological argument for the existence of God, was superfluous.

In the realm of epistemology, delineated in his *A Treatise of Human Nature* (1740), Hume picked up where Berkeley left off by declaring that we have no way of knowing that there is anything independent of us, *including the mind of God*. All we can confidently know is that we exist as individuals. We may know that we exist, but the existence of everything else is subject to serious question. All we have is our mind and our sense impressions (a radical form of empiricism, more accurately denoted as sensorial empiricism). If this is so, *then how do we know there is any reality to anything we experience?* Our minds can concoct some pretty crazy things! In essence, Hume denied any possibility of real knowledge. According to Alistair C. Crombie, "Hume, the 18th-century Ockham [William Occam – J.N.], went even further than Berkeley in claiming that science was irrational and that explanation was strictly speaking impossible."[46] The conundrum of obscurity created by Hume's skeptical skullduggery put the intellectual world in an epistemological crisis. How can we know anything?

Figure 78: David Hume

Hume denied the principle of causality (i.e., cause and effect). According to John Hedley Brooke, for Hume, "the idea of a causal connection had its origin in human psychology rather than in some perceived form of physical necessity."[47] According to 19th century biologist George Romanes (1848-1894):

> Of all philosophical theories of causality the most repugnant to reason must be those of Hume, Kant and Mill which ... attribute the principle of causality to a creation of our own minds, or in other words deny that there is anything objective in the relations of cause and effect, i.e., in the very thing which all physical science is engaged in discovering particular cases of it.[48]

No great scientist of any respect could ever be a *consistent* follower of Hume's logic. Since Hume emphatically denied the law of cause and effect, he concluded that the stated propositions of mathematical physics had no direct relationship to the real world. That it does, in the words of John H. Randall, is a

---

[46]Crombie, 2:330.
[47]John Hedley Brooke, *Science and Religion: Some Historical Perspectives* (Cambridge: Cambridge University Press, 1991), p. 186.
[48]George Romanes, "Notes for a Work on a Candid Examination of Religion to Metaphysics" *Thoughts on Religion*, ed. Charles Gore (London: Longmans Green, 1895), p. 119.

"happy accident."[49] Since mathematics does not give us any valid information about the physical world, Hume concluded that its axioms and theorems, definitions and formulae, are mere tautologies; they are mentally spun out of meanings implicit within themselves.[50]

It is of most importance to note that Hume could not live practically with his philosophy.[51] He had to fudge on his propositions concerning the nature of reality and of causality. He unveiled his irrational schizophrenia when he freely acknowledged that people had to think in terms of cause and effect and had to assume the validity of their perceptions, *or they would go mad.*

Hume's conclusions, although repugnant to most intellectuals, troubled the German idealist philosopher Immanuel Kant (1724-1804). He answered Hume in his celebrated work entitled *Critique of Pure Reason* (1781).

Kant, a great admirer of Newton's mathematical demonstrations that unified the physical phenomena of terrestrial and celestial motion, felt rebuffed by Hume's proposition insinuating the impossibility of mathematical knowledge. How could he reconcile Newton, who showed that the world is intelligible and mathematically ordered, with Hume, who pontificated the impossibility of a strictly speaking mathematical explanation of the world?

Figure 79: Immanuel Kant

Kant's epistemological solution was to conjecture that the mind itself creates this mathematical order. He began by postulating that there may be things external to us (e.g., the motion of the heavenly bodies), but we can have no real knowledge of these things. We do not know that these things exist in themselves or whether the world is mathematically ordered. All we have are our sense impressions of these things. What do we know in truth? Kant posited that we (the subject) and the world that we observe (the object) are at base, one. The world is therefore a part of my sensorial experience (the *phenomenal* realm as he termed it). The world appears ordered only because our minds impose order upon it. He said, "The understanding does not derive its laws (a priori) from, but prescribes them to nature."[52] To Kant, we only have valid knowledge of ourselves and it is the grid of our minds that instructs, or gives order to, the flux of our impressions of the external world. That is, scientific laws and mathematical demonstrations are dictated by

---

[49]John H. Randall, *The Making of the Modern Mind* (New York: Columbia University Press, [1926] 1940), p. 271. See more of the impact of Hume's philosophy on 20th century mathematical thought in section 6.3.

[50]A tautology is a redundant and needless repetition of the same thought in different words.

[51]Man still lives in God's world, not the world of his own autonomous making.

[52]Cited in Paul Carus, ed., *Kant's Prolegomena to Any Future Metaphysics* (La Salle, IL: Open Court, 1955), p. 82.

the structure of our minds. Man never knows these laws except by his ideas of these laws. As Alistair C. Crombie reflects, "Kant found himself able to admit Newtonian science as true only at the price of denying that it had discovered a real world of nature behind the world of appearance."[53] As an important addendum, Kant also denied the possibility of rational knowledge about God (the *noumenal* realm, as he termed it).

Crombie continues, in telling analysis, "It was ... an easy step from Kant's view that theories are read not in but *into* nature, to Auguste Comte's assertion that the real goal of science was and always had been not knowledge at all, but only power."[54] Stanley L. Jaki denoted Kant's theory of knowledge as epistemological geocentrism (or more accurately, mind-centrism). He then unveiled the destructive consequences of Kant's ideas on the scientific method, "If the structure of the mind determines the structure of things that are outside the mind, then the raison d'être [justification – J.N.] for experimenting and observation will hardly ever become a compelling reason."[55] Given Kant's premises, an important question needs to be asked, "Can the structure of my thought be a correct account of the structure of the world?" Kant's answer? "This can never be known." All we know is the phenomenal realm; all we know is things *as they appear to us*. Reality consists of nothing but a world of appearances. Note Alfred North Whitehead's revealing analysis:

> We must not slip into the fallacy of assuming that we are comparing a given world with given perceptions of it. The physical world is, in some general sense of the term, a deduced concept. Our problem is, in fact, to fit the world to our perceptions, and not our perceptions to the world.[56]

Contrary to Whitehead, we must fit our perceptions to the created order, not vice versa. It is God's creation that informs or teaches us (the essence of the experimental method); we do not dictate to it out of our independent, autonomous perceptions. Our reason, intuition, experience, and experimentation must be submitted to a higher authority, one in whom all things hold together – both the created order and the breath of every man (Hebrews 1:3; Colossians 1:17). Note the following confirmation in Job's testimony concerning the created reality:

> But now ask the beasts, and they will teach you; and the birds of the air, and they will tell you; or speak to the earth, and it will teach you; and the fish of the sea will explain to you. Who among all these does not know that the hand of the Lord has done this, in whose hand is the life of every living thing, and the breath of all mankind (Job 12:7-10)?

---

[53]Crombie, 2:330.
[54]*Ibid.*, 2:332.
[55]Jaki, *The Road of Science and the Ways to God*, p. 118.
[56]Alfred North Whitehead, *The Aims of Education and Other Essays* (London: Williams and Norgate, 1929), p. 166.

Rousas J. Rushdoony comments on the position of Kant and of Whitehead:

There is in this position a seeming and deceptive humility which is in actuality a perverse pride. Man's insistence that he has no valid knowledge of reality in itself, his attempts to eliminate causality, order and design while assuming them at every turn, constitute an attempt to resist any interpretation other than that of autonomous man.[57]

As we have seen in Romans 1:18-21, an objective revelation of God has been impressed unmistakably upon man through: (1) the created order and (2) man's own created nature. It is this knowledge that man suppresses in unrighteousness and it is in this context that we should understand this philosophical tomfoolery, "Professing to be wise, they became fools" (Romans 1:22).

Stanley L. Jaki also shows how the thought of Kant has influenced many 20th century seminal theorizers of modern nuclear physics toward oriental mysticism, for example, the 1922 Nobel Prize winner Niels Bohr (1885-1962):

Figure 80: Niels Bohr

... one's knowing is meaningless unless one knows something, that is, unless one's knowledge touches on reality. Elementary as this truth may appear, it has been stolen from Western rationality ever since Kant made his mark. Being heirs to that intellectual larceny, Bohr and his followers tried to understand not reality but only our understanding of reality and in the process Bohr was driven, as Hooker remarked, "toward the twilight zone of mere appearances." A world of appearances is most germane to oriental mysticism, and Bohr's categorical rejection of ontology was rightly seen as a telltale aspect of his basic sympathies with that mysticism.[58]

In summary, note the progression of thought from the Bible to Kant. Genesis 1:1 states that God created everything: the universe and man (including his mind). Kant said that man creates everything; i.e., everything is a part of man and his thinking. The world as we know it is at least relative to our mind if not an outright construct of our mind. The world is something that we have dreamed up; it is our creation. According to Stanley L. Jaki, this is nothing but "the gospel of pure subjectivism,"[59] an "edifice of sheer fantasy."[60] It is instruc-

---

[57]Rushdoony, *By What Standard?*, p. 10.
[58]Jaki, *The Road of Science and the Ways to God*, p. 212.
[59]*Ibid.*, p. 128.
[60]*Ibid.*, p. 115.

tive to note the blind alleys that these highly intelligent men can work themselves into when they reject basic biblical truths of creation. It is similar to the activity of so many followers of Darwinism who devote their entire career to the purpose of proving that there is no purpose to anything.[61] This intellectual schizophrenia is undiluted sophomoricism.[62] Alfred North Whitehead reflected on the folly of all such men, scientists and philosophers (and, by observing his preceding quote, we find that he is certainly subject to self-critique in this regard). He said, "It is a safe rule to apply that, when a mathematical or philosophical author writes with misty profundity, he is talking nonsense"[63] and that such people "constitute an interesting subject for study."[64]

The practical result of the philosophical tomfoolery of Hume and Kant was that, in reference to absolutes, the autonomy of mind's mind (and its imposition of mathematical law on external reality) had replaced the autonomy of the biblical God. Since Kant appreciated the work of Newton, he also approved of the geometry that formed the foundation for it; i.e., Euclidean geometry. To Kant, the matrix of man's mind is geared toward interpreting sense impressions in the context of Euclid's geometry *and no other*. To him, the statements of Euclidean geometry are independent of experience (*a priori*), necessarily true (apodictic), and contain factual content only (synthetic).[65]

5.8 "I SERVE THE LAWS OF NATURE"

The German Carl Friedrich Gauss (1777-1855), nicknamed the "Prince of Mathematicians," is generally regarded as the greatest mathematician of the 19th century. An infant prodigy, at age three he detected an arithmetical error in his father's accounts. At ten years of age, his teacher tried to give his class some "busy work" by telling his pupils to add the numbers from 1 to 100. Almost

---

[61]Or, engaging all their intellectual powers to demonstrate that no intellect is the *sine qua non* of the origin and design of the universe.

[62]*sophos* is Greek for " wise" and *moros* is Greek for "fool."

[63]Cited in Bell, p. xvii.

[64]Alfred North Whitehead, *The Function of Reason* (Princeton: Princeton University Press, 1929), p. 12. For some enlightening and in-depth study of some influential intellectuals, from the French philosopher Jean-Jacques Rousseau (1712-1778) to Bertrand Russell (1872-1970), see Paul Johnson, *Intellectuals* (New York: Harper & Row, 1988).

[65]Kant made a distinction between synthetic and analytic judgments as follows. In an analytic statement the predicate says something about the subject that is already contained in the notion of the subject. Example: "A rainy day is a wet day." The notion of "wet" is already contained in the phrase "rainy day." In contrast, in a synthetic statement the predicate says something about the subject that is *not* already contained in the notion of the subject. Example: "Friday is a wet day." This statement gives us a new fact about Friday. Analytic knowledge is also *a priori* knowledge because this knowledge depends upon the definition of terms and concepts. Synthetic knowledge is *a posteriori* knowledge since it involves observation and experiences through the senses. Kant posited that knowledge – including the mathematical propositions of Euclidean geometry – is both synthetic and *a priori*. By this he meant that the "raw material" of knowledge consists of the outside world perceived by the senses (the synthetic element), but that this is inevitably processed through the intuitive grid of the human mind (the *a priori* element). In perceiving the "raw material" of the outside world, the mind intuitively creates the order that is merely perceived. The mind never perceives the outside world as an independent entity. The mind conditions everything it encounters.

immediately, the young Gauss came forward and presented his slate, face down, before his irked instructor. After all the results were in, only Gauss had the correct answer, 5050. Yet, his slate was clean except for the answer. Gauss had calculated the answer mentally summing the arithmetic progression 1 + 2 + 3 + ... + 98 + 99 + 100 by noting that 100 + 1 = 101, 99 + 2 = 101, 98 + 3 = 101, etc. for 50 such pairs. Hence, the answer is 50 x 101 = 5050. Later in life Gauss used to joke that he could calculate before he could talk.

In addition to his mathematical contributions in the areas of algebra, differential geometry,[66] probability theory,[67] and number theory, Gauss made noteworthy contributions in the areas of astronomy, geodesy, and electromagnetism. In honor of him, the unit of intensity of magnetic fields is today called the *gauss*.

To his credit, Gauss was one of the first to denounce Kant's philosophy and he detested contemporary German philosopher Georg Wilhelm Friedrich Hegel's (1770-1831) mathematical incompetence. Note the biting remarks of Gauss in a letter written in 1844:

You see the same sort of thing [mathematical incompetence – J.N.] in the contemporary philosophers – Schelling, Hegel, Nees von Essenbeck – and their followers. Don't they make your hair stand on end with their definitions? Read in the history of ancient philosophy what the big men of that day – Plato and others (I except Aristotle) – gave in the way of explanations. And even with Kant himself it is often not much better. In my opinion his distinction between analytic and synthetic judgments is one of those things that either peter out in a triviality or are false.[68]

Figure 81: Carl Friedrich Gauss

Concerning philosophical issues, he said:

There are problems to whose solution I would attach an infinitely greater importance than to those of mathematics, for example touching ethics, or our relation to God, or concerning our destiny and our future; but their solution lies wholly beyond us and completely outside the province of science.[69]

---

[66]Differential geometry is the study of those properties of curves and surfaces that vary from point to point and therefore can be understood only with the techniques of the calculus.

[67]The gaussian or normal probability curve is known by most students of probability as the Bell Curve. See Charles Murray, *The Bell Curve*, (New York: The Free Press, 1994).

[68]Letter to H. C. Schumacher, November 1, 1844, in Karl Friedrich Gauss, *Werke* (Göttingen: Königliche Akademie der Wissenschaftern, 1877), 12:63.

[69]Cited in Bell, *Men of Mathematics*, p. 240.

He did believe in God, but apparently his belief in God did not give answers to ethical or teleological problems. Also, to him thoughts about God had nothing to do with mathematics or the search for mathematical laws of nature.[70] Gauss served, not the God of nature, but nature herself as illustrated by the following Shakespearean lines which were his motto, "Thou, nature, are my goddess, to thy laws my services are bound."[71]

With the comments of Gauss fresh in mind and also recalling the incredulous propositions of Hume and Kant, one cannot help but to be drawn as a magnet to the Apostle Paul's letter to the Romans, chapter one. Here Paul describes the adjudication of God that rests upon those who suppress the truth in unrighteousness. Note the specific epistemological judgments in the passage, "And even as they did not like to retain God in their knowledge, God gave them over to a debased mind ... proud, boasters, ... undiscerning [without understanding]..." (Romans 1:28-31).

In a leap of faith, many 19th century mathematicians and scientists chose to reject the revelation of God in the created order and clung to a mechanistic philosophy in which mathematical and scientific knowledge opened the sole gateway to the whole realm of knowledge. In this, we have returned to the old Greek idea that truth must be subject to autonomous human reason.

5.9 MATHEMATICAL IDOLATRY?

Historian Herbert Schlossberg defines idolatry as follows:

> Idolatry in its larger meaning is properly understood as any substitution of what is created for the Creator. People may worship nature, money, mankind, power, history, or social and political systems instead of the God who created them all.[72]

Many mathematicians worship mathematics. The reader is encouraged to carefully reread the above testimonials (starting with Lagrange), this time in the light of the principle of idolatry, for Scripture is most scathing in denouncing it:

> Those who make an image, all of them are useless, and their precious things shall not profit; they are their own witnesses; they neither see nor know, that they may be ashamed.... And the rest of it he makes into a god, his carved image. He falls down before it and worships it; prays to it and says, "Deliver me, for you are my god." They do not know nor understand, for He has shut their eyes, so that they cannot see, and their hearts, so that they cannot understand (Isaiah 44:9, 17, 18).

In light of the definition of idolatry, note carefully the following remarks made by William Spottiswoode (1825-1883) at the Dublin meeting of the British Association in 1878:

---

[70]Kline, *Mathematics: The Loss of Certainty*, p. 73.
[71]Cited in Bell, *Men of Mathematics*, p. 230.
[72]Herbert Schlossberg, *Idols for Destruction* (Nashville: Thomas Nelson, 1983), p. 6.

Coterminous with space and coeval with time is the kingdom of mathematics; within this range her dominion is supreme; otherwise than according to her order nothing can exist, in contradiction of her laws nothing takes place. On her mysterious scroll is to be found written for those who can read it that which has been, that which is, and that which is to come.[73]

Spottiswoode said this just when cracks in the mathematical foundation were beginning to show.[74]

5.10 THE IDOL FALLS

One of the cardinal assumptions accepted uncritically by the majority of mathematicians since the time of the Greeks was the Euclidean nature of physical space. From 300 BC to 1800, the only concept ever brought into serious question about Euclid's *Elements* was the way in which he wrote his fifth, or parallel, postulate. For those who are familiar with high school geometry, this axiom is usually rewritten – following the precedent set by the Scottish physicist and mathematician John Playfair (1748-1819) – as "through a point not on a line, there is exactly one line parallel to the given line." In the diagram below, given line *l* and point A, line *m* is the only line parallel to line *l.*

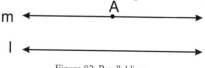

Figure 82: Parallel lines

Originally, Euclid cumbersomely wrote this axiom as follows:

That if a straight line falling on two straight lines makes the interior angles on the same side less than two right angles, the two straight lines, if produced indefinitely, meet on that side on which the angles are less than the two right angles.[75]

Why this complexity? Of all of his axioms, this one, though making intuitive sense, lacked clear description. Even Euclid shied away from using it only until he had to.

Lurking in the shadows of this problem was the thorny issue of infinity. In physical space, do lines continue in a straight direction infinitely? Newton assumed this in his famous first law of motion. But who has yet to travel to the "end of an infinite line" to determine if it is still on a straight course? To the reader, this may sound like unnecessary logic chopping, but since these Euclid-

---

[73]Cited in *British Association Report* (1878), p. 31.
[74]The issue in the analysis to follow (sections 5.10 to 5.16) is *not* good mathematics (of Newton, Kepler, Galileo, etc.) versus bad mathematics (of non-Euclidean geometries, Cantorian set theory, quaternions, matrices, etc.). God did *not* judge good mathematics with bad mathematics; rather, God judged a presuppositional proclivity that bordered on the worship of the creation (i.e., human reason) rather than the Creator.
[75]T. L. Heath, *Euclid: The Thirteen Books of the Elements,* 1:202.

ean axioms had been absolutized as the truth concerning physical space, mathematicians wanted to make sure everything was accounted for. This problem was not just an exercise in intellectual curiosity; it dealt with a conviction about physical space.

In history, mathematicians have tried to corner the fifth axiom of Euclid, either by restating it in simpler terms or by proving this axiom as a theorem from Euclid's other axioms. One, Girolamo Saccheri (1667-1733), a Jesuit priest and professor of mathematics at the University of Pavia, tried to free Euclid from every flaw by proving this axiom as a theorem.[76]

We must now digress and outline some aspects of mathematical method. There are two types of mathematical proof. The most common is called direct proof. From the given axioms, definitions, and other proven theorems, straightforward logic is applied to prove a new theorem. The other less common method is called indirect proof or *reductio ad absurdum*. In this approach, you assume the negation of what you want to prove and reason to a logical contradiction. In this contradiction, you have shown that your assumption is false and what you wanted to prove is therefore true.

Saccheri used indirect proof in his work and it will be illustrated briefly. He had two assumptions to consider: (1) Through a point not on a given line, there exists no line parallel to the given line and (2) through a point not on a given line, there exists more than one line parallel to the given line.

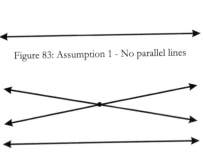

Figure 83: Assumption 1 - No parallel lines

Figure 84: Assumption 2 - More than one parallel line

With Assumption 1, Saccheri reasoned to what seemed to him to be an apparent contradiction (it was apparent because Saccheri, incredulously disbelieving its consistency, arbitrarily contrived a contradiction), but with Assumption 2, the results of his rigidly circumspect logic were strange and intuitively defying, but produced *no* contradictions. An example of one of his conclusions was that the sum of the measures of the angles of a triangle is *less than* 180°! Because the conclusions were so bizarre, Saccheri believed that he had triumphed and that Euclidean geometry was therefore perfect after all. In reality, Saccheri

---

[76]His masterpiece, printed in 1733, was entitled *Euclides ab omni naevo vindicatus* (Euclid cleared of all blemish).

had shown that the fifth postulate was independent of all the others and that logically consistent systems could be constructed based upon Assumption 1 or Assumption 2.

Permission to print Saccheri's *Euclides* was granted by the Jesuit authorities on August 16, 1733 (Saccheri died on October 25 of the same year). This was 48 years *before* Kant's *Critique of Pure Reason* (1781). Had Kant been aware of Saccheri's conclusions, this knowledge would have voided his absolutization of Euclidean geometry. Strangely, its publication *was delayed until 1889*. In other words, for 156 years the book lay buried. Mathematics historian Eric Temple Bell believes that the Jesuit authorities suppressed the book as a matter of conservative policy because of the implications of its conclusions.[77] It was Eugenio Beltrami (1835-1900), a prominent geometer, who resurrected the book in 1889. By that time, non-Euclidean geometries were commonplace.

Early in the 19th century, Carl Friedrich Gauss also noticed what Saccheri had demonstrated; i.e., Gauss could not derive any contradictions starting from either of the two Saccheri assumptions. In 1817, he wrote to his friend Heinrich Olbers (1758-1840):

> I am becoming more and more convinced that the necessity of our geometry [Euclidean – J.N.] cannot be proved, at least not by human reason nor for human reason. Perhaps in another life we will be able to obtain insight into the nature of space, which is now unattainable. Until then we must place geometry not in the same class with arithmetic, which is purely a priori, but with mechanics.[78]

Some mathematics historians think that Gauss performed measurements, calculating the angles formed by three great mountain peaks – Brocken, Hohehagen, and Inselsberg, in the hope of finding out empirically whether space was Euclidean or not.[79] Irrespective of this experimental result, Gauss refuted the Kantian *a priori* proposition about Euclidean space and he concluded that geometries other than Euclid's could be logically constructed in spite of the intuitively defying and bizarre results. Due to the

Figure 85: Nikolai Ivanovich Lobachevsky

---

[77]See Eric Temple Bell, *The Magic of Numbers* (New York: Dover Publications, [1946] 1991), pp. 345-361.

[78]Gauss, 8:177.

[79]The sum exceeded 180 degrees by 14.85 seconds and Gauss attributed it to experimental error. See Kline, *Mathematical Thought from Ancient to Modern Times*, pp. 872-873. According to other sources, Gauss performed this experiment in order to determine whether the shape of the earth (not a perfect sphere because it is flatter at its poles) would affect the accuracy of the calculations of a geodesic survey of Hanover. See Arthur I. Miller, "The Myth of Gauss' Experiment on the Euclidean Nature of Physical Space," *Isis*, 63 (1972), 345-348.

Figure 86: Janos Bolyai

popularity of Euclid, he did not publish for fear of ridicule.

Organized, published presentations were offered simultaneously by two men: the Russian Nikolai Ivanovich Lobachevsky (1793-1856) and the Hungarian Janos Bolyai (1802-1860). Bolyai responded to his findings by writing to his father on November 3, 1823, "I have made such wonderful discoveries that I have been almost overwhelmed by them."[80] Lobachevsky predicted, in response to possible ridicule, that someday the geometry that he developed would be applicable to the real world.[81]

The significance of the discovery of geometries other than Euclid's was that they could be used to adequately model the phenomena of the real world. Let us take the earth as an example. For simplicity, let us assume that it is a perfect sphere. What qualifications would be necessary to define a "straight line" on the "curved surface" of a sphere? Obviously, the Euclidean understanding of a straight line will not do because the perspective of flat, or plane, geometry does not fit in the context of the geometry on a sphere.

Let us wade a little deeper into some mathematical method. Let us define a "straight line" in spherical geometry as a *great circle*, the set of points that are the result of intersecting a flat, plane surface that contains the central point of the sphere with the surface of the sphere (see Figure 87). Some practical examples of a great circle would be the equator line and the lines of meridian.

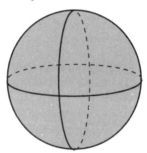

Figure 87: "Lines" on a sphere

Already, we should note that we are developing a geometry that fits the real world. Given "line" *l* on a sphere, the equator, and point A at New York

---

[80]Cited in Roberto Bonola, *Non-Euclidean Geometry* (New York: Dover Publications, [1912] 1955), p. 98.
[81]Cited in D'Ary W. Thompson, *On Growth and Form* (Cambridge: Cambridge University Press, [1948] 1971), pp. 10-11.

City, how many "lines" are there through point A parallel to "line" *l*, the equator (see Figure 88)? Remember that a line is defined as a great circle. Think about it for a minute.

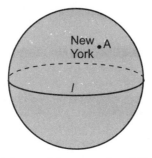

Figure 88: No lines parallel to equator through New York City

Figure 89: George Friedrich Bernhard Riemann

You may have thought to choose the latitude "line" through New York as a "line" parallel to the equator. But, this "line" is *not* a great circle. No matter what point you choose on the sphere that is not on the equator, you will never find a "line" parallel to the given "line." So, in spherical geometry, through a point not on a "line," there exists *no* "lines" parallel to the given "line"! The German mathematician George Friedrich Bernhard Riemann (1826-1866) constructed a type of geometry where the sum of the measures of the angles in a triangle turns out to be *more than* 180°! The important concept about Riemannian geometry is that straight lines, although boundless, are not infinitely long (as in the equator or the lines of the meridian). In 1916, Albert Einstein successfully modeled his general theory of relativity using this concept of Riemannian geometry; i.e., that space could be unbounded without being infinite.[82]

5.11 IMPLICATIONS AND INSIGHTS

The effect of the discovery of geometries other the Euclid's devastated the mathematical community. The obvious question formulated in their minds came out like this: Which geometry is the *true* geometry? For centuries, Euclidean geometry had been viewed as the one, true description of the physical world, and now there was a plurality of geometries from which to choose. Ultimately, mathematicians came to this conclusion: How could any given geome-

---

[82]For a popular, non-technical description of this relationship, see Kline, *Mathematics in Western Culture*, pp. 432-452.

try be considered as true if it and others *contradictory* to it *both* describe physical phenomena? Man's presumably autonomous reason found itself in a quandary.

Actually, if the mathematical context is kept in mind, there are no contradictions among these geometries. The work of the mathematician Felix Klein (1849-1925) proved that the difficulties vanish when the context is recognized.[83] The words "parallel" or "line" can be used in different contexts just as we use the word "court" in the context of an uncovered space surrounded by buildings, wooing a potential marriage partner, or a judicial assembly. In 1871, it was Klein that named the three geometries – that of Lobachevsky and Bolyai, that of Euclid, and that of Riemann – hyperbolic geometry, parabolic geometry, and elliptic geometry, respectively.

Figure 90: Felix Klein

5.11.1 PLURALITY OF GEOMETRIES

Statement	Hyperbolic	Parabolic	Elliptic
Through a point not on a line, there is ...	more than one parallel to the line.	exactly one parallel to the line.	no parallel to the line.
Sums of the measures of the angles in a triangle are ...	< 180°.	= 180°.	> 180°.

In this context, because of the development of non-Euclidean geometries, we must draw a line of demarcation between physical geometry and mathematical geometry. Physical geometry seeks to give a coherent account of the physics of the external world. Each of the three geometries in the above table has proven to be reasonably sufficient, in the physical sense, for certain scientific purposes. Mathematical geometry consists of a system of postulates, and deductions from them, developed independently of the physical world. Each of the three geometries in the above table is internally self-consistent although none is in agreement with another.

How could any of these geometries be considered as true if it and others *contradictory* to it *both* describe physical phenomena? Truth in mathematical geometry is determined by internal consistency. Truth in physical geometry is de-

---

[83]Neumann, 4:2055. Klein, a master teacher at the University of Göttingen, did work in the theory of groups that later became a tool of fundamental importance for quantum mechanics. See Hermann Weyl, *The Theory of Groups and Quantum Mechanics*, trans. H. P. Robertson (London: Methuen, 1931). Group theory initially found application in the description of the symmetry of crystals.

termined by its approximate agreement with observable phenomena. Each of the above geometries is self-consistently true in the abstract, logical, or mathematical sense. At the same time, each is true in the physical sense. Each of the above geometries is a valid geometrical description that describes an appropriate realm of the physical world.

The commentary of Stanley L. Jaki concludes this jaunt into the realm of variegated geometries:

> The concreteness of nature, however, is rich beyond comprehension in aspects and features. This is why even the most bizarre sets of mathematical postulates and geometrical axioms can prove themselves isomorphic with some portion of the observational evidence and useful in systematizing it.[84]

To the minds of most 19th century mathematicians, the discovery of non-Euclidean geometries had seismic implications. Are the foundations of mathematics *really* indubitable and secure? Cracks began to appear all over the place. Not only did the logical foundations of Euclidean geometry appear to be weak, but basic arithmetic and algebra had no logical underpinning at all! And yet, Euclidean geometry, basic arithmetic, and elementary algebra were the roots of the calculus tree, a method that had proven to be so effective in describing and predicting physical phenomena. In fact, the calculus had been built, not on a thoroughly logical basis, as George Berkeley's *The Analyst* (1734) tellingly pointed out, but upon basic intuitive and physical ideas. This understanding provoked Lagrange to conclude that the calculus was successful only because its many logical conundrums offset each other.

To make matters worse, coinciding with the discovery of new geometries came the development of new and strange algebras and arithmetics, all applicable to some aspect of the real world.[85] For example, William Rowan Hamilton (1805-1865), professor of astronomy, poet and close friend of literary giants William Wordsworth (1770-1850) and Samuel Taylor Coleridge (1772-1834), developed a new algebra called quaternions. This algebra made use of complex numbers (of the form $a + bi$ where $i = \sqrt{-1}$) to represent a force, velocity, or any other quantity for which both direction and magnitude are vital; i.e., vector analysis. Hamilton defined a quaternion to be a number of the form $a + bi + cj + dk$ where $i, j,$ and k are *imaginary numbers* (i.e., equal to $\sqrt{-1}$). A property of quaternions is that $i^2 = j^2 = k^2 = -1$. Hamilton went on to show that the multiplication of two quaternions, p and q, is not commutative; i.e., pq ≠ qp. Another example is the work of Arthur Cayley (1821-1895), professor of mathematics at Cambridge University, who invented matrices; i.e., a square or rectangular array of numbers.[86] Like quaternions, the multiplication of two matrices is not commutative. As an addendum, the product of two matrices can be zero

---

[84]Jaki, *The Relevance of Physics*, p. 121.
[85]For a brief survey, see Kline, *Mathematical Thought from Ancient to Modern Times*, pp. 772-794.
[86]In 1925, Werner Karl Heisenberg (1901-1976) realized that what he had set up in his array of atomic spectral terms was nothing but a matrix.

even when neither factor is zero! If the heads of mathematicians were swimming before, they were dizzy now.

5.12 RECONSTRUCTION

In the 19th century, the "faith of the age" among most philosophers, mathematicians, and even theologians rested in an unquestioned conviction concerning the power and ability of human reason to solve all problems. The mathematicians of this era, noting the non-existent logical development of their subject, felt obliged to invoke once again the methodology of the ancient Greek paradigm – axiomatic-deductive logic. These mathematicians were convinced that the cracked foundations of their discipline could be reconstructed logically. The power of man's mind could and certainly would lead them to the "gates of paradise."

The epistemological, metaphysical, and ethical perspectives found in biblical revelation no longer presented viable and ultimate explanations for their perplexing dilemmas. As Morris Kline notes, "Whatever solace these ... men may have derived from the belief that they were uncovering pieces of God's design was nullified by the late-18th-century abandonment of that belief."[87] In mathematical circles, the "age of rigorization" had begun. The goal was to place mathematics on a firm logical and axiomatic base. In spite of its foundational perplexities, the men of this era expressed unshakable confidence that someday a self-consistent, all-embracing mathematical synthesis would be discovered.[88] In striving to reach this goal, 19th century mathematicians did dispense with much of the vague and illogical fog that had inundated the mathematical landscape.[89] But, their vision was ultimately myopic. Morris Kline comments, "What these men did not foresee was that the task of supplying rigorous foundations for all of the existing mathematics would prove to be far more difficult and subtle than any mathematician of 1850 could possibly have envisioned. Nor did they foresee the additional troubles that were to ensue."[90] Since many branches of the mathematical tree could be modeled in number theory, the initial focus of reconstruction pointed to this area.[91]

5.13 INFINITY STRIKES AGAIN

Georg Cantor (1845-1918), born in St. Petersburg, Russia, later moved to Germany. His father was a Danish-Jew who had been converted to Protestantism, and his mother embraced Roman Catholicism. In response to his desire to answer some open questions engendered by the enormously useful infinite se-

---

[87]Kline, *Mathematics: The Loss of Certainty*, p. 170.
[88]The quest for rigor is not wrong; only the absolutization of this quest is in error.
[89]For example, in 1821 Augustin-Louis Cauchy (1789-1857) eliminated all logical inconsistencies from the limit concept of the calculus.
[90]Kline, *Mathematics: The Loss of Certainty*, p. 171.
[91]Rózsa Péter clearly demonstrates how each branch of the mathematics tree finds its roots in number theory in *Playing with Infinity: Mathematical Explorations and Excursions* (New York: Dover Publications, [1961] 1976).

ries called the Fourier series,[92] Cantor introduced a profoundly new theory of numbers, called *Mengenlehre* in German. This theory involved grouping sets of numbers and relating them primarily to the simple, but infinite set of counting numbers; i.e., 1, 2, 3.... Most children, at some time or another, try to see if they can count to and find the last number. No matter how large a number you find all you have to do is add the number one and your discovery is discredited.

Figure 91: Georg Cantor

Cantor founded his study of infinity on the bedrock that nature made use of it everywhere and that the concept of infinity effectively showed the perfection of its Author.[93] In his analysis, the infinite set of counting numbers provided the foundation for his theories. From this basis, he proved many unusual and remarkable theorems about the "arithmetic" of infinite sets. For example, by extending the finite number concept to infinity, he showed that there are two kinds of "infinities." One is the infinity of the set of counting numbers, the cardinality[94] of which he denoted as aleph-null, $\aleph_0$, (aleph is the first letter in the Hebrew alphabet). The other is the infinity of the set of real numbers (the union of rational and irrational numbers), the cardinality of which he denoted $\aleph_1$ or C for the "power of the continuum." He then went on to show that there is no infinite set, of whatever aleph, which cannot be "transcended" by another set of a higher aleph. By that, he meant that there are an infinite number of infinities![95] The Hungarian poet Mihály Babits (1883-1941) called these numbers ($\aleph_0$, $\aleph_1$, $\aleph_2$, $\aleph_3$, $\aleph_4$, ...), towering higher and higher forever, "the towering battlements of infinity." The Argentine writer, Jorge Luis Borges (1899-1986), called them "terrible dynasties."

But the concept of infinity has an uncanny way of exposing the limitations of human thought. In 1895, Cantor's theory gave rise to paradoxical predicaments. For example, let us look at two sets of numbers:

Let A (set of natural or counting numbers): {1, 2, 3, 4, ...}

---

[92]Developed by the French mathematician Joseph Fourier (1768-1830) who formulated a method for analyzing periodic functions based upon his study of the conduction of heat.

[93]Cantor deeply admired the theologians of the Middle Ages who refused to shun the concept of infinity based upon the Scriptural revelation of God as an infinite and eternal Person.

[94]Cardinality signifies the number of elements in a set.

[95]For an interesting summary of Cantor's fascinating life story, see Bell, *Men of Mathematics*, pp. 555-579. For a clear, non-technical discussion of Cantorian set theory, see Martin Gardner, "The Hierarchy of Infinities and the Problems It Spawns," *Mathematics: An Introduction to Its Spirit and Use*, ed. Morris Kline (San Francisco: W. H. Freeman, 1979), pp. 74-78. Cantor conjectured that that there is no aleph between $\aleph_0$ and C. In 1963, a twenty-nine year old mathematician at Stanford University, Paul J. Cohen (1934-), showed that this conjecture, called the "continuum hypothesis," is undecidable; i.e., one can also conjecture that C is not $\aleph_1$ and that there is at least one aleph between $\aleph_0$ and C.

and B (set of even numbers): {2, 4, 6, 8, ...}.

It seems obvious, from inspection, that the number of elements in set B are half as many as the number of elements in set A since set B is missing the elements 1, 3, 5, 7, etc. In mathematical terminology, set B is a proper subset of set A.[96] But Cantor showed that, for each element in set A, there exists one and only one element in set B. That is, you can match elements from each set in a one-to-one correspondence. Therefore, the number of elements in each set is the same! Think about that for awhile! Our common sense tells us that surely there are twice as many elements in set A as there are in set B, but from Cantor's viewpoint, the number of elements in both sets are the same! The Achilles heel in the whole process is that both sets are infinite. In technical terms, the set of even numbers is said to be "denumerable" or "countable." Unlike finite sets, infinite sets can be put in one-to-one correspondence with a *part* of themselves; i.e., their proper subsets. What happens in the realm of the infinite may not always be the same as in the realm of the finite.

The infinite set of primes can also be put into a one-to-one correspondence with the natural numbers:

Set of natural or counting numbers: {1, 2, 3, 4, 5, 6, ...}.

Set of prime numbers: {2, 3, 5, 7, 11, 13, ...}.

5.13.1 AN ODE TO INFINITY

A graduate student at Trinity
Computed the square of infinity.
But it gave him the fidgets
To put down the digits,
So he dropped math and took up divinity.

Anonymous

Although it requires a more complex procedure, the set of rational numbers (i.e., fractions), can also be set in a one-to-one correspondence with the natural or counting numbers. The infinite set of counting numbers, primes, and rational numbers are called transfinite cardinals because they are "countable." The cardinality of real numbers ($\aleph_1$ or C) is larger than the cardinality of the counting numbers and called an "uncountable" infinity. This cardinality is identical with the cardinality of the points on a line, the points in a plane, or the points in any portion of a space of higher dimension. Cantor also showed that a set with $\aleph_0$ elements has exactly $2^{\aleph_0}$ conceivable subsets and that the cardinality of $\aleph_1$ is $2^{\aleph_0}$.[97] This number is what Cantor called the first transfinite number. A set with $\aleph_1$ elements (the real numbers), in turn, has $2^{\aleph_1}$ conceivable subsets.

---

[96] A proper subset of a given set S is any subset of S other than S itself. For example, let S = {1, 2, 3}. S has six proper subsets: {1}, {2}, {3}, {1, 2}, {1, 3}, and {2, 3}.

[97] To conclude this, Cantor used the general formula that states that the number of subsets in a finite set of x elements is 2^x.

$2^{\aleph_1}$ is the next transfinite number and denoted by $\aleph_2$. And, the process of encountering ever larger and larger infinities (the "towering battlements" of transfinite numbers) continues, ad infinitum![98] If the reader feels a bit overwhelmed by all these concepts and wants to drop the mathematics of infinity, then try taking up divinity and unravel the doctrine of the Trinity!

It was Bertrand Russell (1872-1970) who proposed the famous barber paradox in 1918 to dramatize some of the paradoxes found in the study of infinite sets.[99] A barber in a local town puts up a sign saying that he shaves only those people who do not shave themselves. Russell proposed a baffling question, "Who, then, shaves the barber?" If the barber does not shave himself and lets someone else shave him, then he fails to live up to his claim since he should shave himself. If the barber does shave himself, he is likewise failing to live up to his claim since he shaves *only* those people who do not shave themselves. The reader is encouraged to scan the previous sentences again in order to catch their full impact. In the meantime, we will leave our barber poised, with face lathered and razor in hand, mumbling to himself, "To shave or not to shave?" After a few minutes of repeating this Shakespearean dialogue to himself, the barber realizes that the men in the white suits will soon be coming for him. With the specter of straightjackets swirling in his mind, he takes the sign down.

Two other historical paradoxes reflect the same type of logical conundrum and both come from the Greek poet Epimenides (ca. 600 BC).[100] First, the liar paradox: "This statement is not true." If the statement is true, then it is false. If it is false, then it is true. Second, the Cretan liar paradox. If a Cretan says, "The Cretans are always liars, evil beasts, lazy gluttons" (quoted by the Apostle Paul in Titus 1:12), then the statement must include the Cretan who was quoted. In essence, the Cretan is saying, "I am a liar." The statement, "I am lying" is true only if it is false, false only if it is true.

Figure 92: Leopold Kronecker

The formulation of these paradoxes ignited the fires of pure joy in those mathematicians who had suspected Cantor's sanity. Leopold Kronecker (1823-1891), a professor at the University of Berlin when Cantor was there as a stu-

---

[98]How true the remark by the French astronomer Jean Sylvain Bailly (1736-1793), "Infinity is the abyss where our thoughts get lost." Cited in Jan Gullberg, *Mathematics: From the Birth of Numbers* (New York: W. W. Norton, 1997), p. 780.

[99]Russell wanted to illustrate the problem that some set constructions seem to lead to sets that should be members of themselves. For example, the set of all things that are not cats could *not* be a cat. Therefore, *it must be a member of itself*. Consider now the set of all sets that are *not* members of themselves. Is it a member of itself? No matter how you answer, yes or no, you are sure to contradict yourself. The barber paradox illustrates this logical dilemma.

[100]This poet was quoted by the Apostle Paul in Acts 17:28 and Titus 1:12.

dent, called him "a scientific charlatan, a renegade, a 'corrupter of youth'."[101] The Frenchman, Henri Poincaré (1854-1912), said:

> For my part, I think, and I am not the only one, that the important thing is never to introduce entities not completely definable in a finite number of words. Whatever be the cure adopted, we may promise ourselves the joy of the doctor called in to follow a beautiful pathologic case.[102]

Figure 93: Henri Poincaré

The scorn and ridicule heaped upon Cantor by his fellow-mathematicians (especially by Kronecker) eventually broke him. At age 40, he suffered a complete mental breakdown and for the rest of his life these breakdowns recurred with varying degrees of intensity. His mental instability obliterated his mathematical creativity as evidenced by the fact that he published only three more mathematical papers between 1885 and his death. He died in a psychiatric hospital.

Others were impressed with the power of set theory, albeit too late to save Cantor's sanity. David Hilbert (1862-1943), one of the leading mathematicians of the 20th century, declared emphatically that "no one shall drive us out of the paradise which Cantor has created for us."[103] He eulogized Cantor's method as "the most astonishing product of mathematical thought, one of the most beautiful realizations of human activity in the domain of the purely intelligible."[104] Bertrand Russell believed Cantor to be "one of the greatest intellects of the nineteenth century."[105]

5.14 DIVERGENT SCHOOLS

As noted above, not all mathematicians in the early 20th century waved the same flag. Disagreements abounded, so much so, that three schools of divergent mathematical thought arose. Each school claimed to have the last word on mathematics. Only one factor unified each: All still placed indubitable faith in the power of human reason to build a sufficient and complete foundation for mathematics.

---

[101]Cited in A. Schoenflies, "Die Krisis in Cantor's Mathematischen Schaffen," *Acta Mathematica*, 50 (1927), 2.

[102]Henri Poincaré, *The Foundations of Science*, trans. George Bruce Halsted (New York and Garrison: The Science Press, 1913), p. 382.

[103]David Hilbert, "On the Infinite," *Philosophy of Mathematics: Selected Readings*, ed. Paul Benacerraf and Hilary Putnam (Cambridge: Cambridge University Press, [1964] 1983), p. 191.

[104]*Ibid.*, 167.

[105]Bertrand Russell, *The Autobiography of Bertrand Russell* (London: George Allen and Unwin, 1969), 1:217.

The three schools were logicism, intuitionism, and formalism. Detailed inspections of each will be waived. Only pertinent quotes and analyses from each school will be given in order to reveal their epistemological base.

5.14.1 LOGICISM

Bertrand Russell and Alfred North Whitehead founded the logistic school. In the early 20ᵗʰ century, logic was understood and accepted as a body of truth. If mathematics could be shown to be a subset of logic, then it too would be under the umbrella of truth. And, since truth cannot be contradictory, the structure of mathematics would be necessarily consistent – *sine qua non*.

In their work Russell and Whitehead attempted to prove that logic transcended all other pathways to knowledge. Absolute and indubitable knowledge could be achieved through the procedures delineated in their three volume work entitled *Principia Mathematica* (1910-1913), in which it took the authors *362 pages* to establish the proposition that 1 + 1 = 2. According to mathematicians Philip J. Davis (1927-) and Reuben Hersh (1923-), this extensive edifice is "an outstanding example of an unreadable masterpiece."[106]

Note that this compilation of logical propositions, according the Yale University Professor of Philosophy Carl G. Hempel (1905-1997), "conveyed no information whatever on any empirical subject matter."[107] It was an explanation and reduction of mathematics in the context of logic alone. According to this analytical opus, the theorems of mathematics are derived only from human thought; they have no affinity with the physical world. Hempel confirms this incongruity as he continues his commentary on the logicism school:

> This result seems to be irreconcilable with the fact that after all mathematics has proved to be eminently applicable to empirical subject

Figure 94: Alfred North Whitehead

> matter, and that indeed the greater part of present-day scientific knowledge has been reached only through continual reliance on and application of the propositions of mathematics.[108]

5.14.2 INTUITIONISM

The intuitionist school[109] finds its roots in Leopold Kronecker, who made famous the remark, "God made the integers, but all else is the work of man."[110]

---

[106]Philip J. Davis and Reuben Hersh, *The Mathematical Experience* (Boston: Houghton Mifflin, 1981), p. 138.
[107]Carl G. Hempel, "On the Nature of Mathematical Truth," *The World of Mathematics*, ed. James R. Newman (New York: Simon and Schuster, 1956), 3:1631.
[108]*Ibid.*

In response, mathematician Karl Weierstrass retorted, "… all those who up to now have labored to establish the theory of functions are sinners before the Lord."[111]

Figure 95: Luitzen Brouwer

Mathematician Arend Heyting (1898-1980) defines the working vision of the proponents of this school, "The intuitionist mathematician proposes to do mathematics as a natural function of his intellect, as a free, vital activity of thought. For him, mathematics is a production of the human mind."[112] In the early years of the 20th century, Luitzen Brouwer (1882-1966) took up the torch of intuitionism and systematized it. In Brouwer's opinion, the only source of mathematical knowledge is man's intuition.[113] According to Morris Kline, Brouwer and his followers believed that "the intuition, not experience or logic, determines the soundness and acceptability of ideas."[114] It is important to note that Brouwer understood intuitionism as a resurrection of Kantian thought:

> In Kant, we find an old form of intuitionism, now almost completely abandoned, in which time and space are taken to be forms of conception inherent in human reason. For Kant the axioms of arithmetic and geometry were synthetic a priori judgments, i.e., judgments independent of experience and not capable of analytical demonstration; and this explained their apodictic exactness in the world of experience as well as in abstracto. For Kant, therefore, the possibility of disproving arithmetical and geometrical laws experimentally was not only excluded by a firm belief, but it was entirely unthinkable."[115]

Because of its Kantian underpinnings, this school presupposes that truth in mathematics can be known explicitly in the intuitive capabilities of man's mind.

---

[109]Since a mathematical proposition is only valid if it can be intuitively "constructed," this school of thought is also referred to as constructionist (although the two ideas are *not* coextensive).

[110]Cited in Moritz, p. 269. Kronecker believed this sincerely. Negative numbers, fractions, imaginary numbers, and especially irrational numbers were anathema to him and he advocated banishing them completely.

[111]Letter to Sonja Kowalewsky, March 24, 1855, in G. Mittag-Leffler, "Weierstrasss et Sonja Kowalewsky," *Acta Mathematica*, 39 (1923), 194.

[112]Arend Heyting, "The Intuitionist Foundations of Mathematics," *Philosophy of Mathematics: Selected Readings*, ed. Paul Benacerraf and Hilary Putnam (Cambridge: Cambridge University Press, [1964] 1983), p. 52.

[113]Richard Von Mises, "Mathematical Postulates and Human Understanding," *The World of Mathematics*, ed. James R. Newman (New York: Simon and Schuster, 1956), 3:1747.

[114]Kline, *Mathematical Thought from Ancient to Modern Times*, p. 1200.

[115]Luitzen E. Brouwer, "Intuitionism and Formalism," *Philosophy of Mathematics: Selected Readings*, ed. Paul Benacerraf and Hilary Putnam (Cambridge: Cambridge University Press, [1964] 1983), p. 78.

That mathematics reflects and expresses the laws inherent in the pre-established and ordered patterns of the universe is of no importance to the discussion. Hence, the question as to why mathematics works, i.e., why it describes the workings of the physical world so accurately, is left open and unanswered.

5.14.3 FORMALISM

Figure 96: David Hilbert

Cantorian set theory naturally led to the rise of the formalist school led by David Hilbert. Using the counting numbers as a model, this school relied strongly on axiomatics in which mere symbols (Hilbert called them "meaningless marks on paper") are formulated according to set rules without any reference to meaning, context, or interpretation. In 1912, Luitzen Brouwer compared intuitionism with formalism, "The question where mathematical exactness does exist, is answered differently by the two sides; the intuitionist says: in the human intellect, the formalist says: on paper."[116] It was very easy, therefore, for mathematicians of this school to say good-bye to physical reality and make mathematics into a formal, intellectual game.

Hilbert made some staggering claims for his method. In 1930, he said, "Ultimate certainty can be achieved in number theory, that fundamental branch of math, by formalizing it in a superior form of math, or metamath."[117] He recognized the need to "carefully investigate fruitful deductions and deductive methods" and the necessity to "nurse them, strengthen them, and make them useful."[118] In order to drive out the contradictions and paradoxes he had to first corral the infinite. He said:

> We must establish throughout mathematics the same certitude for our deductions as exists in ordinary elementary number theory, which no one doubts and where contradictions and paradoxes arise only through our own carelessness. Obviously these goals can be attained only after we have fully elucidated *the nature of the infinite*.[119]

Confidence radiated from him as indicated by his following remarks:

> ... we have resolved a problem which has plagued mathematicians for a long time, viz., the problem of proving the consistency of the axioms of arithmetic.... What we have twice experienced, once with the paradoxes of

---

[116]*Ibid.*
[117]David Hilbert, "Die Grundlagen der elementaren Zahlenlehre," *Gesammelte Abhandlungen,* 3 (1930), 193.
[118]Hilbert, "On the Infinite," p. 191.
[119]*Ibid.*

the infinitesimal calculus, and once more with the paradoxes of set theory, will not be experienced a third time, nor ever again.... The theory of proof ... is capable of providing a solid basis for the foundations of mathematics.[120]

In 1930, Hilbert retired from his university post in Göttingen, Germany. He entitled his farewell address "The Understanding of Nature and Logic." His final six words were, "Wir mussen wissen, wir werden wissen" (We must know, we shall know).[121] Absolute certainty of mathematical knowledge was the goal of the formalist school.

5.15 THE HIROSHIMA OF MATHEMATICS

A few months after Hilbert's farewell address, a young mathematician at the University of Vienna, Kurt Gödel (1906-1978), published an article that literally dropped an atomic bomb on the mathematical foundations postulated by these schools.[122]

Essentially, Gödel demonstrated, in a deeply involved process of reasoning, that the "axiomatic method has certain inherent limitations, which rule out the possibility that even the ordinary arithmetic of integers can ever be fully axiomatized."[123] In other words, the soundness, acceptability, and completeness of any mathematical system, even the arithmetic of the counting numbers, *cannot* be established by logical principles.

In the words of Ernest Nagel (1901-1985) and James R. Newman (1907-1966), Gödel's work showed that "it is impossible to establish the internal consistency of a very large class of deductive

Figure 97: Kurt Gödel

systems – elementary arithmetic, for example – unless one adopts principles of reasoning so complex that their internal consistency is as open to doubt as that of the systems themselves."[124] Furthermore, Gödel showed that any formal system that attempted to embrace even the theory of the counting numbers would, of necessity, be incomplete. Nagel and Newman clarify the meaning of incompleteness: "Gödel's paper is a proof of the impossibility of demonstrating cer-

---

[120]*Ibid.*, p. 200.
[121]Cited in Irving Kaplansky, "David Hilbert," *The New Encyclopaedia Britannica*, 1976 ed. These words are carved onto Hilbert's tombstone in Göttingen.
[122]See Kurt Gödel, *On Formally Undecidable Propositions of Principia Mathematica and Related Systems* (New York: Dover Publications, [1931, 1962] 1992). For a clear, non-technical demonstration of Gödel's reasoning see Péter, pp. 255-265.
[123]Ernest Nagel and James R. Newman, *Gödel's Proof* (London: Routledge and Kegal Paul, 1958), p. 6.
[124]*Ibid.*

tain important propositions in arithmetic."[125] That is, there would exist some true statement that could not be logically proven from the given and acceptable axioms.[126]

According to Morris Kline, Gödel's results "dealt a death blow to comprehensive axiomization."[127] Nagel and Newman comment, "No final systematization of many important areas of mathematics is attainable, and no absolutely impeccable guarantee can be given that many significant branches of mathematical thought are entirely free from internal contradiction."[128] The axiomatic-deductive method, so highly acclaimed since the time of the Greeks, was not capable of producing a perfect and complete system of thought.

Hilbert's goal is therefore unattainable, a shattering realization to many mathematicians, especially those who had placed mistaken faith in rationalism. In the words of Stanley L. Jaki:

> For one thing, Gödel's theorem casts light on the immense superiority of the human brain over such of its products as the most advanced forms of computers. Clearly, none of these machines can ever yield an answer comparable in its breadth and depth to Gödel's theorem. For another, despair can grow only in a soil where a rigid rationalism has already killed off the seeds of intellectual humility. Such a soil cannot nurture the recognition that there is no escape from admitting that in mathematics and *a fortiori* in physics certainty is not the fruit of a "pure rationalistic" procedure alone.[129]

Man's reason is not autonomous; man is not omniscient. As Stanley L. Jaki reiterates, "Only a limited range of the full reality of things can ever be accommodated in the molds of mathematics, advanced and esoteric as these might be."[130] To the biblical Christian, Gödel's results confirmed the truth that Scrip-

---

[125]*Ibid.*, p. 10.

[126]Two examples of such "unprovable" statements in number theory:

(1) In 1742, Christian Goldbach (1690-1764), a German mathematics teacher, wrote a casual letter to Leonhard Euler asking him if he could supply the proof of the following observation. Goldbach noted that every *even* counting number greater than 2 could be written as the sum of two prime numbers. For example, $4 = 2 + 2$; $6 = 3 + 3$; $8 = 5 + 3$; $16 = 13 + 3$; $18 = 11 + 7$; $48 = 29 + 19$; $100 = 97 + 3$. Could Euler prove this for *all* even counting numbers? He could not. A computer has shown that Goldbach's "conjecture" is true up to 2,000,000, but no machine will ever prove that Goldbach's conjecture is true by testing *all* even numbers greater than 2. In the context of Gödel's result, the inability of mathematicians for almost 250 years to prove this conjecture to be true suggests that it is probably either unprovably true or else clearly false (just one exception to the rule could disprove it). This dilemma pains mathematicians like a decaying tooth for as more time passes, the *less certain* it becomes whether the conjecture's unwillingness to submit to proof signifies a greater probability that it is false or that it is unprovably true.

(2) "Twin" primes of the form p, p + 2 (e.g., 3, 5; 11, 13; 29, 31) keep turning up in the longest series of primes that have yet to be listed. Will they continue to recur indefinitely? Is their number infinite? It seems probable that there are infinitely many pairs of twin primes, but no one has been able to prove it.

[127]Kline, *Mathematical Thought from Ancient to Modern Times*, p. 1207.

[128]Nagel and Newman, p. 6.

[129]Jaki, *The Relevance of Physics*, pp. 129-130.

[130]*Ibid.* p. 135.

ture had always proclaimed. Autonomy belongs to the biblical God *alone*. Whenever man tries to construct any system of thought without reference to this God, it will ultimately fall short.

5.16 TRUE FOUNDATIONS

Morris Kline continues to comment on the implications of Gödel's work by observing that "no system of axioms is adequate to encompass, not only all of mathematics, but even any one significant branch of mathematics...."[131] If the reason of man is not sufficiently powerful to place mathematics on a secure foundation, then what is?

Nagel and Newman remark, "Most of the postulate systems that constitute the foundations of important branches of mathematics cannot be mirrored in finite models."[132] Hermann Weyl observes, "If the game of mathematics is actually consistent then the formula of consistency cannot be proved within this game."[133] By that observation, Weyl meant that in order to prove this consistency one needs another class of games, called metamathematics. To prove the consistency of metamathematics one needs a game superior to it, a super theory of metamathematics. To prove that this super theory of metamathematics is consistent, one needs a super super theory of metamathematics, *ad infinitum*. According to Stanley L. Jaki:

> One is, in fact, caught in a process of endless regression when trying to formalize a metamathematical theory of proof as a set of symbols manipulated according to specified rules. Each set of rules points beyond itself for its proof of consistency. This is why one has to consider dim the prospect of mathematics ever becoming established as the system of "absolute truths."[134]

Gödel's theorems are like a fork in the road for the epistemological foundations of mathematics. One path leads to the possibility that the formulation of mathematical consistency within itself is not yet wholly known. The other path leads to the realization that a rigorously definitive and absolute proof of mathematical consistency is outside the reach of man's mind. Both paths imply that the *ultimate* foundation for truth in mathematics must, of necessity, *lie outside the system of mathematics* and *outside the reach of man's mind*. It was this conclusion that caused Hermann Weyl to make the remarks quoted in Chapter 1:

> The questions of the ultimate foundations and the ultimate meaning in mathematics remain an open problem; we do not know in what direction it

---

[131]Kline, *Mathematical Thought from Ancient to Modern Times*, p. 1207.
[132]Nagel and Newman, p. 22.
[133]Hermann Weyl, "A Half-Century of Mathematics," *American Mathematical Monthly*, 58 (1951), 553.
[134]Jaki, *The Relevance of Physics*, p. 128.

will find its solution nor even whether a final objective answer can be expected at all.[135]

To Weyl, the final solution to the ultimate foundations of mathematics remains a mystery.[136] Morris Kline reveals the source of all the difficulties that creates remarks like these:

Figure 98: Hermann Weyl

> All of the developments since 1930 leave open two major problems; to prove the consistency of unrestricted classical analysis and set theory, and to build mathematics on a strictly intuitionistic basis or to determine the limits of this approach. The source of the difficulties in both of these problems is infinity as used in infinite sets and infinite processes. This concept, which created problems for the Greeks in connection with irrational numbers and which they evaded … has been a subject of contention ever since and prompted Weyl to remark that mathematics is the science of infinity.[137]

As David Hilbert once remarked, "From time immemorial, the infinite has stirred men's *emotions* more than any other question. Hardly any other *idea* has stimulated the mind so fruitfully."[138] He went on to say, "In a certain sense, mathematical analysis is a symphony of the infinite."[139] Like Weyl, Hilbert could not escape the undeniable truism that infinity and mathematical analysis were indelibly linked.

What, in Scripture, does the mathematical concept of infinity give echo to? The echo comes, not from a what, but from a Who. The direction in which the question of the ultimate foundations and the ultimate meaning in mathematics finds its final solution is in the infinite, personal God revealed in Holy Scripture. The structure of mathematics must be built upon a biblical worldview bedrock, the cornerstone of which is the person of the Lord Jesus Christ, or it will crumble into the dust of meaninglessness, mystery, and uncertainty.

The reason that mathematics is such a beautiful and effective system is not because of man's efforts to autonomously create a logically foolproof foundation. Mathematics is beautiful and effective because the biblical God, the creator of the real world with its mathematical properties and the human mind with

---

[135]Weyl, *Philosophy of Mathematics*, p. 219.
[136]Weyl, "A Half-Century of Mathematics," 553.
[137]Kline, *Mathematical Thought from Ancient to Modern Times*, p. 1209.
[138]Hilbert, "On the Infinite," p. 185.
[139]*Ibid.*, p. 187.

its mathematical capabilities, upholds and sustains it, along with everything else, by the "word of His power" (Hebrews 1:3).[140]

QUESTIONS FOR REVIEW, FURTHER RESEARCH, AND DISCUSSION

1. Comment on Aquinas and the influence of his thinking on mathematics.
2. What role did mathematics play in the philosophy of Descartes?
3. Explain the shift in thinking that 18th century mathematicians made and its influence on mid-20th century mathematics education.
4. Distinguish between pure and applied mathematics. What is the danger in absolutizing pure mathematics? What is the danger in absolutizing applied mathematics?
5. What constitutes *mathematical* idolatry?
6. What is "non-Euclidean" geometry?
7. Relate the development of non-Euclidean geometry with the judgment of God on mathematical idolatry.
8. Explain the most serious implication of the development of non-Euclidean geometry on the 19th century mathematical community.
9. Explain the historical context and motivation for the mathematical "age of rigorization."
10. Explain the paradoxical predicaments of Cantor's theory of infinite sets.
11. Explain the factors that divided and united the three foundational schools of mathematical thought.
12. Relate the convictions of Hilbert with the demonstrations of Gödel.
13. Relate Gödel's theorem to God's exhaustive knowledge and man's limited knowledge.
14. Present a case that demonstrates the power and perspective that biblical revelation gives to the foundational crisis in mathematics.

---

[140]This conclusion does not infer that mathematicians are to wholly forsake reasoning (rigor and proof) and empirical confirmation in their work. What it does mean is that mathematicians are to continue in their work with one very important proviso: do not expect absolute certainty and exhaustive consistency in mathematics to be the fruit of a "pure rationalistic" procedure alone.

6: WHY DOES MATHEMATICS WORK?

The eternal mystery of the world is its comprehensibility.[1]

Albert Einstein

In this chapter, the author would like to show that, when a person begins with the idea that the reason of man is autonomous in mathematics, then, when it comes to the comprehensible applicability of mathematics to reality, words like "mystery" and "incredible" appear all over the place.

6.1 THE MYTH OF NEUTRALITY

Early in the 20th century, Bertrand Russell reacted to the paradoxes of set theory with this statement: "Mathematics is the subject in which we never know what we are talking about, nor whether what we are saying is true."[2] Then, using the medium of logic, Russell and Whitehead tried to build a secure and indubitable foundation for mathematics. Gödel's results stopped them dead in their tracks. Toward the end of his life, Russell evaluated his efforts:

Figure 99: Bertrand Russell

> I wanted certainty in the kind of way in which people want religious faith. I thought certainty is more likely to be found in mathematics than elsewhere. But I discovered that many mathematical demonstrations, which my teachers expected me to accept, were full of fallacies, and that, if certainty were indeed discoverable in mathematics, it would be in a new field of mathematics, with more solid foundations than those that had hitherto been thought secure. But as the work proceeded, I was continually

---

[1]Albert Einstein, *Out of My Later Years* (New York: Citadel Press, [1950, 1956, 1984] 1991), p. 61.
[2]Bertrand Russell, "Recent Work on the Principles of Mathematics," *The International Monthly*, 4 (1901), 84.

reminded of the fable about the elephant and the tortoise. Having constructed an elephant upon which the mathematical world could rest, I
found the elephant tottering, and proceeded to construct a tortoise to keep
the elephant from falling. But the tortoise was no more secure than the elephant, and after some twenty years of very arduous toil, I came to the conclusion that there was nothing more that I could do in the way of making
mathematical knowledge indubitable.[3]

Russell's perspective of life and mathematics revealed a clear presuppositional stance:

Man is the product of causes which had no prevision of the end they were
achieving; that his origin, his growth, his hopes and fears, his loves and his
beliefs, are but the outcome of accidental collocations of atoms; that no
fire, no heroism, no intensity of thought and feeling, can preserve an individual life beyond the grave; that all the labor of the ages, all the devotion,
all the noonday brightness of human genius, are destined to extinction in
the vast death of the solar system, and that the whole temple of man's
achievement must inevitably be buried beneath the debris of a universe in
ruins – all these things, if not quite beyond dispute, are yet so nearly certain, that no philosophy which rejects them can hope to stand. Only within
the scaffolding of these truths, only on the firm foundation of unyielding
despair, can the soul's habitation henceforth be safely built.[4]

As we have surveyed the history of mathematics, we have noticed that
what an individual or culture believes concerning the origin, purpose, and destiny of the cosmos affects the way mathematics is viewed and ultimately, the
way mathematics progresses. In the words of Oswald Spengler, "The style of
any mathematic which comes into being ... depends wholly on the culture in
which it is rooted, the sort of mankind it is that ponders it."[5]

Mathematics is not a neutral discipline; it is always linked with presuppositions. In fact, in its presuppositional base, mathematics either thrives or dies. In
the civilizations of antiquity, we have seen that mathematics progressed for a
few centuries, then stagnated due to a false worldview. We have also noted the
great creative mathematical stirrings that took place in a culture steeped in the
biblical worldview. Today, the majority of mathematicians and scientists philosophically deny the worldview that birthed modern science. Given this stance,
what is the present condition of mathematics and where will it end? Will it
eventually exhaust itself?

Francis Schaeffer observes that "the world view determines the direction
such creative stirrings will take, and how – and whether the stirrings will con-

---

[3]Russell, *The Autobiography of Bertrand Russell*, 3:220.
[4]Bertrand Russell, "A Free Man's Worship" (1903) *Mysticism and Logic, and Other Essays* (London:
Longmans, Green & Co., 1921), pp. 47-48.
[5]Spengler, 4:2318.

tinue or dry up."[6] It is a fact that, in the 20th century, more work has been done in mathematics than all other centuries combined.[7] If mathematics should exhaust itself in a culture steeped in humanism, then why all this activity? Schaeffer explains, "Later, when the Christian base was lost, a tradition and momentum had been set in motion, and the pragmatic necessity of technology, and even control by the state, drives science on, but ... with a subtle yet important change in emphasis."[8]

That change in emphasis, from Christianity to humanism, has serious ramifications. Stanley L. Jaki makes this clear as he comments on the philosophical movements in the 18th and 19th centuries:

> The next two centuries saw the rise of philosophical movements, all hostile to natural theology. Whatever their lip service to science, they all posed a threat to it. The blows they aimed at man's knowledge of God were as many blows at knowledge, at science, and at the rationality of the universe.[9]

It is extremely important to note that many scientists today believe in a cosmology that posits an "oscillating" universe. Through observational astronomy, distant galaxies are known to be receding from ours, the Milky Way, at great speed. In fact, the farther these galaxies are away from ours, the greater is their speed away from us.[10] To many astronomers, this indicates that the universe is "breathing out." If the universe is exhaling, there had to be a point in time when this "breathing" began. At one time in our distant past, say these astronomers, all the matter of the universe found itself concentrated in one extremely dense atom. Then an explosion occurred, called the Big Bang, and the elements of the universe galloped off into space. Now, and it is essential to note this, these astronomers do not conclude their theory with a universe "breathing" out. If the universe is now exhaling, it will need to "inhale" again. That is, the elements of the universe will eventually "breathe in" coming together again

---

[6]Schaeffer, *How Should We Then Live? The Rise and Decline of Western Thought and Culture*, p. 133.

[7]See Keith Devlin, *Mathematics: The New Golden Age* (London: Penguin Books, 1988). Devlin points out that in the year 1900, all the world's mathematical knowledge could be compiled in about eighty books. One hundred years later, one hundred thousand books would be needed to store that knowledge. The validity of this mathematical storehouse of knowledge to the realm of physical science is questionable, however. Like the accomplishments of the Alexandrian Greeks, this repository may just be a "mass of detail without focus." See the chapter entitled "The Isolation of Mathematics" in Kline, *The Loss of Certainty*, pp. 278-306. In the words of John von Neumann (see section 5.5), this mathematical knowledge, due to abstract inbreeding, may just be a "multitude of insignificant branches ... a disorganized mass of details and complexities."

[8]Schaeffer, *How Should We Then Live? The Rise and Decline of Western Thought and Culture*, p. 134.

[9]Jaki, *The Road of Science and the Ways to God*, p. 160.

[10]These ideas are based upon the theory of spectroscopic red-shift. This is a controversial topic in the scientific community and it illustrates how scientists read their cosmogonical and teleological preconceptions into their observations. For further detail, see Paul M. Steidl, *The Earth, the Stars and the Bible* (Phillipsburg: Presbyterian and Reformed, 1979), pp. 211-218. For an interesting and fascinating attempt to analyze spectroscopic red-shift from a biblical cosmogonic perspective, see D. Russell Humphreys, *Starlight and Time: Solving the Puzzle of Distant Starlight in a Young Universe* (Colorado Springs: Master Books, 1994).

into one extremely dense atom. This is what "oscillating" universe means. This idea is a mirror image of the ancient eternal cycle theory.

Stanley L. Jaki sees the theory of the oscillating universe as the gravest perplexity of the modern, scientific world. He says:

> The very roots of that perplexity form a mirror-image of the age-old need to make a choice between two ultimate alternatives: faith in the Creator and in a creation once-and-for-all, or surrender to the treadmill of eternal cycles. Such should indeed be the case, as the present is always a child of the past. The present and past of scientific history tell the very same lesson. It is the indispensability of a firm faith in the only lasting source of rationality and confidence, the Maker of heaven and earth, of all things visible and invisible.[11]

Figure 100: Aleksandr Solzhenitsyn

If Western man continues to aim his metaphysical, epistemological, and ethical blows at the revelation of God in His word and in His works, Western civilization will eventually stagnate and die just like the cultures of antiquity. This fact has been graphically portrayed to the West through the articulate and heart-rending words of the Russian exile and one-time mathematics/physics teacher, Aleksandr Solzhenitsyn (1918-).[12] In 1946, responding to the appalling devastation (nuclear and otherwise) of World War II, Albert Einstein (1879-1955) reflected, "It is easier to denature plutonium than it is to denature the evil spirit of man."[13] In 1948, Omar Bradley (1893-1981), famed World War II general, scrutinized both the scientific advances and ethical forfeitures of the 20th century: "We have grasped the mystery of the atom and we rejected the Sermon on the Mount. Ours is a world of nuclear giants and ethical infants."[14] For Bertrand Russell, it was the likely prospect of nuclear holocaust that caused him to add a disclaimer to his scientific ethical manifesto – his "firm foundation of unyielding despair centered on a accidental collocations of atoms" – made fifty years earlier. In 1950, he reluctantly admitted:

---

[11]Jaki, *Science and Creation: From Eternal Cycles to an Oscillating Universe*, p. 357.
[12]See Alexandr Solzhenitsyn, "Men Have Forgotten God," The Templeton Address (1983), *National Review*, July 22, 1983. See also his commencement address given at Harvard University on June 8, 1978 entitled "A World Split Apart" (available from many sources).
[13]Albert Einstein, from an interview with Michael Amrine, *The New York Times Magazine*, June 23, 1946, p. 42.
[14]General Bradley said this in Boston on November 10, 1948. Cited in Jaki, *The Road of Science and the Ways to God*, p. 304.

The root of the matter is a very simple and old-fashioned thing, a thing so simple that I am almost ashamed to mention it, for fear of the derisive smile with which wise cynics will greet my words. The thing I mean – please forgive me for mentioning it, is love, Christian love or compassion. If you feel this, you have a motive for existence, a guide for action, a reason for courage, an imperative necessity for intellectual honesty.[15]

In spite of its metaphysical, epistemological, and ethical trauma, modern science and mathematics drive on, either in ignorance or in blunt rejection of the biblical foundations that engendered its viable birth. The low-octane fuel of pragmatic necessity fuels science. Mathematics is driven by the ivory tower of abstract analysis.[16] In spite of the imposing discoveries made by modern mathematicians and in spite of the voluminous mathematical activity in the past one hundred years, the price tag of modernity's ethical relativism will be paid. In part, the price is being paid now if we discerningly "understand the times" (I Chronicles 12:32). Stanley L. Jaki comments on these themes:

To cultivate a science which has grown, in virtue of a viable birth, into a robust being, an explicit faith in Creation is not necessary. But since any such being lives in terms of the logic of its conception and birth, scientific blind alleys immersed in philosophical darkness will be in store for those who chart, intentionally or not, avenues whose sense is diametrically opposed to the most creative innovation in human thought, the Christian doctrine of creation of all out of nothing in the Beginning.[17]

Returning to presuppositions, in the philosophy of mathematics, they can be based either upon the autonomy of man or in the biblical revelation of the sovereign, Creator God. A world of difference separates the two. One believes that all things happen by chance, the other by design. One, using the words of Bertrand Russell, is the philosophy of empty despair and the other the dynamic of living certainty. These two presuppositional camps will be delineated as we inspect the statements made by 20th century mathematicians and scientists.

6.2 THE STATE OF THE ART

6.2.1 PREESTABLISHED HARMONY

Sir James Jeans (1877-1946), renowned British mathematician and scientist, said in 1930, "The universe shows evidence of a designing or controlling power that has something in common with our own mathematical minds ... the ten-

---

[15]Bertrand Russell, *The Impact of Science on Society* (New York: Columbia University Press, 1951), p. 59.

[16]That some of this abstract analysis in mathematics may sometime, in the future, relate to the physical world is not denied. When this happens, it happens in spite of the philosophy that motivates this type of study, not because of it (and due to God's common grace). The author does not relegate pure mathematics to some Satan-inspired realm of darkness. The danger in absolutizing pure mathematics is in separating it from its life-blood; i.e., God's variegated creation.

[17]Jaki, *Science and Creation: From Eternal Cycles to an Oscillating Universe*, pp. 366-367.

dency to think in a way which, for want of a better word, we describe as mathematical."[18] He adds:

> The essential fact is simply that all the pictures which science now draws of nature, and which alone seem capable of according with observational fact, are *mathematical pictures*.... Nature seems very conversant with the rules of pure mathematics.... In any event it can hardly be disputed that nature and our conscious mathematical minds work according to the same laws.[19]

Since Jeans perceived an ingrained harmony between mathematics and nature, he concluded that only a mathematician could understand the essential workings of the universe.

Figure 101: Sir James Jeans

It is this alliance of mathematics and nature that inspired him to say, "... the Great Architect of the Universe now begins to appear as a pure mathematician."[20]

Pierre Duhem agreed with Jeans by observing that "it is impossible for us to believe that this order and this organization produced by theory are not the reflected image of a real order and organization."[21]

Referring to the givens of nature, Hermann Weyl saw in mathematics "a wonderful harmony between the given on one hand and reason on the other."[22]

Max Planck (1858-1947), a German theoretical physicist, laid the foundations for the development of the quantum theory, a theory which revolutionized physics. Near the end of his life, he said:

> What has led me to science and made me since youth enthusiastic for it is the not at all obvious fact that the laws of our thoughts coincide with the regularity of the flow of impressions which we receive from the external world, [and] that it is

Figure 102: Max Planck

---

[18]James Jeans, *The Mysterious Universe* (New York: Macmillan, 1930), p. 149.
[19]*Ibid.*, p. 127.
[20]*Ibid.*, p. 144. For an analysis of this statement, "the Great Architect of the Universe appears to be a pure mathematician," see section 6.10.
[21]Pierre Duhem, *The Aim and Structure of Physical Theory*, trans. Philip P. Weiner (Princeton: Princeton University Press, 1954), p. 26.
[22]Weyl, *Philosophy of Mathematics*, p. 69.

therefore possible for man to reach conclusions through pure speculation about those regularities. Here it is of essential significance that the external world represents something independent of us, something absolute which we confront, and the search for the laws valid for this absolute appeared to me the most beautiful scientific task in life.[23]

Albert Einstein (1879-1955) refined Newton's model of the universe with his special and general relativity theories. Some understand Einstein's relativity theory as a total rejection of the ordered time and space of Newtonian mechanics.[24] Remember Alexander Pope's epitaph on Newton,

Nature and Nature's laws lay hid in night;
God said, "Let Newton be!" and all was light.

In response to Einstein's discoveries, Sir John Collings Squire added,

It did not last: the Devil howling "Ho!
Let Einstein be!" restored the status quo.

Experimentally, Einstein's theories were limited in application to the realms of the very fast (e.g., the speed of light) and the very large (e.g., the entire universe). Everywhere else, including flying in an airplane and sending men to the moon, Einstein's theories paralleled those forecasted by Newtonian mechanics.

Although Einstein talked about a "God who does not play dice," he remained an agnostic most of his life.[25] Stanley L. Jaki remarks that in Einstein's "cosmic religion" there was "no room for creation or Creator."[26] Einstein himself defined his conception of God as follows:

Certain it is that a conviction, akin to religious feeling, of the rationality or intelligibility of the world lies behind all scientific work of a higher order. This firm belief, a belief bound up with deep feeling, in a superior mind that reveals itself in the world of experience, represents my conception of God.[27]

He then admitted that his conception of God, the superior mind, was pantheistic in nature.[28] When it came to his scientific and mathematical work, he

---

[23]Max Planck, "Wissenschaftliche Selbstbiographie," *Physikalische Abhandlungen*, 3 (1948), 374.

[24]Einstein replaced the metaphysical absolutes of time and space (they are relative to one's position) with the material absolute of the speed of light. For a study of the impact of Einstein's thought (more indirectly than directly) on culture, see the chapter entitled "A Relativistic World" in Paul Johnson, *Modern Times: The World from the Twenties to the Nineties*, pp. 1-48.

[25]Einstein may have talked about God, but biblical revelation did not provide him with any direction and purpose in life. To him, "to ponder interminably over the reason for one's own existence or the meaning of life in general seems to me, from an objective point of view, to be sheer folly." Albert Einstein, *What I Believe* (London: George Allen & Unwin, 1966), p. 27.

[26]Stanley L. Jaki, *Cosmos and Creator*, p. 4.

[27]Einstein, *Essays in Science*, p. 11.

[28]*Ibid.* That Einstein, a Jew, would freely embrace pantheism is explainable only when one recognizes the difference between Christian monotheism and Jewish (also Muslim) monotheism. In Christian monotheism, Christ is the only-begotten (the *monogenes*) of the Father (see section 3.3 and John 1:14). In Judaism (as in Islam) there is no place for the *monogenes* to reside in a person (i.e., Je-

placed his faith in, using the words of Leibniz, a "preestablished harmony in the universe."[29] To him, as illustrated by the above quotation, God was simply a name for this principle of harmony in the universe.

Einstein's passion for order is revealed in the following legend told by Otto Neugebauer.[30] When Einstein was a youth, his parents were worried about him because he did not speak at all. Finally, one day at supper, he broke the silence saying, "Die Suppe ist zu heiss" (The soup is too hot). Greatly relieved, his parents asked him why he had not spoken up to that time. He replied, "Bisher war alles in ordnung" (Until now everything was in order).

According to Stanley L. Jaki "... the mathematicians and especially the geometry that the scientists of Galileo's time held in such high esteem was not considered by them a free creation of mind but rather a pattern to be learned from observation of the actual contours of nature."[31] This statement reflects the viewpoints of all of the above scientists; a viewpoint that presupposes mathematics to be a splendid tool that enables man to discover order in a preestablished universe. This assumption is in agreement with biblical revelation, even though most, if not all, of these men would not overtly align themselves with the Christian faith. Unfortunately, their belief is the minority opinion because most modern mathematicians and scientists understand mathematics to be, not a tool, but a divining rod.

6.2.2 "THE FOOTPRINT IS OUR OWN"

John W. N. Sullivan (1886-1937), who wrote many interpretive works on science, expressed the majority opinion by saying:

> And it seems that the mathematician, in creating his art, is exhibiting that movement of our minds that has created the spatio-temporal material universe we know ... The significance of mathematics resides precisely in the fact that it is an art; by informing us of the nature of our own minds it informs us of much that depends on our minds. It does not enable us to explore some remote region of the eternally existent; it helps to show us how far what exists depends upon the way in which we exist. We are the lawgivers of the universe; it is even possible that we can experience nothing

---

sus Christ). Hence, it is natural and logical for a scientist believing in one God (but not in His only-begotten Son, Jesus Christ), whether personal or impersonal (as in the case of Einstein), to slip into pantheism when faced with explaining the order of the universe (called *monogenes* by the Greeks and which the Greeks attributed to an impersonal *logos*). Another prominent Jewish example of this "slip into pantheism" is Baruch Spinoza (1632-1677). As for the Muslims, it is enough to think of Averroës (1126-1198) and his followers.

[29]*Ibid.*, p. 4.
[30]Cited in Davis and Hersh, *The Mathematical Experience*, p. 172.
[31]Jaki, *The Relevance of Physics,* p. 101.

but what we have created and that the greatest of our mathematical creations is the material universe itself."[32]

Percy W. Bridgman (1882-1961), a 1946 Nobel Prize winner in physics, said, "It is the merest truism, evident at once to unsophisticated observation, that mathematics is a human invention."[33]

Sir Arthur Stanley Eddington (1882-1944), British astronomer, graphically explained the origin and originator of all things:

> We have found that where science has progressed the farthest, the mind has but regained from nature that which the mind has put into nature. We have found a strange foot-print on the shores of the unknown. We have devised profound theories, one after another, to account for its origin. At last, we have succeeded in reconstructing the creature that made the foot-print. And Lo! it is our own.[34]

Figure 103: Sir Arthur Stanley Eddington

Morris Kline gives affirmation:

> It may be that man has introduced limited and even artificial concepts and only in this way has managed to institute some order in nature. Man's mathematics may be no more than a workable scheme. Nature itself may be far more complex or have no inherent design.[35]

To the majority of mathematicians today, the universe does not reveal a preestablished harmony. Hence, to these men, mathematics, as a method, does not reveal a harmonious order established by the biblical God; it enables man to create order out of a multiverse of assumed chaos. The presuppositions of these men have blinded them to the realization that the basic function of mathematics is interpretive. Mathematics is a tool of quantification; it cannot create what is quantifies. Mathematics is a reporter of the external, objective, and pre-established world.

The difference in perspectives is clear. Man is either a discoverer or an autonomous creator. Since both viewpoints posit explanations concerning the

---

[32]John W. N. Sullivan, "Mathematics as an Art," *The World of Mathematics*, ed. James R. Newman (New York: Simon and Schuster, 1956), 3:2021. Note the distinct Kantian impress in his remarks.

[33]Percy W. Bridgman, *The Logic of Modern Physics* (New York: Macmillan, 1927), p. 60.

[34]Arthur Stanley Eddington, *Space, Time and Gravitation: An Outline of the General Theory of Relativity* (Cambridge: Cambridge University Press, 1920), p. 201.

[35]Kline, *Mathematics: The Loss of Certainty*, p. 350. Even the complexity of the so-called "Chaos" theory reveals an underlying structure and order. James Gleick notes that many in this field of study have "discovered suggestions of structure amid seemingly random behavior." See *Chaos: Making a New Science* (New York: Penguin Books, 1987), p. 44.

origin, purpose, and destiny of the cosmos, then *both are integrally religious in nature.* As Rousas J. Rushdoony points out:

> ... mathematics is not the means of denying the idea of God's pre-established world in order to play god and create our own cosmos, but rather is a means whereby we can think God's thoughts after Him. It is a means towards furthering our knowledge of God's creation and towards establishing our dominion over it under God. The issue in mathematics today is root and branch a religious one."[36]

6.3 A QUANTUM LEAP

Figure 104: Werner Karl Heisenberg

To some, this talk about harmonious order and its underlying ontological commitment to causality (cause and effect relationships) is irrelevant given the "uncertainty principle" first conceptualized by Werner Karl Heisenberg (1901-1976) in 1927.[37] In the realm of quantum mechanics, this principle states that you cannot measure momentum and position of an atomic particle at the same time. That is, there is an inherent limitation to the accuracy that can be achieved in measurements and this limitation is *only* noticeable on the atomic level (physicists have to resort to statistical averages rather than exact measurement). From this, Heisenberg, using a Humean springboard, immediately jumped to the conclusion that causality thereby had definitively been disproved. In 1929, he argued that "the resolution of the paradoxes of atomic physics can be accomplished only by further renunciation of old and cherished ideas. Most important of these is the idea that natural phenomena obey exact laws – the principle of causality."[38]

It is important to note that Heisenberg *assumed* that causality in a physical process depended upon its exact measurability. This assumption cannot be faulted for it was the fruit of the philosophical barrenness that had been growing for centuries in the Western mind. During these centuries the scientific mind had come under subjugation to a purely mechanistic cosmology due to the quantitative successes of measurement in physics. Because of this absolutization of a quantitative and mechanistic picture of the world, Heisenberg made

---

[36]Rushdoony, *The Philosophy of the Christian Curriculum*, p. 58.
[37]Werner Heisenberg, *The Physical Principles of the Quantum Theory*, trans. C. Eckart and F. C. Hoyt (Chicago: University of Chicago Press, 1930). This principle is also called the indeterminacy principle.
[38]*Ibid.* p. 62. Scientific indeterminacy is, in reality, much more than using statistical probabilities to understand the quantum matrix. The denial of causality embraces a metaphysic of a blind, impersonal, and purposeless force or energy.

the mistake that the *inability of the physicists to measure nature exactly* showed the *inability of nature to act exactly*. The seeming truth of the foregoing statement depended on taking the word "exactly" in two different meanings: one operational (quantifiable), the other ontological (metaphysical). The philosophical conclusion that Heisenberg drew from his uncertainty principle is a classic example of falling prey to the logical fallacy of equivocation, the fallacy of mixing up unequal and heterogeneous quantities.[39]

It may be that behind the physics of quantum mechanics lies a higher degree of unity and harmony that our current instrumentation cannot yet measure. The wisdom and logic of the quantum realm may be so complex that we may never be able to unravel it. The only instrumentality that we have to help us describe this realm now are the wonderful tools that mathematics gives us.

6.4 A DEEPER LOOK INTO MORRIS KLINE

Perhaps no one has been more prolific in writing about mathematics in the late 20th century than Morris Kline (1908-1992). Obviously well qualified, talented, and articulate, his erudite works on the history and scientific applications of mathematics have had beneficial influence worldwide.

But, at the same time, he was an apparent thorn in the flesh to the professional mathematical community. In 1973, he published a critical appraisal of the new mathematics curriculum of the 1960s entitled *Why Johnny Can't Add: The Failure of the New Mathematics*. After shooting down the methodology of pre-university mathematics, he next took aim at the

Figure 105: Morris Kline

university professors of mathematics. In 1977, his publication *Why the Professor Can't Teach: Mathematics and the Dilemma of University Education* certainly did not win him too many friends in the higher circles of the educational elite. Finally, in 1980, the publication of *Mathematics: The Loss of Certainty* unveiled a most thorough indictment of modern mathematics. We will leave it to the mathematics professionals to quibble over Dr. Kline's bombastic exposures. In the mean time, we will take note of some of Kline's revealing conclusions concerning mathematics.

In his introduction to *Mathematics: The Loss of Certainty*, he states, "It behooves us therefore to learn why, despite its uncertain foundations, and despite

---

[39]That Heisenberg and other physicists did not and do not recognize this fallacy reflects on the calamitous consequences resulting from the dearth of unity between departments of the modern university (or, more accurately, multiversity – see section 3.13); i.e., the logic department has nothing to say to the physics department. For more analysis, see the chapter entitled "Turtles and Tunnels" in Jaki, *God and the Cosmologists*, pp. 117-147.

the conflicting theories of mathematicians, mathematics has proved to be so incredibly effective."[40]

In his preface to *Mathematics and the Physical World*, he reflects:

Finally, a study of mathematics and its contributions to the sciences exposes a deep question. Mathematics is man-made. The concepts, the broad ideas, the logical standards and methods of reasoning, and the ideals which have been steadfastly pursued for over two thousand years were fashioned by human beings. Yet with this product of his fallible mind man has surveyed spaces too vast for his imagination to encompass; he has predicted and shown how to control radio waves which none of our senses can perceive; and he has discovered particles too small to be seen with the most powerful microscope. Cold symbols and formulas completely at the disposition of man have enabled him to secure a portentous grip on the universe. Some explanation of this marvelous power is called for.[41]

Kline is not alone in this acute cry for an explanation. Richard Courant (1888-1972), formerly head of the mathematics department at the pre-Hitler world's center for mathematics, the University of Göttingen, and then head of the Courant Institute of Mathematical Sciences of New York University, remarks, "That mathematics, an emanation of the human mind, should serve so effectively for the description and understanding of the physical world is a challenging fact that has rightly attracted the concern of philosophers."[42]

Richard E. Von Mises (1883-1953), who was born in Austria and later became a lecturer at Harvard University, agrees with Courant by stating that "the coordination between mathematics ... and reality cannot be reached by a mathematicized doctrine...."[43] He goes on to remark:

None of the three forms of the foundation of mathematics, the intuitionist, the formalist, or the logistic, is capable of completely rationalizing the relation between tautological systems and (extramathematical) experiences, which is its very purpose, i.e., to make this relation a part of the mathematical system itself.[44]

Norman Campbell, British physicist and philosopher of science, queries about the remarkable power of mathematics in prediction:

Why do they predict? We return once again to the question which we cannot avoid. The final answer that I must give is that I do not know, that nobody knows, and that probably nobody ever will know.[45]

---

[40]Kline, *Mathematics: The Loss of Certainty*, p. 8.
[41]Kline, *Mathematics and the Physical World*, p. ix.
[42]Richard Courant, "Mathematics in the Modern World," *Scientific American*, 211 (1964), 48-49.
[43]Von Mises, 3:1752-1753.
[44]*Ibid.*, 3:1754.
[45]Norman Campbell, *What is Science?* (New York: Dover Publications, 1952), p. 71.

Mathematics historian Salomon Bochner (1899-1982) confesses:

What makes mathematics so effective when it enters science is a mystery of mysteries, and the present book wants to achieve no more than to explicate how deep this mystery is.[46]

All of these men begin with the explicit assumption that mathematics is merely and only a creation of the human mind. Given this premise, they are unable to completely explain the marvelous power of mathematics, the power of describing and predicting the workings of the physical world. In finality, they must consign themselves, as Bochner and Kline do, to the use of the expression, "It is a mystery."[47]

6.5 THE CONFESSIONS OF A NOBEL PRIZE WINNER

In 1963, Eugene Wigner (1902-1995) won the Nobel Prize in physics for his research in quantum mechanics. In 1960, he wrote an article with a revealing title: "The Unreasonable Effectiveness of Mathematics in the Natural Sciences." To begin his discussion, he quoted the philosopher Charles Sanders Peirce (1839-1914): "It is probable that there is some secret here which remains to be discovered."[48] Then he presented his thesis, "The enormous usefulness of mathematics in the natural sciences is something bordering on the mysterious and ... there is no rational explanation for it."[49] He explains these successes using an interesting metaphor:

Figure 106: Eugene Wigner

We are in a position similar to that of a man who was provided with a bunch of keys and who, having to open several doors in succession, always hit on the right key on the first or second trial. He became skeptical concerning the uniqueness of the coordination of keys and doors.[50]

Wigner continues to express his bafflement over the fact that "it is not at all natural that 'laws of nature' exist, much less that man is able to discern them."[51] Concerning the effectiveness of Newton's law of universal gravitation, he says that it "has proved accurate beyond all reasonable expectations."[52]

---

[46]Salomon Bochner, *The Role of Mathematics in the Rise of Science* (Princeton: Princeton University Press, 1966), p. v.
[47]Kline, *Mathematics: The Loss of Certainty*, p. 7.
[48]Cited in Eugene Wigner, *Symmetries and Reflections: Scientific Essays* (Cambridge and London: The MIT Press, 1970), p. 222.
[49]*Ibid.*, p. 223.
[50]*Ibid.*
[51]*Ibid.*, p. 227.
[52]*Ibid.*, p. 231.

He continues to illustrate the mysterious usefulness of mathematics by citing the application of imaginary numbers (e.g., $\sqrt{-1}$) in the laws of quantum mechanics. First, he observes that "the use of complex numbers is in this case not a calculational trick of applied mathematics but comes close to being a necessity in the formulation of the laws of quantum mechanics."[53] Given this fact, he responds with this amazing remark, "It is difficult to avoid the impression that a miracle confronts us here."[54]

6.5.1 THE STRUCTURE OF THE ATOM IN QUANTUM MECHANICS

Figure 107: Quantum Atomic Structure

In the hydrogen atom, electrons are pictured as clouds that orbit the positive nucleus. The darker section of this time-average view shows where the electron is most probably found.

Finally, he concludes, "Fundamentally, we do not know why our theories work so well."[55] And to this, he concludes:

The miracle of appropriateness of the language of mathematics for the formulation of the laws of physics is a wonderful gift which we neither understand nor deserve. We should be grateful for it and hope that it will extend, for better or for worse, to our pleasure even though perhaps also to our bafflement, to wide branches of learning.[56]

6.6 MORE MIRACLES, MYSTERY, AND WONDER

Erwin Schrödinger (1887-1961) developed the famous wave equation of quantum mechanics that includes the "miraculous" imaginary number mentioned by Wigner.[57] He affirms Wigner in observing that humanity's power to

---

[53] *Ibid.*, p. 229.
[54] *Ibid.*
[55] *Ibid.*, p. 237.
[56] *Ibid.*
[57] The time dependent Schrödinger equation where $\psi(t)$ is a function giving both the space and time of the matter wave associated with a particle of mass m moving in a region where the potential energy is U(x) is:

$$\frac{\partial^2 \psi(x,t)}{\partial x^2} = -\frac{8\pi^2 m}{h^2}\left[\frac{ih}{2\pi}\frac{\partial \psi(x,t)}{\partial t} - U(x)\psi(x,t)\right]$$ where h = Planck's constant and $i = \sqrt{-1}$.

discover the laws of nature is beyond human understanding.[58] The same miracle appears again as he reflects on the atomic structure of genes:

> How can we, from the point of view of statis-
> tical physics, reconcile the facts that the gene
> structure seems to involve only a compara-
> tively small number of atoms (of the order of
> 1000 and possibly less), and that nevertheless
> it displays a most regular and lawful activity –
> with a durability or permanence that borders
> upon the miraculous.[59]

In 1980, Richard W. Hamming (1915-1998), university professor, tried to explain the mystery proposed by Dr. Wigner. He introduced his treatise, "We must begin somewhere and sometime to explain the phenomenon that the world seems to be organized in a logical pattern that parallels much

Figure 108: Erwin Schrödinger

of mathematics."[60] After several pages of discourse, he came to this conclusion, "From all of this I am forced to conclude both that mathematics is unreasonably effective and that all of the explanations I have given when added together simply are not enough to explain what I set out to account for."[61]

Dr. Remo J. Ruffini, physicist at Princeton University, reacted to the successful landing of men on the moon:

> How a mathematical structure can correspond to nature is a mystery. One
> way out is just to say that the language in which nature speaks is the lan-
> guage of mathematics. This begs the question. Often we are both shocked
> and surprised by the correspondence between mathematics and nature, es-
> pecially when the experiment confirms that our mathematical model de-
> scribes nature perfectly.[62]

Albert Einstein once remarked concerning this issue, "The eternal mystery of the world is its comprehensibility."[63] One of his friends, Maurice Solovine, asked Einstein to clarify this remark. Einstein replied:

---

[58]Erwin Schrödinger, *What is Life? The Physical Aspects of the Living Cell* (Cambridge: Cambridge University Press, 1945), p. 31.

[59]*Ibid.*, p. 46.

[60]Richard W. Hamming, "The Unreasonable Effectiveness of Mathematics," *American Mathematical Monthly*, 87 (1980), 81.

[61]*Ibid.*, 90.

[62]Remo J. Ruffini, "The Princeton Galaxy," interviews by Florence Heltizer, *Intellectual Digest*, 3 (1973), 27.

[63]Einstein, *Out of My Later Years*, p. 61.

You find it surprising that I think of the comprehensibility of the world ... as a miracle or an eternal mystery. But surely, a priori, one should expect the world to be chaotic, not to be grasped by thought in any way. One might (indeed one *should*) expect that the world evidence itself as lawful only so far as we grasp it in an orderly fashion. This would be a sort of order like the alphabetical order of words of a language. On the other hand, the kind of order created, for example, by Newton's gravitational theory is of a very different character. Even if the axioms of the theory are posited by man, the success of such a procedure supposes in the objective world a high degree of order which we are in no way entitled to expect a priori. Therein lies the "miracle" which becomes more and more evident as our knowledge develops.... And here is the weak point of positivists [true knowledge is that which can *only* be verified by the senses or experience - J.N.] and of professional atheists, who feel happy because they think that they have not only pre-empted the world of the divine, but also of the miraculous. Curiously, we have to be resigned to recognizing the "miracle" without having any legitimate way of getting any further. I have to add the last point explicitly, lest you think that, weakened by age, I have fallen into the hands of priests.[64]

To Einstein, there is no "legitimate" way to get around recognizing the miracle. To him, to explain the miracle in terms of the "divine" would be "falling into the hands of priests" and therefore, in accordance with his convictions, sacrilegious. Stanley L. Jaki exposes the obvious by remarking that Einstein "perceived that such a train of thought was not only a road of science but it also came dangerously close to turning at the end into a way to God."[65] Dr. Ruffini is another scientist who openly admitted, after his testimony above, that the mystery of mathematical effectiveness can be solved by positing the biblical God. But, as Einstein, he considered this explanation to be unacceptable. According to

Figure 109: Albert Einstein

Rousas J. Rushdoony, Ruffini "prefers to deny the theoretical possibility of a correlation and meaning than to admit the reality of the Creator God."[66]

Most scientists, however, run away from this problem and do what Morris Kline describes, "Indeed, faced with so many natural mysteries, the scientist is only too glad to bury them under a weight of mathematical symbols, bury them

---

[64]Albert Einstein, *Lettres À Maurice Solovine* (Paris: Gauthier-Villars, 1956), pp. 114-115 and cited in Jaki, *The Road of Science and the Ways to God*, pp. 192-193.

[65]Jaki, *The Road of Science and the Ways to God*, p. 193.

[66]Rushdoony, *The Philosophy of the Christian Curriculum*, p. 102.

so thoroughly that many generations of workers fail to notice the conceal-ment."[67]

6.7 THE REAL ISSUE

Why bury and conceal? Is the mathematician running away from an issue that he does not want to confront? Yes. Using the words of Herbert Schloss-berg, "Scientific scabbards fall away to reveal ideological swords."[68] Morris Kline summarizes the attitude of most mathematicians today, "Many mathema-ticians are happy to accept the remarkable applicability of mathematics but con-fess that they are unable to explain it."[69]

Willem Kuyk, professor of mathematics at the University of Antwerp in Belgium, explains why mathematicians do not want to explain:

> The question whether it is possible to make some kind of ontology the ba-sis of modern mathematics is left open by most people working in mathe-matical fields. Fearing to introduce into mathematics arguments of a meta-physical nature, the philosophically minded mathematician will avoid as much as possible reference to mathematical existence independent of hu-man thought. In general it can be said that under the impact of the prag-matist attitude, for the philosopher of mathematics the workability of mathematical systems rather than their interpretability has become a central point of view. Reflections of an epistemological nature as well as reflec-tions regarding for example mathematical truth are not readily undertaken by mathematicians of the pragmatistic type.[70]

Most mathematicians today would rather hide in the dark closet of pragma-tism than come out into the bright light of biblical revelation.

6.8 THE WONDERS OF CREATION

The structure of the honeycomb is a series of interlocking regular hexago-nal prisms. Through differential calculus, one can determine that this design is the most efficient possible. It wastes no space at all and is the most effective structure for strength against collapsing.

Creation is revelatory of God's attributes. According to Kepler, when one contemplates God's created order, he "immediately takes hold of God."[71] What is made reflects something about the maker. In the case of the honeycomb, we see the wisdom of the infinite Creator.

---

[67]Morris Kline, *Mathematics and the Search for Knowledge* (New York: Oxford University Press, 1985), p. 146.
[68]Schlossberg, p. 145.
[69]Kline, *Mathematics and the Search for Knowledge*, p. 224.
[70]Willem Kuyk, "The Irreducibility of the Number Concept," *Philosophia Reformata*, 31 (1966), 37.
[71]Caspar, p. 374.

Figure 110: Honeycomb

6.8.1 THE MATHEMATICS OF THE HONEYCOMB

The honeybee constructs the honeycomb by tessellating a series of regular hexagonal prisms with one end open and the other end a pointed trihedral apex (see Figure 111). The bee fabricates the displacement, x (see Figure 112), *so that the waxed surface area is minimal* (the bee also does less work for the same space when compared to the construction of square or triangular tessellations). Using the differential calculus, we can prove that God has designed the bee to construct a comb using the *minimal* surface area (i.e., the least amount of wax).

Let A = surface area,
a = length of each side of the hexagon,
h = height of the hexagonal prism, and
x = displacement.

If x = 0, then the base is flat. But, A (surface area) is *not* minimal. We can find the minimal surface area by taking the first derivative of x given the formula for the surface area. Then, we will set this derivative equal to zero and solve for x. Setting a = 1, the value of x thus calculated will give us the minimal surface area.

apex

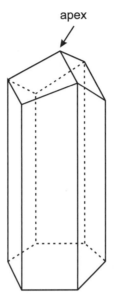

Figure 111: Trihedral apex of honeycomb cell

apex

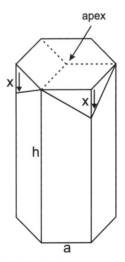

Figure 112: Regular hexagonal prism

$$A = 6(ah - \frac{ax}{2}) + 3a\sqrt{3}\sqrt{x^2 + \frac{a^2}{4}}$$

Take the derivative of A with respect to x; i.e. $\frac{d}{dx}A$

$$\frac{d}{dx}A = \frac{d}{dx}6ah - 3a\frac{d}{dx}x + 3a\sqrt{3}\frac{d}{dx}\left(x^2 + \frac{a^2}{4}\right)^{\frac{1}{2}}$$

$$\frac{d}{dx}A = 0 - 3a + 3a\sqrt{3}\left[\frac{1}{2}\left(x^2 + \frac{a^2}{4}\right)^{-\frac{1}{2}}2x\right]$$

$$\frac{d}{dx}A = -3a + 3a\sqrt{3}x\left(x^2 + \frac{a^2}{4}\right)^{-\frac{1}{2}}$$

Now, set $\frac{d}{dx}A = 0$ *and solve for x :*

$$-3a + 3a\sqrt{3}x\left(x^2 + \frac{a^2}{4}\right)^{-\frac{1}{2}} = 0$$

$$3a\sqrt{3}x\left(x^2 + \frac{a^2}{4}\right)^{-\frac{1}{2}} = 3a$$

$$x\left(x^2 + \frac{a^2}{4}\right)^{-\frac{1}{2}} = \frac{1}{\sqrt{3}} \Rightarrow \frac{\sqrt{x^2 + \frac{a^2}{4}}}{x} = \sqrt{3} \Rightarrow \frac{x^2 + \frac{a^2}{4}}{x^2} = 3 \Rightarrow 1 + \frac{a^2}{4x^2} = 3 \Rightarrow$$

$$\frac{a^2}{4x^2} = 2 \Rightarrow \frac{4x^2}{a^2} = \frac{1}{2} \Rightarrow 4x^2 = \frac{a^2}{2} \Rightarrow x^2 = \frac{a^2}{8}.$$

Hence, $x = \sqrt{\frac{a^2}{8}} = \frac{a}{\sqrt{8}} = \frac{a}{2\sqrt{2}} = \frac{a\sqrt{2}}{4}.$

If $a = 1$, *then* $x = .35355339$

The graph in Figure 113 plots the independent variable x (displacement) versus the dependent variable A (surface area) showing the minimal surface area at x ≈ .35

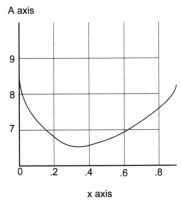

Figure 113: Minimal surface area plot of honeycomb cell

6.8.2 THE PATH OF LIGHT THROUGH A WATER DROPLET

The path of light can be determined by the laws of geometric optics. Each time that the beam strikes the surface of an individual droplet, part of the light is reflected and part is refracted. The following laws govern the degree of reflection/refraction (see Figure 114):

Let I = the angle of incidence,
R = the angle of reflection,
r = the angle of refraction,
V_1 = the velocity of light in the incident medium (air in this example),
and
V_2 = the velocity of light in the refraction medium (water in this example).
According to the law of reflection, I = R.
According to the law of refraction, $\dfrac{\sin I}{\sin r} = \dfrac{V_1}{V_2}$.

Rays that are directly reflected from the surface are designated as TYPE 1 rays. Those rays transmitted directly through the droplet are classified as TYPE 2. TYPE 3 rays emerge from the droplet after one internal reflection. These rays produce the primary rainbow. A much fainter bow, called the secondary rainbow, is sometimes seen behind the primary rainbow. This bow is made up of TYPE 4 rays, which have undergone two internal reflections.

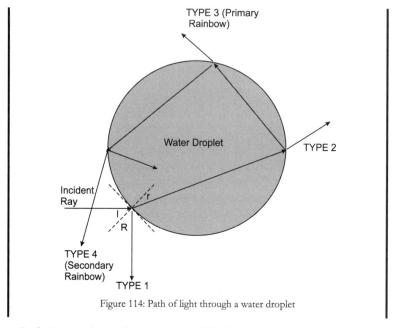

Figure 114: Path of light through a water droplet

In Scripture, the rainbow is a sign of God's covenant mercies (Genesis 9:8-17). The structure of a rainbow is complex and intricate. Detailing it could fill a large book. Carl Boyer extols the rainbow's properties:

> The rainbow is just about the most subtle phenomenon that everyday nature presents us; the creation of the familiar sun and rain, it is nevertheless as unapproachable as a spirit.[72]

H. E. Huntley reflects:

> The light of day is reflected, chromatically refracted, reflected again and dispersed by gently falling water spheres into a thousand hues, conforming the while to lovely theorems of mathematics so simple in some respects that the schoolboy may understand, so complex in others as to defy analysis.[73]

According to Larry Zimmerman, Christian educator, "the knowledge of mathematics unveils not only vistas of beauty and power unsuspected before

---

[72]Carl Boyer, *The Rainbow: From Myth to Mathematics* (New York: Thomas Yoseloff, 1959), p. 20.
[73]H. E. Huntley, *The Divine Proportion: A Study in Mathematical Beauty* (New York: Dover Publications, 1970), p. 11. Huntley is citing from another book written by him, H. E. Huntley, *The Faith of a Physicist* (Bles, 1960), p. 12. For more detail on the mathematics of the rainbow, see H. Moyses Nussenzveig, "The Theory of the Rainbow," *Scientific American*, 236 (1977), 116-127.

but also an order, symmetry and infinitude which stuns and awes the beholder."[74]

We have seen that not only is mathematics useful in helping man to fulfill the dominion mandate of Scripture (Genesis 1:26-28), it also uniquely reveals certain attributes of the Creator God (for further detailed discussion, see sections 7.5 and 7.6). After God finished creating the heavens and the earth, He pronounced everything as "good." Although made by the word of His power and designed by His infinite wisdom, the works of God ultimately reflect the goodness of God (Genesis 1:31). The heart of goodness is generosity. In its essence, goodness is the desire to *do* good, to create a medium through which one can communicate freely and extravagantly. Being infinite, the biblical God communicates Himself to the degree of infinity! The voice of the good God is everywhere, waiting to be heard by those who have ears to hear.

The whole of creation is for man's services, food, and enjoyment. There is nothing in creation that does not contribute something either (1) to our welfare in the provision of our means, our health, our clothing, our service, or (2) to our sheer and unmitigated delight.

Mathematics serves as a unique method of describing the arrangement of God's good creation. In this arrangement, we see God's great and gracious concern to bless man. Over and above the mathematical formulae describing matter, motion, and forces, there is a message conveyed through the loving touch of personality. Through the manifold works of God, from the variegated rainbow to the delicate rose, the language of God's goodness reaches our hearts.

6.9 GUILTILY BLIND

We must expect that humanistic mathematicians will miss the whole point of the place of mathematics in the purposes of God. Not willing to submit their lives to their Maker, they are guiltily blind to the glory of God reflected in the unique mirror of mathematics. Because of this willful denial and suppression of evident truth, the mathematical structure of creation will be misunderstood and ultimately perverted.

Yet, if their practical day to day work is to be effective, scientists and applied mathematicians must make biblical assumptions about the physical world that are contrary to their voiced humanistic presuppositions. In the words of Albert Einstein, "Don't listen to their words, fix your attention on their deeds."[75] He continues:

Without the belief that it is possible to grasp the reality with our theoretical constructions, without the belief in the inner harmony of our world, there could be no science.[76]

---

[74]Zimmerman, 2:3, 2.
[75]Einstein, *Essays in Science*, p. 12.
[76]Albert Einstein and Leopold Infeld, *The Evolution of Physics: The Growth of Ideas from the Early Concepts to Relativity and Quanta* (Cambridge: Cambridge University Press, 1938), p. 312.

Mathematicians and scientists today are living on borrowed capital; they are earning interest off a deposit that they no longer acknowledge or recognize to be genuine. As Stanley L. Jaki states, "Science is now in possession of such a vast interconnection of data, laws, and instruments as to continue its progress even if no attention is paid any longer to that faith which played an indispensable role in its rise."[77]

In essence, unbelieving scientists can do science *only because they operate secretly on Christian premises while denying that faith*. What they do in their scientific work expresses a biblical faith that contradicts their profession of unbelief. We must always remember that it is a gift of God's common grace that enables the unbelieving scientist of this age (or any age) to formulate and discover valid mathematical laws. It is Christ that "gives light to every man" (John 1:9) and it is God in Christ that "teaches man knowledge" (Job 32:8-9). According to Rousas J. Rushdoony:

> The unbeliever is thus able to think and work only on the basis of a practical reason which presupposes the Christian frame of things.... On his own premises, he can know nothing; on borrowed premises, he is able to think and work, but for all his results, he remains in the paradoxical position of the cattle rustler.... He has no knowledge on the basis of his own principles, he has valid knowledge only as a thief possesses stolen goods.[78]

Cornelius Van Til is direct and to the point:

> Sinners use the principle of Chance back of all things and the idea of exhaustive rationalization as the legitimate aim of science. If the universe were actually what these men assume it to be according to their principle, there would be no science. Science is possible and actual only because the non-believer's principle is not true and the believer's principle is true. Only because God has created the universe and does control it by His providence, is there such a thing as science at all.[79]

The rationalist John W. N. Sullivan echoed Van Til's remark by querying, "Why the external should obey the laws of logic; why, in fact, science should be possible, is not at all an easy question to answer."[80] Reflecting on this link between the laws of logic and the external world, the French physicist Louis-Victor de Broglie (1892-1987) remarked that the progress of science "has revealed to us a certain agreement between our thought and things, a certain possibility of grasping, with the assistance of the resources of our intelligence and the rules of reason, the profound relations existing between phenomena. We

[77]Jaki, *Cosmos and Creator*, p. 139.
[78]Rushdoony, *By What Standard?*, pp. 61-62.
[79]Cornelius Van Til, *A Christian Theory of Knowledge* (Philadelphia: Presbyterian and Reformed, 1953), p. 193 and cited in Rushdoony, *The One and the Many: Studies in the Philosophy of Order and Ultimacy*, p. 15.
[80]Sullivan, 3:2020.

are not sufficiently astonished by the fact that any science may be possible."[81] Scientists must accept objective coherence in a *uni*verse of God's making, not a *multi*verse of man's construction, if there is to be any such thing as real science. If not, using the words of Stanley L. Jaki, "any analysis of knowledge becomes a celebration of incoherence."[82]

6.10 MATHEMATICS: METAPHYSICAL AND EPISTEMOLOGICAL REALITY

According to Max Caspar, Kepler's view of the world and his doctrine of knowledge were as follows:

> What supported and gave wings to Kepler in the execution of his program was a refreshing optimism about cognition. "Man, stretch thy reason hither, so that thou mayest comprehend these things"; that was the call which rang out for him from the material world. While he accepted this call with open ear, with the complete and unreserved readiness of a young mind, he believed in the reality of the things outside us and in the possibility of being able to comprehend them in their essence, order and meaning. What the eye brought him was that which he saw, and was in reality as he saw it; and the mind repeated the thoughts which God had materialized in His Creation. He did not start with doubt, as another soon did [Descartes – J.N.], but with an unquestioned faith in *ratio*. He did not limit himself to the framework of immanent thought, but became intoxicated with the contemplation of a transcendental truth. He had not yet fallen into the abyss of relativity, but was deeply convinced that there is an absolute truth. Admittedly our mind never can completely grasp this truth, but it is the noble task of scientific and philosophical research to draw nearer to it.[83]

James Clerk Maxwell (1831-1879) developed mathematical equations that enabled scientists to accomplish wonders with electrical and magnetic phenomena.[84] Not only are these equations profound, comprehensive, and effective, they are also extremely beautiful and symmetric. According to Norman Campbell, these equations illustrate "the marvellous power of pure thought, aiming only at the satisfaction of intellectual desires (e.g., beauty, order, symmetry), to con-

Figure 115: James Clerk Maxwell

[81]Louis-Victor de Broglie, *Physics and Microphysics*, trans. M. Davidson (New York: Pantheon Books, 1955), pp. 208-209.
[82]Jaki, *Cosmos and Creator*, p. 97.
[83]Caspar, p. 377.
[84]For a description of Maxwell's method, see Kline, *Mathematics and the Search for Knowledge*, pp. 126-147, 166-167 and Kline, *Mathematics in Western Culture*, pp. 304-321.

trol the external world."[85] According to Stanley L. Jaki, Maxwell's electromagnetic equations are "possibly the most beautiful equations until then formulated in theoretical physics."[86]

When Heinrich Hertz (1857-1894) discovered the existence of radio waves in space, he verified all of Maxwell's equations and ensuing predictions. The response of Hertz is informative:

> One cannot escape the feeling that these mathematical formulas have an independent existence and an intelligence of their own, that they are wiser than we are, wiser even than their discoverers, that we get more out of them than was originally put into them."[87]

Philip E. B. Jourdain (1879-1914), mathematician and son of a Derbyshire vicar, observed that:

> ... the nature of Mathematics is independent of us personally and of the world outside, and we can feel that our own discoveries and views do not affect the Truth itself, but only the extent to which we or others see it .... Some philosophers have reached the startling conclusion that Truth is made by men, and that Mathematics is created by mathematicians, and that Columbus created America....[88]

Godfrey H. Hardy (1877-1947) considered God to be his personal enemy and prided himself in the claim that none of his mathematics would ever apply to any aspect of the real world. He finally had to confess that "mathematical reality lies outside of us. Our function is to discover, or observe it, and that the theorems which we describe grandiloquently as our creations are simply notes on our observations."[89] After Hardy's death, his mathematical results were applied to the physical world.[90]

Figure 116: Godfrey H. Hardy

---

[85]Campbell, p. 156. For Maxwell, the motivation for his scientific studies stemmed from his belief in the Genesis account of the creation of man in God's image and the command to man to subdue the earth. See Lewis Campbell and William Garnet, *The Life and Times of James Clerk Maxwell* (London: Macmillan and Company, 1882), p. 323 (http://www.sonnetusa.com/bio/maxbio.pdf, p. 160).

[86]Jaki, *Cosmos and Creator*, pp. 31-32.

[87]Cited in Bell, *Men of Mathematics*, p. 16.

[88]Philip E. B. Jourdain, "The Nature of Mathematics," *The World of Mathematics*, ed. James R. Newman (New York: Simon and Schuster, 1956), 1:71.

[89]Godfrey H. Hardy, *A Mathematician's Apology* (Cambridge: Cambridge University Press, 1967), pp. 123-124.

[90]How were Hardy's results applied? One of his conclusions has been applied to genetics and it has turned out to be known as Hardy's law, a law "of central importance in the study of Rh-blood groups and the treatment of haemolytic disease of the newborn." E. C. Titchmarsh, "Obituary of G. H. Hardy," *The Journal of the London Mathematical Society* (April 1950), p. 83. His work in number

Albert Einstein confessed in 1934, "To him who is a discoverer in this field the products of his imagination appear so necessary and natural that he regards them, and would like to have them regarded by others, not as creations of thought but as given realities."[91]

Nicholas Bourbaki, a collective pseudonym for a group of French mathematicians, said in 1950:

> That there is an intimate connection between experimental phenomena and mathematical structures, seems to be fully confirmed in the most unexpected manner by the recent discoveries of contemporary physics ... but we are completely ignorant as to the underlying reasons for this fact ... and we shall perhaps always remain ignorant of them.[92]

Bourbaki concludes:

> Mathematics appears thus as a storehouse of abstract forms ... and it so happens – without our knowing why – that certain aspects of empirical reality fit themselves into these forms, as if through a kind of preadaptation.[93]

6.11 THE UNIFYING FACTOR

Sir Oliver Graham Sutton remarks, "How can the manipulation of symbols which *we* have invented, according to rules which *we* alone make (and sometimes break), reveal that which lies beyond our senses?"[94] To him, this question "is one which is unlikely to receive a satisfactory answer...."[95] Then, he makes a remarkable and accurate observation:

> The universe, both as a whole and in its microstructure, suggests that in neither aspect can it be treated merely as an enlarged or diminished version of the world which we know through our senses. The ultimate secrets of nature are written in a language which we cannot yet read. Mathematics provides a commentary on the text, sometimes a close translation, but in words we can read because they are our own.[96]

Why does mathematics work? Why does it fit the real world? What is the reason for this mysterious coherence between mathematical thought and empirical reality? What is the ultimate metaphysical "language of the universe" that mathematics gives dim commentary to? If mathematics is just a product of

theory has been used extensively in the investigation of the temperature of furnaces. See John B. S. Haldene, *Everything Has a History* (London, 1951), p. 240.

[91]Einstein, *Essays in Science*, p. 12.

[92]Nicholas Bourbaki, "The Architecture of Mathematics," *American Mathematical Monthly*, 57 (1950), 231.

[93]*Ibid.*

[94]Sutton, p. 3.

[95]*Ibid.*, p. 4.

[96]*Ibid.*, p. 23.

man's autonomous reason, then the answers to these questions will forever remain a mystery.

Early in the 20th century, Philip E. B. Jourdain remarked that mathematics "really occupies a place sometimes reserved for an even more sacred Being."[97] According to Max Caspar's understanding of Kepler's epistemology, "The fact that the world and man's mind are images of God in their manner, makes knowledge possible, a knowledge which not only is certain but also carries sense and value in itself."[98]

Henry Morris (1918-), a pioneer in the field of creation science, observes:

> The more intensively and thoroughly man probes the universe – whether the submicroscopic universe of the atomic nucleus or the tremendous metagalactic universe of astronomy – the more amazingly intricate and grand are God's reservoirs of power revealed to be.[99]

For Larry Zimmerman, mathematics is more than just the free creation of the human mind. He says:

> It is possible, that mathematics is an entity which always exists in the mind of God, and which is for us the universal expression of His creative and sustaining word of power.... So we would expect the deepest scientific probes into the micro- or macro-cosmos to reveal a language fabric in which are woven the forces and relationships governing the tangible creation. This language fabric should itself be suggestive of an intellectual antecedent, an orderly, powerful, infinitude of thought, a "terra incognita of pure reasoning" which "casts a chill on human glory."[100]

Vern S. Poythress agrees with Zimmerman and reflects upon the linguistic character of God's creation:

> The created world, *as* result of God's speech, bears within it from top to bottom a kind of quasilinguistic character ... through God's act of creation, things in the world themselves become wordless voices to the praise of God.[101]

Edward Everett (1794-1865), president of Harvard, was the first American to earn a doctor's degree at the University of Göttingen. In the autumn of 1863 at Gettysburg, Pennsylvania, he took the stage and spoke for two hours as part of a tribute ceremony for those who fought and for those who died in the Battle of Gettysburg (July 1-3, 1863). What he said no one remembers or cares to remember. It was what the succeeding speaker said in a short, two to three mi-

---

[97]Jourdain, 1:71.
[98]Caspar, p. 381.
[99]Henry Morris, *The Biblical Basis for Modern Science* (Grand Rapids: Baker, 1984), p. 52.
[100]Zimmerman, 2:2, 1.
[101]Poythress, p. 5.

nute address that everyone remembers. That speaker was Abraham Lincoln. Note Everett's remarkable observation about the nature of mathematical truth:

> The great truths with which it [mathematics – J.N.] deals, are clothed with austere grandeur, far above all purposes of immediate convenience or profit. It is in them that our limited understandings approach nearest to the conception of that absolute and infinite…. In the pure mathematics we contemplate truths, which existed in the divine mind before the morning stars sang together, and which will continue to exist there, when the last of the radiant host shall have fallen from heaven. They existed not merely in metaphysical possibility, but in the actual contemplation of supreme reason. The pen of inspiration, ranging all nature and life for imagery, to set forth the Creator's power and wisdom, finds them best symbolized in the skill of the surveyor.[102]

Figure 117: Edward Everett

Whether pure mathematics can be *equated* with the divine mind is open to serious debate. Can we say with assurance, in the words of James Jeans, that God, the Great Architect of the universe, is a pure mathematician? In other words, can we *equate* the pure mathematics of today with the very mind of God? This statement is an example of univocal reasoning (see section 2.5) and according to Isaiah 55:8-9, this type of reasoning is impossible. To equate pure mathematics with the mind of God is to lift the finite into the infinite. It is the infinite that penetrates the finite, not vice versa. It is God who teaches man, not man who teaches God (Job 32:8-9; Job 38:2; Job 38:36; Psalm 94:10; Isaiah 40:13-14). Whatever man knows rightly, it is a gift of God in Christ (John 1:9). This gift of grace sometimes enthralls man, sometimes humbles man, and, sometimes confuses man in his disobedience (we see all three of these aspects in the mathematical realm). Man's mathematical knowledge will never exhaust the infinite panorama of God's knowledge (I Samuel 2:3; Psalm 147:5). At best, man is gifted with an infinitesimal subset of God's exhaustive wisdom and knowledge (Psalm 104:24). At best and in accordance with Kepler's convictions, the propositions of mathematics dimly reflect that ultimate reality of God's sustaining word of power. Every atom of the universe reflects His order and conforms to His decree. The mind of man, created in His image, can grasp a semblance of this order through the propositions of mathematics. This understanding can never be exhaustive (it can never encompass comprehensive intelligibility), but it can approximate truth. As Stanley L. Jaki remarks, "Only the kernel of scientific truth will become better defined as time goes on."[103]

---

[102]Edward Everett, *Orations and Speeches* (Boston, 1870), 3:514.
[103]Jaki, *The Relevance of Physics*, p. 137.

As a Fellow of the Royal Society and Regius professor of mathematics at the University of St. Andrews in Scotland, Herbert Westren Turnbull (1885-1961) was well known for his research in algebra (determinants, matrices, and theory of equations). At the conclusion of his biographical history of mathematics, his thoughtful summarization reflects a unique balance of simplicity and elegance:

Figure 118: Herbert Westren Turnbull

The story has now been told of a few among many whose admirable genius has composed the lofty themes which go to form our present-day heritage ... if this little book perhaps may bring to some, whose acquaintance with mathematics is full of toil and drudgery, a knowledge of those great spirits who have found in it an inspiration and delight, the story has not been told in vain. There is a largeness about mathematics that transcends race and time: mathematics may humbly help in the market-place, but it also reaches to the stars. To one, mathematics is a game (but what a game!) and to another it is the handmaiden of theology. The greatest mathematics has the simplicity and inevitableness of supreme poetry and music, standing on the borderland of all that is wonderful in Science, and all that is beautiful in Art. Mathematics transfigures the fortuitous concourse of atoms into the tracery of the finger of God.[104]

Figure 119: Paul A. M. Dirac

What is made reflects the maker. Creation is a showcase of God's splendor, cunning, and power; i.e., creation is the tracery of the finger of God. Man, made in the analogical image of God, has been, for millennia, using his powers of ingenuity, in the words of Paul A. M. Dirac (1902-1984) – 1933 Nobel Prize winner in atomic theory – in "developing fundamental physical laws ... in terms of a mathematical theory of great beauty and power."[105] In probing the creation, man has discovered and formulated relationships that reflect the language fabric of the "word of God's power" and in so doing has exposed the ingenuity of the Creator.

---

[104]Herbert Westren Turnbull, *The Great Mathematicians* (New York: Barnes & Noble, [1929] 1993), p. 141.

[105]Paul A. M. Dirac, "The Evolution of the Physicist's Picture of Nature," *Scientific American*, 208 (1963), 53.

According to Stanley L. Jaki, the universe "has supreme coherence from the very small to the very large.... It is beautifully proportioned into layers of dimensions and yet all of them are in perfect interaction."[106] God, the author of this wonderful, marvelous, and coherent display, has gifted the mind of man with the capabilities of grasping it.

Charles Hermite (1822-1901), a mathematical analyst who proved the transcendence of *e* in 1873, offered this explanation for the agreement between mathematics and the physical world:

> There exists, if I am not deceived, a world which is the collection of mathematical truths, to which we have access only through our intellects, just as there is the world of physical reality; the one and the other independent of us, both of divine creation, which appear distinct because of the weakness of our minds, but for a more powerful mode of thinking are one and the same thing. The synthesis of the two is revealed partially in the marvelous correspondence between abstract mathematics on the one hand and all the branches of physics on the other.[107]

Figure 120: Charles Hermite

In *For the Beauty of the Earth,* a hymn of grateful praise written by Folliott S. Pierpoint in 1864, the poet thanked God for several marvelous truths: the beauty of the earth, the wonder of the day and night, the joy of human love, the church. In stanza five, using classical Keplerian ethos, he gave thanks to God for Hermite's "marvelous correspondence" with enchanting prose, "For the joy of ear and eye, for the heart and mind's delight, for the mystic harmony linking sense to sound and sight."

The Author of that mystic harmony linking sense to sound and sight is the biblical God. The mind of man, with its mathematical capabilities, and the physical world, with its observable mathematical order, *cohere* because of a common Creator. Einstein's eternal mystery has a solution. The biblical revelation of the Creator God is the unifying factor that reconciles what is irreconcilable in the humanistic context.

Men who follow in the train of Sullivan, Bridgman, Eddington, and Kline will never know the delight of this divinely orchestrated accord. Every time man boasts that the "footprint is his own," he is denigrating the gift of Christ (John 1:9). Men who claim the self-autonomy which seeks the measure of all things in man, men who pursue, with Promethean stubbornness, a reliance on

---

[106]Jaki, *Cosmos and Creator*, p. 42.
[107]Cited in Kline, *Mathematics: The Loss of Certainty*, p. 345.

their own reasoning and their own abilities, in the end only fracture themselves, their science, and their mathematics with their impotent strikes at the Almighty. Instead of a science reflective of the aesthetic-metaphysical beauty of the Creator God, we now have a science of brute factuality. To the ethos of the noble Kepler we must return. We need scientists and mathematicians who boldly confess, "How great is the Creator who has made both the mind and nature so compatible!" We need scientists and mathematicians who see the universe, not as a mere mass of mechanistic and impersonal laws, but as the handiwork of God, and delight themselves therein. May God in His mercy add to this tribe and may the reader of this book be one of them.

QUESTIONS FOR REVIEW, FURTHER RESEARCH, AND DISCUSSION

1. Explain why the notion of "neutrality" in mathematics is mythological.
2. Explain the implications of a humanistic worldview in relationship to the current status and the future of mathematics and science.
3. Explain the only two assumptions that one can make about mathematics and its relationship to the physical world.
4. Carefully explain the following statement, "Only Christian monotheism can prevent one from venturing into the forest of pantheism."
5. Explain the statement, "The issue in mathematics today is root and branch a religious one."
6. Explain the dilemma that humanistic mathematicians face when they assume that mathematics is merely and only an invention of the human mind.
7. Explain how Heisenberg's denial of causality based upon his theory of quantum mechanics is an example of the logical fallacy of equivocation.
8. Comment on Einstein's reasons for claiming that the discourse of God does *not* belong to "natural philosophy."
9. Explain the schizophrenic attitude that humanistic mathematicians and scientists must have in order for their practical day to day work to be effective.
10. Explain why a biblical worldview necessitates a *uni*verse and why a humanistic worldview necessitates a *multi*verse.
11. Mathematics, as a language, serves as a faint "echo" of what metaphysical reality?
12. Carefully explain the following statement, "To equate the abstract logic of pure mathematics with the very mind of God is an example of autonomous univocal reasoning."
13. Carefully explain the following statement, "The validity of man's mathematical knowledge does not rest on it being exhaustively equal to the mind of God, but on it being reinterpretively analogous with the revelation of God."
14. Explain the biblical answer to the question, "Why does mathematics work?"

Part II: How Should We Then Teach?

7: OBJECTIVES

In exploring mathematics one is exploring the nature of God's rule over the universe; i.e. one is exploring the nature of God Himself.[1]

Vern S. Poythress

The next two chapters deal with mathematics education. Some readers, especially those who are not mathematics teachers, may decide to forgo this information, but the author would encourage everyone to realize that these chapters are a necessary and strategic outgrowth of the preceding six chapters. To point out what is wrong with humanistic mathematical philosophy and education is not enough. As biblical Christians, we must give direction for reform.

7.1 THE BIBLICAL WORLDVIEW AND MATHEMATICS

Worldview and mathematics ... never the two shall meet? Is it true that a person's worldview is totally irrelevant to the understanding and doing of mathematics? From the foregoing study of mathematics history, the reader should answer both questions with a resounding no. Before we begin an analysis of the objectives of a biblical Christian approach to mathematics, we need to summarize and carefully articulate its worldview underpinnings from a biblical Christian perspective. As we noted in section 1.5, there are three fundamental components of a worldview. The first deals with the nature of reality, the second with how we know what we know, and the third with how we live our lives.[2]

---

[1] Vern S. Poythress, "A Biblical View of Mathematics," *Foundations of Christian Scholarship*, ed. Gary North (Vallecito, CA: Ross House, 1976), p. 184.

[2] The author credits the seminal work of Vern S. Poythress in the analysis that follows.

7.1.1 METAPHYSICS

Let us investigate two approaches to the nature of reality: radical monism and radical pluralism. An ancient example of radical monism can be found in the Greek philosopher Parmenides (see section 2.10). He posited that it appears that things change, but they really do not. Reality for him was a single, permanent substance that is uncreated, indestructible, and unchangeable, and that reality is to be found only in reason. A modern example of radical monism is Eastern Hinduism. This philosophy posits that there is no plurality and no change; it looks like there is plurality and change but it is a mere illusion.

The problem with this position is that if everything is one, then there can be *no* science and mathematics. For mathematics to work you must have a philosophy of plurality; i.e., that there are many things. As long as one attempts to explain everything in terms of one principle (e.g., the logicism of Russell and Whitehead), then any remaining diversity will be a problem.

An ancient example of radical pluralism is the Greek philosopher Democritus (see section 2.12). He embraced an atomistic theory of matter in which all things are composed of minute, invisible, indestructible particles of pure matter, which move about eternally in infinite empty space. A modern example is the nominalism or empiricism of Western civilization where only individual objects or particulars have real existence. You can experience five dollars or five soft drinks, but you cannot "abstract" from these experiences the universal concept of "fiveness."

The problem with this position can be illustrated as follows. First, with a pencil, write "5" on a piece of paper. Next, ask yourself, "Is this the number five?" According to the radical pluralism, you must answer "yes." Now, erase "5" from your paper. Hence, "5" no longer exists! We all know that "V," "////," or "5" are numerals (symbols or names), not numbers. If *only* particulars exist, then the abstract concept of five does not exist. Has anyone ever experienced five, not five dollars or five soft drinks, but the abstract reality of five? Five is a different kind of metaphysical reality. With radical pluralism there is no concept of number, sets, classes, universals, relationships, coherence (between man's mind, laws of logic, and the physical world) or laws. Hence, you *cannot* do mathematics.

To summarize, if all is one (radical monism), then there are no particulars and no numbers. Hence, you *cannot* do mathematics. If all is many (radical pluralism), then there are no universals and no number. Hence, you *cannot* do mathematics. According to Cornelius Van Til:

> ... if one begins with an ultimate plurality in the world, or we may say regarding plurality as ultimate, there is no way of ever coming to an equally fundamental unity. On the other hand, if one should begin with the assumption of an ultimate abstract, impersonal unity, one cannot account for the fact of plurality. No system of thought can escape this dilemma. No system of thought has escaped this dilemma. Many systems of thought

have denied one of the horns of the dilemma, but all that they have accomplished by doing this is to find relief in the policy of the ostrich.[3]

In order to do mathematics you must resolve the thorny metaphysical question of the "one and the many." This tension is resolved and answered in the nature of the ontological trinity, the eternal *one and the many* (see section 7.5.2). We can do mathematics *only* because the triune God exists.

Only biblical Christianity can *account* for the ability to count.

7.1.2 EPISTEMOLOGY

Epistemology deals with the nature and limits, the grounds, of human knowledge. It deals with questions like: How are mathematical truths justified or proven? How do we know what is truth in mathematics?

These questions have been answered historically in one of three ways. First, the *a priori* approach. Independent of any experience, certain mathematical truths exist. For example, the equation $2 + 2 = 4$ is a universal, eternal truth. We have already seen the main problem with this approach (see section 5.14). If we assume that we can prove the truths in mathematics *a priori*, then why does mathematics work? No answer is forthcoming from the logicism, intuitionism, or formalism schools.

Second, the *a posteriori* approach. This position states that truth is based upon experience. For example, to prove that the Corvette two blocks down the street is blue, I must walk the two blocks and look at it myself. If we assume that we can prove the truths in mathematics *a posteriori*, then the following questions cannot be answered:

1. On what basis do you generalize from experienced cases to inexperienced cases? Has anyone experienced $2,646,123 + 10,126,484$?
2. Has anyone ever experienced complex numbers or quaternions?
3. Why can we universalize the truths of mathematics? Could there be one exception? Does $2 + 2 = 4$ in the Andromeda Galaxy? If we can only say, "As far as we know, $2 + 2 = 4$," then there is no such thing as *truth* in mathematics.
4. Why can you generalize? If your answer is, "That's just the way the human mind works," you are back to the *a priori* position.

The third position, recognizing the impasse of the first two approaches, basically states that there are no truths in mathematics. The propositions of mathematics are just linguistic conventions; e.g., $2 + 2 = 4$ because, in our language, we have agreed to use 2, +, =, and 4 in this way. This is just a subtle rehash of the *a priori* argument. If mathematics is just a linguistic convention, then why are these conventions so useful in the real world? If the answer is that these conventions are chosen because they are useful, then mathematics is true because it is useful (the utilitarian approach). Again, this is a rehash of the *a posteriori* argument.

---

[3]Van Til, *A Survey of Christian Epistemology*, p. 47.

How, then, do we know what we know about the truths of mathematics? Before we answer this question, we need to make an important distinction. There is a difference between the ability to do something and the ability to explain what you are doing (practice and theory). Everybody can count, but not everyone can account for the ability to count. Unbelievers can do mathematics but they cannot offer a theory that accounts for their practice of doing mathematics. Can biblical Christianity *account*?

How do we know? Do we know *a priori* or *a posteriori*? For the Christian, God's unchanging character guarantees 2 + 2 = 4. The entire universe reflects His faithfulness. As we noted in section 5.16, the biblical God is the creator of the human mind with its mathematical capabilities and the physical world with its mathematical properties. Biblical faith unites *a priori* and *a posteriori* under the umbrella of the biblical God as the true source of knowledge and revelation. The workings of man's mind and the laws of the physical world cohere because of a common Creator. The *a priori* capabilities of the human mind correspond to the *a posteriori* properties of the external world by *prearranged design*.

What about errors in reasoning? What about incompleteness or mistakes in scientific and mathematical thought? Humans err because they are *finite* and *sinful* creatures made in the image of the *infinite*, all-wise, and holy God. What about things that we cannot prove? What about Goldbach's conjecture? What about the continuum hypothesis? Does the decimal expansion of π ever end? God, in His infinite knowledge, knows all and we can rest in that assurance.

Insights in mathematics come ultimately from God. Job 32:8-9 states that "… the breath of the Almighty gives him understanding." In Job 38:36, God challenges Job with a rhetorical question, "Who has put wisdom in the mind? Or who has given understanding to the heart?" According to Psalm 94:10, God is the one "who teaches man knowledge." Christ is the light that enlightens every man (John 1:9). If we want know mathematics aright, then we must acknowledge God in all of our mathematical ways (Proverbs 3:6). When man rejects these epistemological underpinnings, when man spurns reverence to God as the substructure of all knowledge, when man refuses to retain God in his knowledge, then God confuses man; his epistemology is shipwrecked (Psalm 111:10; Proverbs 1:7; Proverbs 9:10; Romans 1:20-28). The history of non-Christian approaches to epistemology amply exemplifies this conundrum. The Christian and the non-believer both discover and share mathematical truths, but the non-believer comes to know mathematical truths *contrary to their worldview*. It is because Christianity is true, *because* God is who He is, *because* man is the image of God, that the non-Christian knows anything. As Cornelius Van Til states:

> If the Christian position with respect to creation, that is, with respect to the idea of the origin of both the subject and the object of human knowledge is true, there is and must be objective knowledge. In that case the world of objects was made in order that the subject of knowledge, namely man, should interpret it under God. Without the interpretation of the universe by man to the glory of God the whole world would be meaningless.

The subject and the object are therefore adapted to one another. On the other hand if the Christian theory of creation by God is not true then we hold that there cannot be objective knowledge of anything.[4]

Only biblical Christianity can *account* for the ability to know mathematical truths.

7.1.3 ETHICS

Since God is truth (Deuteronomy 32:4), He is the author and standard of all truth. The biblical Christian ethic commands man to love God with the mind (Mark 12:29-31). God has gifted the mind with the ability to discover the mathematical truths that God has ordained for this world. In obedience to the "love your neighbor" ethic, man applies these mathematical truths for the betterment of mankind; i.e., for the healing for the nations (Revelation 22:2).

God has always purposed for man to take dominion over the earth (Genesis 1:28) by understanding, observing, classifying, and taking pleasure in God's works (Genesis 2:18-21; Psalm 111:2). This process of categorizing God's creation is the essence of naming something in the biblical sense. To name something means you know it so well that you can use it effectively; i.e., you have dominion over it.

Figure 121: Richard P. Feynman

The underlying goal of the biblical Christian ethic is to unfold the glory and the beauty of God (Psalm 104, Psalm 145, Psalm 148, Romans 11:36). According to Richard P. Feynman (1918-1988), noted American physicist, gifted instructor, and 1965 Nobel Prize winner in quantum electrodynamics, "To those who do not know mathematics it is difficult to get across a real feeling as to the beauty, the deepest beauty, of nature."[5] The biblical ethic leads to genuine aesthetical appreciation because it is *only* the biblical Christian who can glorify God for (1) the beauty and usefulness found in the mathematical description of nature, (2) the incomprehensible nature of God that this mathematical description displays, and (3) the human mind that God has gifted to understand His creation mathematically.

Only biblical Christianity can *justify* the ethic that undergirds the understanding and doing of mathematics.

---

[4]Cornelius Van Til, *The Defense of the Faith* (Phillipsburg: Presbyterian and Reformed, [1955, 1963] 1967), p. 43.

[5]Richard P. Feynman, *The Character of Physical Law* (Cambridge, MA: The MIT Press, 1967), p. 58. Feynman was a master teacher of physics but is perhaps best known for solving the mystery of the Challenger Space Shuttle disaster (1986). With a simple physics demonstration, he showed that the Shuttle's O-rings, though environmentally safe, were functionally defective (they failed under freezing conditions). The Environmental Protection Agency had banned the original O-rings (which were functionally safe) because they were environmentally defective (made of asbestos).

7.2 THREE CHRISTIAN APPROACHES TO MATHEMATICS EDUCATION

When it comes to the actual teaching of mathematics, there are several approaches. The most common is the so-called "neutral" approach. Here, mathematical facts are interpreted in and of themselves. In the religious institutions of secular humanism, man deals with knowledge without reference to its true foundation, the biblical God.[6]

When Christians enter the field of education, they deal with mathematics in one of three possible ways. First, the most common way merely duplicates what happens in the government schools and mathematics is taught "as usual." Yes, the Bible is taught, but the biblical worldview framework has nothing to do with mathematical understanding or pedagogy. Prayer, Bible classes, and Chapel are "sacred" subjects and mathematics is a "secular" subject.

Second, other Christians feel that the secular must be "baptized" with the spiritual. In the teaching of mathematics, Scriptures are artificially "tacked on" and the students are encouraged to learn mathematics "for God's glory."

Very few Christians approach mathematics in the third way. In this approach, God is seen as the foundation of all knowledge, not just "spiritual" knowledge. He knows everything that there is to know, and He knows everything thoroughly and completely. To God, every item of His creation, invisible and visible, reflects back to Him the beauty, wonder, and infinity of His attributes. In the words of the Netherlands' Confession, "Before our eyes as a beautiful book, in which all created things, large or small, are as letters showing the invisible things of God."[7] Since mathematics is a unique, "alphabetical" description of God's creation, we must expect to find, upon reading it, the invisible things of God.

In the words of Larry Zimmerman:

> If mathematics is the basis of creation, its nature is revelatory of God and its purpose is to glorify God; the Christian teacher must desecularize mathematics for his students. That is, he must polish away the patina of secularization with which mathematics has become encrusted so its true, God-reflecting nature shines through.[8]

7.3 FOUR CURRICULUM OBJECTIVES

For mathematics education to be biblically Christian, the bull's-eye on the target must be the Living God of Scripture, His attributes, His word, and His acts. Although God's enscripted word is His primary means of self-disclosure, what He has made also reveals the essence of His being:

---

[6]Although government schools have tried to excise God from their premises in the guise of church/state separation, an anonymous but sagacious observer has noted, "As long as mathematics is taught in these schools, prayer to God will persist – especially on test day."
[7]Hooykaas, p. 105.
[8]Zimmerman, 2:3, 2.

For since the creation of the world, His invisible attributes are clearly seen, being understood by the things that are made, even His eternal power and Godhead, so that they are without excuse (Romans 1:20).

The heavens declare the glory of God; and the firmament shows His handiwork (Psalm 19:1).

Here we see that creation reveals the glory of God. Glory, in Hebrew, means "to be heavy." God is not a man, so this does not mean that He is over-weight. In response to a significant fact, the beatniks of the 1950s used to say, "Man, that's heavy." God is "heavy" with significance and His creation reveals that. Mathematics, as a description of God's creation, plays a unique role in showing forth the significance of the Creator.

From the above statements, we can formulate the first two objectives of a biblical Christian mathematics curriculum.

1. Biblical Christian mathematics describes the wonders of God's creation, and in so doing,
2. Biblical Christian mathematics reveals the invisible attributes of God.

If we would stop here, we would be left with just thought, wondrous as it is. The ability to think is a gift of God. God wants us to use our thought proc-esses to further His purposes on the earth. A thorough study of Genesis 1-3 re-veals that God created man to have dominion over, or subdue, the earth. God gave man his brain for the purpose of taking control over creation. With sin came distortion, in man and in creation. Yet God's dominion mandate holds. Tainted by sin, man will tend to misinterpret and misuse God's creation. Knowing this, God gave man a promise that a deliverer would come who would redeem man and thereby encompass him within a liberating epistemo-logical and ethical enclosure.[9] It is within these limits that man can be construc-tive in his dominion activities. Redemption is always linked with a view toward godly dominion.

Now, we can complete our objectives. Mathematical thought is meant to further God's purposes of redemption and dominion.

3. Biblical Christian mathematics serves to aid man in fulfilling God's mandate of dominion,
and
4. Biblical Christian mathematics serves to assist God's people in fulfilling God's mandate of worldwide evangelism.

As we survey each objective, we must not put each into its own isolated compartment. These four objectives overlap and integrate with each other. For example, as we observe God's creation we can formulate associated mathemati-cal relationships. In these relationships, we will be able to not only behold and marvel in unfolding vistas of beauty, power, order, symmetry, and infinitude, but also use these relationships in practical ways. This usage will usually involve technology in some manner or form. This discovered technology can then be

---

[9]These limitations are, in reality, pathways to true liberty. They are like railroad tracks for a train and they are twofold: (1) The objective standards of the Word of God and (2) the illuminating and ena-bling power of the Holy Spirit to understand and apply those objective standards to life and culture.

applied to either take control of some aspect of creation or to assist, in some way, God's evangelism program. Each objective compliments and serves the other three.

7.4 THE WONDERS OF GOD'S CREATION

Note Max Caspar's commentary on Keplerian thought:

> The entire work of Creation which God has prepared for man is for Kepler a richly laid table. And just as nature sees to it, that the living creature never lacks food, so, he supposes, we can also say that the diversity in the phenomena of nature is so great only so that the human intellect never lacks fresh nourishment nor experiences weariness of mind in age nor comes to rest, but far rather, that in this world, a workshop for the exercise of man's mind always remains open.[10]

Galileo Galilei spoke of the great "book of nature." The writing of this book is mathematical; the symbols are triangles, circles, and other geometrical figures. He believed that it is impossible to comprehend a single word of this book unless one has the aid of mathematics.[11] Jacob W. A. Young, educator, observed in 1907, "Little can be understood of even the simplest phenomena of nature without some knowledge of mathematics, and the attempt to penetrate deeper into the mysteries of nature compels simultaneous development of the mathematical processes."[12] Sir James Jeans confessed, "So true is it that no one except a mathematician can hope fully to understand those branches of science which try to unravel the fundamental nature of the universe."[13] Albert Einstein adds, "I am convinced that we can discover by means of purely mathematical constructions the concepts and the laws connecting them with each other, which furnish the key to the understanding of natural phenomena."[14] Carl Hempel remarks:

> ... the greater part of present-day scientific knowledge has been reached only through continual reliance on and application of the propositions of mathematics .... The majority of the more far-reaching theories in empirical science ... are stated with the help of mathematical concepts.[15]

Hermann Weyl echoed the aspirations of both Pythagoras and Kepler in these words: "Reason, ... in spite of all disappointments and errors, ... is able to follow the intelligence which has planned the world.... Our ears have caught a few of the fundamental chords from the harmony of the spheres...."[16] Larry

---

[10]Caspar, p. 374.
[11]Galileo Galilei, *Discoveries and Opinions of Galileo*, p. 238.
[12]Jacob W. A. Young, *The Teaching of Mathematics* (New York, 1907), p. 16.
[13]Jeans, *The Mysterious Universe*, p. 136.
[14]Einstein, *Essays in Science*, p. 17.
[15]Hempel, 3:1631, 1634.
[16]Hermann Weyl, *Space-Time-Matter*, trans. Henry L. Brose (London: Methuen, 1922), p. 312.

Zimmerman links the knowledge of mathematics with effective dominion action:

> Without the knowledge of the patterns of God's speech in the creation, you are powerless to replenish the earth and will instead be subdued by it. Now it is true that the melody of nature's song can be enjoyed without a knowledge of the underlying mathematical structure ... but the lyric of the "music of the spheres" is mathematics....[17]

Mathematics is a disciplined thought structure that describes both the numerical and spatial aspects of the structure of God's creation. Using these two aspects as a guideline, let us explore some of the many wonders of God's creation.

7.4.1 MUSIC, TRIGONOMETRY, AND WAVE MOTION

At the creation, the "morning stars sang together" (Job 38:7). Music is universal and its structure is thoroughly mathematical. Whoever participates in music is really counting without being aware of counting.

7.4.1.1 MIDDLE OCTAVE FREQUENCIES

c	261.6	f	349.2	a	440.0
c#	277.2	f#	370.0	a#	466.2
d	293.6	g	392.0	b	493.9
d#	311.1	g#	415.3	c'	523.2
e	329.6				

7.4.1.2 THE MATHEMATICAL HARMONY OF MUSIC

Interval	Frequency Ratio
Unison	1:1
Octave	2:1
Fifth	3:2
Fourth	4:3
Major Third	5:4
Major Sixth	5:3
Minor Third	6:5
Minor Sixth	8:5
Second	9:8

Every note of the musical scale has an exact frequency; each note vibrates a certain number of times per second. For example, the frequency of middle C is about 261.6 cycles per second. Originally conceptualized by Pythagoras, the

---

[17]Zimmerman, 2:3, 2.

frequency doubles every octave and musical intervals depend upon simple arithmetical ratios. For example, the ratio of frequencies between the key of C and the key of G is 2:3.

The ratio between successive frequencies in an octave (13 notes) is determined by the formula:

$$r_{n+1} = \sqrt[12]{2}\, r_n \text{ where } r = given\ frequency$$

In order to produce a certain pitch (frequency) in a pipe organ, the length of the pipe in feet is determined by the mathematical formula:

$$L = \frac{512}{P} \text{ where } L = length\ and\ P = pitch$$

The human voice and musical sounds can be changed into electrical current through the medium of a microphone. If this microphone is connected to a special instrument, called an oscilloscope, you will see a picture as illustrated in a simple manner in Figure 122.

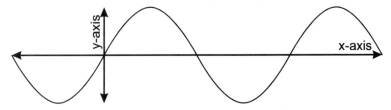

Figure 122: Sine curve

Sounds that are pleasing to the ear will display a picture (or graph) that reflects an order and regularity. This periodicity can be described mathematically as follows:

$$y = D\ sin\ 2(ft)$$

By definition, y represents the displacement, D the amplitude, f the frequency and t the time. This equation describes very simple sounds, like those of a tuning fork. For more complex sounds, the equation still holds, but is expanded in the terms of a summation of a mathematical series, called the Fourier series after its originator, Joseph Fourier (1768-1830). The work of this Frenchman in the analysis of heat provided the foundation for the mathematical study of music, a study that has enabled man in the 20th century to construct musical electronic gadgets like the synthesizer.[18]

Using the Fourier series, middle C on the piano can be described mathematically as follows:

---

[18]Joseph Fourier, *The Analytical Theory of Heat*, trans. Alexander Freeman (Cambridge: Cambridge University Press, 1878).

y = sin 2(512t) + 0.2 sin 2(1024t) + 0.25 sin 2(1536t) + 0.1 sin 2(2048t) + 0.1 sin 2(2560t).

Every sound that is pleasing to the ear can be described mathematically as the summation of what are called sinusoidal functions. The order and harmony of true music will create order and harmony in those who listen to it and play it. Those sounds that are not pleasing to the ear we call noise. Noise and dissonant music do not display regularity and cannot be described in terms of the mathematics above. The disorder of noise, and much of the popular "rock music" of today could be proved mathematically to be noise, will create disorder in those who listen to it and play it.

Figure 123: Joseph Fourier

Note the wondrous order and complexity in music! To fully comprehend it involves a thorough knowledge of trigonometry. The foundation of trigonometry is the simple right triangle. In the right triangle (see Figure 124), the rudimentary trigonometric ratios are defined:

$$\sin A = \frac{a}{h}$$

$$\cos A = \frac{b}{h}$$

$$\tan A = \frac{a}{b}$$

The reader should carefully note the progression from a simple right triangle to the complex order of music. Amazing! In fact, sinusoidal functions not only perfectly describe sound waves, but they also completely describe the distinct, wavelike motion of visible light and in fact, the entire electromagnetic spectrum. Who is the originator of light? God.

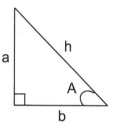

Figure 124: Right-angle triangle

Any wavelike, or regular, motion – e.g., the path of meandering rivers, ocean tides, the crest and trough of ocean waves, and the majestic rotation of galaxies – can be described mathematically in terms of trigonometric functions.

7.4.2 THE FIBONACCI SEQUENCE

Figure 125: The Fibonacci Sequence and the flower kingdom

Another fascinating numerical property found in God's creation is called the Fibonacci sequence. Leonardo of Pisa popularized it in the 13th century.[19] The sequence looks as follows:

1,1,2,3,5,8,13,21,34,55,89,144,233,377,...

Note that the third term is the sum of the first two, the fourth the sum of the second and third, and so on. These numbers are so common in creation that in 1963, *The Fibonacci Quarterly* began to be published by an organization called "The Fibonacci Association."[20] The sole purpose of this publication is to document the occurrence of this sequence in nature!

The Fibonacci numbers occur repeatedly in the petal arrangement of flowers. For example, two petals are found on the Enchanter's Nightshade, three on the Trilium, Lily, and Iris, five on the Wild Geranium, Spring Beauty, and Yel-

---

[19]The French mathematician Edouard Lucas (1842-1891) attached Fibonacci's name to this sequence in the 19th century.

[20]See <www.sdstate.edu/wcsc/http/fibhome.html> for current issues of this periodical and subscription information. Back issues of *The Fibonacci Quarterly* are available in microfilm or hard copy from *University Microfilms International* and reprints from *UMI Clearing House*, both at 300 North Zeeb Road, Dept. P.R., Ann Arbor MI 48106.

low Violet, eight on the Lesser Celandine, Sticktight, and Delphinium, thirteen on the Corn Marigold, Mayweed, and Ragwort, twenty-one on the Chicory, Aster, and Helenium, thirty-four on the Plantain, Ox-eye Daisy, and Pyrethrum, fifty-five on the Field Daisy, Helenium, Michaelmas, and Daisy, and eighty-nine on the Michaelmas and Daisy.[21]

Figure 126: Pine cone (top view)

Figure 127: Sunflower

These numbers are found in the spiral arrangement of petals, pine cones, and pineapples. In the pine cone spiral (see Figure 126) there are five spirals one way and eight the other. In pineapple spirals, the pair one way and the other can be five and eight, or eight and thirteen. In the sunflower spirals (see Figure 127), the combination can be eight and thirteen, twenty-one and thirty-four, thirty-four and fifty-five, fifty-five and eighty-nine, and eighty-nine and one hundred forty-four.

Figure 128: Pine cone (side view)

---

[21]LeRoy C. Dalton, *Algebra in the Real World* (Palo Alto: Dale Seymour Publications, 1983), p. 75.

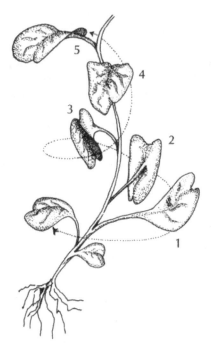

Figure 129: Phyllotaxis

The phyllotaxis of trees exhibits the Fibonacci numbers in a unique way. The spiraling pattern is seen in the turning of leaves about the stem (see Figure 129). Starting from a given leaf at a specific position, the number of turns required to find another leaf in the same position is a Fibonacci number. Also, the number of leaves between and within those turns is a Fibonacci number! For example, in the Basswood and Elm, the ratio of turns to leaves is 1:2, in the Hazel and Beech, the ratio is 1:3, in the Apricot, Cherry and Oak it is 2:5, in the Pear and Poplar it is 3:8, and in the Almond and Pussy Willow it is 5:13.[22]

The Fibonacci number also finds its home in the mathematics of the quantum matrix, of rabbit populations, and of the genealogy of male bees.[23]

7.4.2.1 THE GENEALOGY OF THE MALE BEE

The male bee has only one parent, the female bee. The female bee has two parents, the male bee and female bee. In Figure 130, the one male bee has one mother, two grandparents, three great-grandparents, five great-

---

[22]*Ibid.*

[23]Huntley, pp 156-161. See also Rochelle Newman & Matha Boles, *Universal Patterns* (Bradford, MA: Pythagorean Press, [1983] 1992), pp. 169-196.

great grandparents, eight great-great-great grandparents, and thirteen great-great-great-great grandparents.

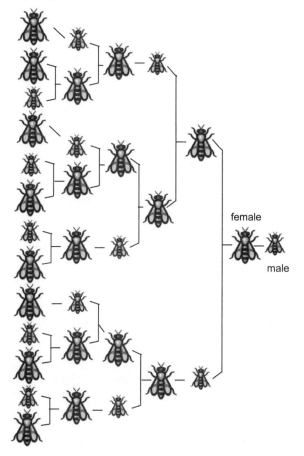

Figure 130: The Genealogy of the Male Bee

Figure 131: Piano keyboard

Returning to music for a moment, the basic structure of a piano keyboard consists of an octave of *thirteen* notes, *eight* of which are white and *five* of which are black. The black keys come in groups of *two* and *three!*

We have yet to scratch the surface of the Fibonacci numbers. Starting with the second and third terms of the sequence (1, 2), we can calculate the ratio between terms and come up with the following:

2:1 =	2.0	34:21 =	1.6190476
3:2 =	1.5	55:34 =	1.617647
5:3 =	1.67	89:55 =	1.6181818
8:5 =	1.6	144:89 =	1.6179775
13:8 =	1.625	233:144 =	1.6180555
21:13 =	1.6153846	377:233 =	1.6180257

We can see that the sequence of ratios approaches the number 1.618. Using mathematical terms, the limit of the sequence of ratios in the sequence of Fibonacci numbers is 1.618. This number is called φ, the Greek letter phi, which is the first letter of the name of the Greek sculptor Phidias (ca. 490-430 BC) who consciously made use of this ratio in his work.

What is astonishing about this ratio is that it *always* appears when applying the following algorithm. Take *any* two numbers at random, say 4 and 9. Add them to get 13. Add 13 and 9 to get 22. Add 22 and 13 to get 35. Keep this process going and observe what happens to the ratios. The ratios approach φ as a limit! The reader is encouraged to verify this result inductively by choosing a wide variety of number pairs as starting points. For those who want a challenge, try proving the results of this algorithm deductively.

9:4 =	2.25	149:92 =	1.6195652
13:9 =	1.444	241:149 =	1.6174497
22:13 =	1.692	390:241 =	1.6182573
35:22 =	1.590	631:390 =	1.6179487
57:35 =	1.629	1021:631 =	1.6180666
92:57 =	1.61403	1652:1021 =	1.6180215

Phi (φ) is also equal to the division of a line segment into its extreme and mean ratio. According to Euclid (Book VI, Definition 3), "A straight line is said to have been cut in extreme and mean ratio when, as the whole line is to the greater segments, so is the greater to the less."[24] That is, given a line segment below:

---

[24]Heath, *Euclid: The Thirteen Books of the Elements,* 2:188.

The extreme and mean ratio is: $\dfrac{x}{1} = \dfrac{x+1}{x}$

Cross-multiplying, we get: $x^2 = x + 1$

Solving for x by applying the quadratic formula,[25] we get: $x = \dfrac{1 \pm \sqrt{5}}{2}$

Solving for positive x, x $\approx$ 1.618033989 = φ!

This number φ has been called the golden ratio. Since it has so many applications in God's creation, scientists like Kepler called it the "Divine Proportion."[26]

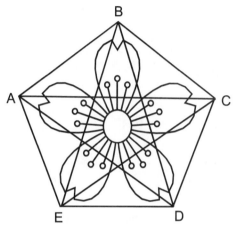

Figure 132: Pentagram and the Golden Ratio

The Divine Proportion finds acute application in the pentagram. In Figure 132, the ratio of AC to AB is equal to φ! God's creation is replete with five-petaled flowers. A few are the Hibiscus, Wild Geranium, Spring Beauty, Wood Sorrel, Yellow Violet, Plumeria, Saxifrage, Wild Rose, Pipsissewas, Bellflower, Filaree, Hoya plant, Columbine, Cinquefoil, Chickweed, and Passion flower.[27]

---

[25]The general form of a quadratic equation is ax² + bx + c = 0. Solving this general equation for x, we get the quadratic formula (dutifully memorized by most high school algebra students):

$$x = \frac{-b \pm \sqrt{b^2 - 4ac}}{2a}$$

[26]The Italian friar Luca Pacioli (ca. 1445-ca. 1509) wrote *De Divina proportione* (1509). In this book, he identified φ as the "Divine Proportion" (*sectio divina* in Latin). Pacioli is often called the father of double-entry bookkeeping, not because he invented the system, but because he was the first to put its principles in print in the 1494 work entitled *Summa de arithmetica, geometrica, proportioni et proportionalita*.

[27]Dalton, p. 70.

Figure 133: Starfish   Figure 134: Sand dollar

Also, the starfish, the sand dollar, the cross section of an apple core, and a wide variety of marine animals all reflect the pentagramic form of the Divine Proportion.

7.4.2.2 SOME FASCINATING MATHEMATICAL PROPERTIES OF φ (PHI)

Given: $\varphi = \dfrac{1+\sqrt{5}}{2}$, then $\dfrac{1}{\varphi} = \dfrac{2}{1+\sqrt{5}}$.

Rounding to five decimal places, $\dfrac{1}{\varphi} = .61803$.

Note that $1.61803 - 1 = .61803$ or $\varphi - 1 = \dfrac{1}{\varphi}$!

Phi is the only number that becomes its own reciprocal when one is subtracted from it.

Now, let $\varphi' = \dfrac{1-\sqrt{5}}{2}$ (the negative x solution of the mean/extreme ratio). We read φ' as "phi prime."

Multiplying $\varphi\varphi'$, we get:

$$\varphi\varphi' = \left(\frac{1+\sqrt{5}}{2}\right)\left(\frac{1-\sqrt{5}}{2}\right) = \frac{1-5}{4} = \frac{-4}{4} = -1.$$

This means that φ and φ' are negative reciprocals since their product is -1.

Adding $\varphi+\varphi'$, we get: $\varphi+\varphi' = \left(\dfrac{1+\sqrt{5}}{2}\right)+\left(\dfrac{1-\sqrt{5}}{2}\right) = \dfrac{2}{2} = 1.$

So, $\varphi\varphi' = -1$ and $\varphi+\varphi' = 1$!

Let us consider the geometric sequence, $1, \varphi, \varphi^2, \varphi^3, \varphi^4, ..., \varphi^n, ...$

In a geometric sequence, the next term is calculated by multiplying the previous term by the same number, in this case, φ.

$1\varphi = \varphi$ (*Nota bene*: $\varphi^0 = 1$ and $\varphi^1 = \varphi$)

$\varphi\varphi = \varphi^2$

$\varphi^2\varphi = \varphi^3$

$\varphi^n\varphi = \varphi^{n+1}$

Here is another example of a geometric sequence: 1, 2, 4, 8, 16, 32, 64, 128. In this sequence, the next term is calculated by multiplying the previous term by 2.

Now, back to the φ geometric sequence. If we analyze the terms more carefully, we discover that each term is also the sum of the two preceding terms. Thus, this sequence is at the same time both a geometric sequence and an arithmetic or summation sequence.

Note, $\varphi^2 - \varphi - 1 = 0$ (Let the reader do the math!). Hence,

$\varphi^2 = 1 + \varphi$ or $\varphi^2 = \varphi^0 + \varphi^1$.

Now, $\varphi^3 = \varphi^2\varphi = (1 + \varphi)\varphi = \varphi + \varphi^2 = \varphi^1 + \varphi^2$ and

$\varphi^4 = \varphi^3\varphi = (\varphi + \varphi^2)\varphi = \varphi^2 + \varphi^3$, and, in general,

$\varphi^n = \varphi^{n-2} + \varphi^{n-1}$ or $\varphi^n + \varphi^{n+1} = \varphi^{n+2}$. This is the general formula for determining the next term in the Fibonacci sequence.

Finally, the following elegant formula of continued fractions unveils the delightful φ:

$$\varphi = 1 + \cfrac{1}{1 + \cfrac{1}{1 + \cfrac{1}{1 + \ldots}}}$$

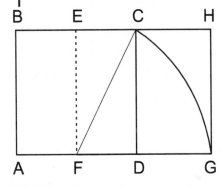

Figure 135: Golden rectangle

In the golden rectangle (also called golden section), the ratio of the length to the width, BH/BA, is φ (see Figure 135). To construct such a rectangle, begin with a square ABCD. Divide the square into two equal parts by the dotted line EF. Point F is the center of a circle whose radius is the diagonal FC. Draw an arc of the circle (CG) and extend the base line AD to intersect it. This becomes the base, AG, of the rectangle. Draw the new side, HG, at right angles to the new base, bringing the line BC to meet it at H. The resultant golden rectangle has one unusual property: If the original square is taken away, what remains will still be a golden rectangle! The golden rectangle is said to be one of the most visually satisfying of all geometric forms; it appears repeat-edly in art and architecture.

The geometric shape of the chambered nautilus and the ammonite is called the logarithmic, or equiangular, spiral. This shape can be constructed using the golden ratio (see Figure 136). First, draw a golden rectangle as described above. Divide the rectangle into a series of

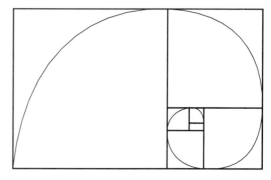

Figure 136: Golden rectangle and the equiangular spiral

squares as shown in the figure. Draw an arc that is a quarter circle from a corner of each square, starting with the largest square, to form a continuous curve. The resulting curve is close to the shape of the chambered nautilus!

7.4.3 SPIRALS

Figure 137: Chambered Nautilus

Figure 138: Storm clouds of a hurricane

We have already seen examples of this spiral in the botanical structure of the pine cone, the pineapple, and the sunflower. Other botanical spirals are found in the growth of creeping vines, the Honey-suckle, and the Bindweed. Also, the vortex of a whirlpool, the motion of a tor-nado, the storm clouds of a hurricane, ocean waves, and certain shorelines all reveal the spiral pattern.[28]

---

[28]For over 400 pages devoted to the spiral, see Theodore A. Cook, *The Curves of Life* (New York: Dover Publications, [1914] 1979). See also Newman and Boles, pp. 197-217.

In the animal kingdom, we find the spiral appearing in the horns of wild sheep, seashells (chambered nautilus and snails), claws, elephant tusks, beaks, teeth, fangs, and spider webs.[29]

A quick glance at the ridges on our fingertips will reveal varied spiral patterns. Many internal organs (e.g., the cochlea of the human ear) are spirals. The human umbilical cord is a triple spiral helix.

Figure 139: Cochlea of the human ear

The genetic structure of all living things is the spiral helix of the DNA molecule. In the words of Erwin Schrödinger, the gene structure "displays a most regular and lawful activity – with a durability or permanence that borders upon the miraculous."[30]

Figure 140: Spiral galaxy

As we look to the heavens, we see the spiral in the structure of the tail of a comet curving away from the sun and in many galaxies. Modern astrophysicists have posited a fascinating thesis with regards to the creation of these galaxies. They believe that creative forces (assumed by these scientists to be self-generating) released compound ripples, or oscillatory patterns, causing violent, abrupt changes in the pressure and density of the galactic mass. They refer to these oscillatory patterns as galactic "sonic booms." These whirling sound waves produced a spin in the galactic cloud and stars were born within the interior regions set up by this spin. The Bible reveals that God created the stars in these galaxies by His voice, "Let there be" (see Genesis 1:14-19). The first chapter of the Gospel of John describes the eternal Word (*logos*) of God as the agent through whom all things (including the galaxies and stars) were made (John 1:1-3). It is possible that astrophysicists, albeit governed by evolutionary presuppositions, have simply restated the biblical image of creation through the Word (or sound) of God. Has science, in its affirmation that visible stars and galaxies are spiral blast patterns, inadvertently confirmed the biblical message of creation? When we view these galaxies, we are observing the residual imprints of standing shock waves generated by the thundering and creative voice of Almighty God.

---

[29]Dalton, p. 98.
[30]Schrödinger, p. 46.

7.4.4 CONIC SECTIONS

The conic sections have intrigued mankind for ages. The Greeks carefully analyzed all four of the basic conics: the circle, the ellipse, the parabola, and the hyperbola. Unless we view it from an airplane, we never see the rainbow completely (it is a circle).[31] What we see from ground level is a half circle. Kepler showed that planetary motion follows an elliptical path. The rings of Saturn are ellipses. The path of a comet and the arc of spouting water from a waterfall are parabolic and in quantum mechanics, the path of an alpha particle can be described using the hyperbola.

7.4.5 HEXAGONS

Figure 141: Snowflakes

The hexagonal pattern is not only found in the honeycomb, but also in the atomic structure of honey! Job speaks of the treasures of the snow (Job 38:22).[32] Our minds boggle at the hexagonal patterns revealed in snowflakes; patterns infinitely rich in variety and intricacy. The hexagonal tessellation also appears in the chemical structure of adenovious, a common cold virus, and the endothelium layer of human and animal cornea.

7.4.6 CRYSTALLOGRAPHY

Crystallography demonstrates in the mineral and chemical realm a geometric beauty, clarity, and perfection beyond description. Stanley L. Jaki observes:

> Photographs taken of distant galaxies turned into common knowledge a magnificent variety of patterns that are not a whit less beautiful than snowflakes and crystals under magnifying glasses ... the models of atoms and molecules ... give a glimpse of the intricate beauty of the world of atoms.[33]

---

[31]See Elizabeth A. Wood, *Science from Your Airplane Window* (New York: Dover Publications, [1968] 1975), pp. 59-79.

[32]See W. A. Bentley and W. J. Humphreys, *Snow Crystals* (New York: Dover Publications, [1931] 1962).

[33]Jaki, *Cosmos and Creator*, p. 29.

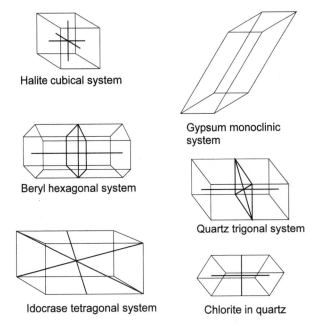

Figure 142: Crystalline structures

7.4.7 CONGRUENCE, SIMILARITY, SYMMETRY, ETC.

Other basic shapes in creation reflect basic mathematical properties. The comparable shapes of animals, plants, and stars give rise to the principles of congruence and similarity. Beautiful and balanced symmetry is reflected in the human body, the animal body, plants, crystal structure, etc. The surface of lakes illustrates the plane. Tree trunks are cylindrical. At certain locations, riverbanks reflect the property of parallel lines.

7.4.8 TRIGONOMETRY AND THE HEAVENLY EXPANSE

Through basic principles of triangulation, the trigonometric functions serve to expand our understanding of the vast outreaches of space. Yet, the heaven and the heaven of heavens cannot contain the infinite, personal, Creator God!

7.5 THE "ONE AND THE MANY"

7.5.1 THE PROBLEM

Throughout the history of philosophical thought, autonomous man has attempted to answer a recurring and oftentimes difficult dilemma. The problem is stated in terms of nature versus grace, the many versus the one, or the particulars versus the universal.

NATURE vs. GRACE
MANY vs. ONE
PARTICULARS vs. UNIVERSAL

Man tends to make one or the other absolute. Today, for example, "scientific" man generally tends to reject anything that has to do with grace or the universal, making nature or the particulars absolute. The result of this materialistic mindset is that the particular facts under study are unrelated and meaningless. They just stand by themselves, the chaotic result of eons of evolutionary change. Herbert Schlossberg remarks about the ramifications of this position:

> Materialism and pantheism, whatever their differences, share immanence, the identification of ultimacy with the creation. They must therefore end up deterministic, fatalistic, and pessimistic, or else cover up their logical conclusions with mystification.[34]

In Chapter 6, we noted that humanistic mathematicians are doing exactly what Schlossberg describes. Man *cannot* continue to believe in the ultimacy of materialism. If he does, he will end up with a suicidal mentality looking in despair at everything around him saying, "Is that *all* there is?" It is noteworthy to see that, in recent years, many top intellectuals, those making key contributions in thought and discovery, are returning to a belief in God.[35]

This belief in God recognizes a return to the metaphysical and epistemological underpinnings of science and mathematics, underpinnings that derive their subsistence from the biblical worldview. We have been brainwashed by modernity into believing a lie; a lie that boldly states that the scientist is a man of reason and certainty and the person who believes in the Bible is a man of faith and uncertainty.[36] Note what K. Lonsdale, a leading expert on crystal structure, confessed, "The scientist, as well as the man of religion lives by faith and not by certainty."[37] In this context, we must note that it is very easy for the scientist of modernity to become so focused on his work that he neglects some of the basic assumptions upon which his own science is founded. Indeed, his focus can become so concentrated that he forgets *that there are basic assumptions*.

The following are some of those important assumptions or metaphysical and epistemological presuppositions of science and mathematics; all of which are articles of *faith*. First, the scientist must have faith in the existence of an external, objective world. Second, the scientist must believe in the orderliness of the external, objective world. To detect that order, the scientific method is engaged. Underpinning the scientific method are the following articles of faith: (1) faith that abstract mathematical modeling can both explain and predict the workings of the objective world, (2) faith that things behave in the same way

---

[34]Schlossberg, p. 175.
[35]See Roy Abraham Varghese, ed., *The Intellectuals Speak Out About God* (Chicago: Regnery Gateway, 1984).
[36]As a by-product, this faith is deemed both blind and irrational.
[37]K. Lonsdale, in the foreword to Max Born, *Physics and Politics* (Edinburgh: Oliver, 1962), p. v.

whether they are being observed or not, (3) faith that memory is trustworthy, (4) faith in the reports of other scientists, and (4) faith in the reality of distinctions (the distinction between you, it, and I; the distinction between likeness and unlikeness; the distinction between unity and diversity).

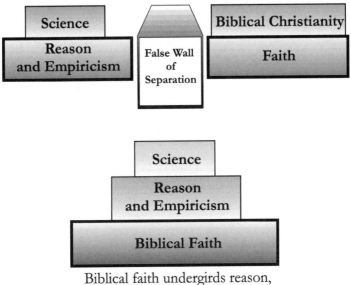

Biblical faith undergirds reason,
empiricism, and science

Figure 143: Faith, reason, empiricism, and science

Universals can also be absolutized. Before the Reformation, many Aristotelian academic leaders of Catholicism tended to do just that. Hence, when a person like Galileo challenged the system, he experienced virtual excommunication. At one point, Galileo invited these leaders of his day to peer through his telescope and see the plain (particular) facts, but they even refused to look! When a universal is absolutized, individual facts tend to be invalidated and therefore unessential.

7.5.2 THE SOLUTION IN THE ONTOLOGICAL TRINITY

Particular facts must have a reference point, and that reference point must be big enough to provide sufficient motivation and meaning to life. All facts are not isolated and unrelated. Every fact of the universe is a created fact; created by and understood truly only in terms of the infinite, personal, and triune God of Scripture.

God being infinite implies that the reference point is big enough. God being personal implies that the particular created facts are meaningful. God being triune implies a solution to the problem of the one and the many.

When we think of trinity, reason must kneel before love and faith. We cannot thoroughly comprehend the essence of God. For example, try thinking about a person who is unbeginning. Try thinking about a person who is infinite (boundless and limitless). Try thinking about a person who is everywhere all at once. This is who God is, and our minds "strike at the wind."

The Westminster Confession of Faith (1646) defines the triune nature of God as follows:

> In the unity of the Godhead there be three persons, of one substance, power, and eternity: God the Father, God the Son, and God the Holy Ghost: the Father is of none, neither begotten, nor proceeding; the Son is eternally begotten of the Father; the Holy Ghost eternally proceeding from the Father and the Son.[38]

There is only one God but, in the unity of the Godhead,[39] there are three persons, coeternal and coequal, all three fully God, but each distinct in existence. In the Godhead, the three distinct persons are the Father, the Son, and the Holy Spirit.[40] We might call each, using philosophical terms, particulars. All the "particulars" of the Godhead are related to the "universal" and the "universal" is fully expressed in the "particulars." The Father is fully God, the Son is fully God, and the Holy Spirit is fully God. In this sense we can say that the "plurality and the unity of the Godhead are both equally ultimate."[41]

When the triune God creates, He externalizes His very nature in what He makes. Creation is a temporal revelation of the eternal "One and the Many." First, the Living God gives meaning to every facet of creation. No particular fact is a fact apart from God. No fact can be interpreted truly in isolation. Second, creation is made up of an almost infinite variety of "particulars" that includes the mathematical concepts of number and space. These particular facts give expression to the nature of the triune God in that particulars relate to a universal (e.g., patterns or laws) and a universal is fully expressed in the particulars.[42] Like their creator, but on a temporal scale, the plurality and the unity of the created order are both equally ultimate. As Rousas J. Rushdoony summarizes, "Christian philosophy is thus able to give a comprehensive and unified picture of reality without doing any injustice either to unity or particularity."[43]

---

[38]*The Westminster Confession of Faith,* chapter II, article III (Norcross, GA: Great Commission Publications, 1994), p. 6.

[39]Deuteronomy 6:4 states that "God is one." The Hebrew word translated as "one" is transliterated *echad.* This word does not refer to mathematical singularity (as the Hebrew word transliterated *yachid* does); it denotes a compound or collective unity (i.e., a unity of persons).

[40]This doctrine is contrary to the ancient and modern heresy of modalism or Sabellianism (God is a mathematical one and expresses Himself historically in modes: Father, Son, and Spirit). This denial of distinction between the persons of the Godhead is not only philosophically feeble (the universal is absolutized), but ecclesiastically irrelevant and culturally impotent.

[41]Rushdoony, *The One and the Many: Studies in the Philosophy of Order and Ultimacy,* p. 356.

[42]In other words, scientific laws are but generalizations of God's way of making the particulars of His creation cohere or consist by the sustaining word of His power (Colossians 1:17; Hebrews 1:3).

[43]Rushdoony, *The One and the Many: Studies in the Philosophy of Order and Ultimacy,* p. 356.

Polish-born English humanist and mathematician Jacob Bronowski (1908-1974) states, "Science is nothing else than the search to discover unity in the wild variety of nature; ... beauty is unity in variety."[44] God is the author of this beautiful "unity in variety." He is the creator of the real world with its inherent harmony and relationships. To pontificate otherwise, as modern science does, is to make a pretentious metaphysical claim. As Rousas J. Rushdoony remarks, "To say that science makes no pronouncement about the ontological trinity is to ascribe to science a tremendous pronouncement, one which makes brute factuality the ultimate reality."[45]

Let Cornelius Van Til summarize how the biblical revelation of the triune God resolves once and for all the problem of the one and the many. Digest the following carefully and slowly:

> In the first place we are conscious of having as our foundation the *metaphysical* presupposition of Christianity as it is expressed in the creation doctrine. This means that in God as an absolutely self-contained being, in God as an absolute personality, who exists as the triune God, we have the solution of the one and the many problem. The persons of the trinity are mutually exhaustive. This means that there is no remnant of unconsciousness of potentiality in the being of God. Thus there cannot be anything unknown to God that springs from his own nature. Then too there was nothing existing beyond this God before the creation of the universe. *Hence the time-space world cannot be a source of independent particularity.* The space-time universe cannot even be a universe of exclusive particularity. It is brought forth by the creative act of God, and this means in accordance with the plan of the universal God. Hence there must be in this world universals as well as particulars. *Moreover they can never exist in independence of one another.* They must be equally ultimate which means in this case they are both derivative.[46]

7.5.3 THE SOLUTION APPLIED

What is number? Each number is a unity. For example, the number one expresses the unity of "oneness." This unity has a variety of expressions: one tree, one cow, one moon, one river, etc.

What about the basic mathematical operations of addition, subtraction, multiplication and division? In base ten, $1+1=2$, $3-2=1$, $2\times3=6$, and $8\div2=4$.

---

[44]Jacob Bronowski, *Science and Human Values* (Harmondsworth Middlesex: Pelican, 1964), p. 27.
[45]Rushdoony, *By What Standard?*, p. 61.
[46]Cornelius Van Til, *Psychology of Religion*, Syllabus (Philadelphia: Westminster Theological Seminary, 1935), p. 49f. and cited in Rushdoony, *The One and the Many: Studies in the Philosophy of Order and Ultimacy*, p. 355. The temporal one and the many (universals and particulars; unity and diversity; *a priori* and *a posteriori* knowledge; reason and empiricism; induction and deduction) are means to proximate and needed knowledge; they are never the means to ultimate or absolute knowledge. Neither the temporal one nor the temporal many is to be seen as the absolute pathway to proximate knowledge (both are equally ultimate; i.e., they interact and feed off each other) and both depend upon God as the foundation and source of all knowledge.

The unified principle that each fact represents can be seen in a variety of real world applications.

We have already noticed, in our survey of the wonders of God's creation, many examples of "unity in diversity." Basic trigonometric functions relate to the right triangle, music, and the electromagnetic spectrum. The Fibonacci sequence unifies many diverse aspects of creation. The spiral, the hexagon, and the pentagram are just a few of the other beautiful and inherent relationships in creation.

Certain numbers appear again and again in a diversity of situations. The Greek letter pi (π) represents the ratio of the circumference of a circle to its diameter. This number is another transcendental number like e and it finds its home in trigonometry, the study of music, the geometry of a spherical raindrop, probability theory, the sum of an infinite series, etc.

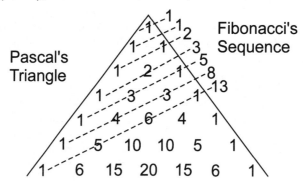

Figure 144: Pascal's triangle

Pascal's triangle, developed by Blaise Pascal (1623-1662) in 1644, unites several diverse aspects: The Fibonacci sequence, the binomial formula developed by Isaac Newton, and the laws of probability.

Binomial formula:

$(x+y)^0 = 1$

$(x+y)^1 = 1x + 1y$

$(x+y)^2 = 1x^2 + 2xy + 1y^2$

$(x+y)^3 = 1x^3 + 3x^2y + 3xy^2 + 1y^3$

$(x+y)^4 = 1x^4 + 4x^3y + 6x^2y^2 + 4xy^3 + 1y^4$

Probability:

When four coins are tossed in the air, the possible combinations of heads and tails are:

4 heads: 1 combination (HHHH)

3 heads and 1 tail: 4 combinations (HHHT, HHTH, HTHH, THHH)

2 heads and 2 tails: 6 combinations (HHTT, HTHT, THHT, HTTH, THTH, TTHH)

1 head and 3 tails: 4 combinations (HTTT, THTH, TTHT, TTTH)
4 tails: 1 combination (TTTT)
The sum of the fourth row of Pascal's triangle represents the total number of combinations (16). The probability of tossing 2 heads and 2 tails is 6/16 or 3/8 (37.5%).

When we speak of the "laws of nature" we must remember that science really studies the footprints of God in creation. These laws are man's attempt to quantitatively describe the evidence of the faithfulness of God in His power which ever preserves the creation. According to Vern S. Poythress, "God is so consistent and regular about it [His rule - J.N.] that we can plot His activity with mathematical equations."[47] Science may tell us the laws that explain the formation of dew in the morning, the motion of the planets, the force of gravitation, the structure of the atom, or the process of photosynthesis. But, what caused these causes? If we just speak of the "laws of nature," then we have removed ourselves from God by at least one step.

With this in mind, Isaac Newton's "law" of universal gravitation has also been called the inverse square law. The force of gravitation between two objects varies inversely with the square of the distance between them (see section 4.6.5.2). The inverse square law is another wonderful example of unity in diversity; it also finds application in the study of the intensity of light, the volume of sound, the uniform motion of heat in homogenous media, and the forces of electricity and magnetism.

Why do our mathematical constructions relate to physical reality? We have already established (section 6.11) that these two diverse aspects, the working of man's mind and the working of the physical world, are in union, or cohere, because of a common Creator.

Within the discipline of mathematics, called number theory, there exists a wondrous and almost endless variety of internal, unifying relationships. For an elementary example, let us look at one number pattern called casting out nines.[48] If we multiply any number by nine and add the digits of the answer, we will always end up with the number nine!

---

[47]Poythress, "Science as Allegory-Mathematics as Rhyme," p. 4. This does not imply that plotting His activity is easy. The plotting activity requires hard work, many revisions, regrouping from false starts, testing of hypotheses, etc. Anyone who studies the scientific and mathematical labors of a Kepler, Newton, or Einstein will heartily concur. Regularities in God's created order are easy to locate in some areas, extremely complex and possibly even beyond absolute understanding in others (e.g., the physics of the rainbow and the mathematics of the quantum matrix). Because we are unable to completely quantify a certain area of God's creation does not immediately imply that the belief in a patterned creation is a chimera. God's creation might be complex in certain areas, but it is never chaotic. We must remember that our models of that pattern are just that – models (not reality). They are always subject to revision and correction when anomalies are noted. The detection of anomalies in a model often serves as a springboard to the development of a better or more comprehensive model (e.g., what Einstein did with Newton's model).
[48]The Arabs used this method as early as the 9th century. See Smith, 2:151-152.

$2 \cdot 9 = 18$ $(1 + 8 = 9)$; $31 \cdot 9 = 279$ $(2 + 7 + 9 = 18$ & $1 + 8 = 9)$; $235 \cdot 9 = 2115$ $(2 + 1 + 1 + 5 = 9)$

7.6 THE ATTRIBUTES OF THE LIVING GOD

7.6.1 INFINITE, ETERNAL, AND TRIUNE

God is not subject to any limitations. In *all* that we understand Him to be, He is the *unlimited* one (Genesis 21:23; Deuteronomy 32:27; Psalm 90:2). As finite humans, we have difficulty with this concept. Our limited minds cannot thoroughly grasp the unlimited.

God is without bounds. You cannot add to God or take away from God. The concepts of addition and subtraction do not apply to the infinite. In all that He is, He is unlimited and unbounded.

Mathematics is a unique model of the idea of infinity. To review sections 5.13 through 5.16, Hermann Weyl defined mathematics as "the science of the infinite."[49] David Hilbert made these remarks about the subject, "From time immemorial, the infinite has stirred men's *emotions* more than any other question. Hardly any other *idea* has stimulated the mind so fruitfully."[50]

The counting numbers are without end. The integers have neither beginning nor end. Rational numbers are those numbers that can be written as the ratio of two integers (e.g., 3/4, 8/7, etc.) where the denominator cannot be equal to zero. On the number line, you can find two rational numbers as close together as you wish (e.g., 135/233 and 136/233), but there will always be another rational number between them. In fact, there will be a limitless number of rational numbers between any two rational numbers! Furthermore, the entire set of rational numbers leaves infinitely many gaps, called irrational numbers, in any interval on the number line. The set of real numbers includes all the rational numbers and fills all the "gaps" as well, forming the idea of continuity or the continuous number line.[51]

The method of the calculus finds its foundation in the concept of infinity. With this method, man is able to explain and predict, to a marvelous degree of accuracy, the workings of God's creation.

The study of infinite sets provides an amazing analogy for the doctrine of the Trinity. In considering this, we can gain insight into how the concept of the "One and the Many" can be possible.

Consider four infinite sets:

N = 1, 2, 3, 4, 5, 6, 7, 8, 9,10,11,12, …
A = 1, 4, 7,10,13,16,19,22,25,28,31,34, …
B = 2, 5, 8,11,14,17,20,23,26,29,32,35, …
C = 3, 6, 9,12,15,18,21,24,27,30,33,36, …

[49]Weyl, *Philosophy of Mathematics*, p. 66.
[50]Hilbert, "On the Infinite," p. 136.
[51]See Péter for an excellent and clear discussion of how number theory relates to infinity.

The sets A, B, and C each consist of distinct elements and are called disjoint sets. Each of these sets can be placed in a one-to-one correspondence with the set N, the counting numbers. According to Cantor's theory of infinite sets, this means that all four sets have the same cardinality, or the same number of elements. Finally, the set N is the union of the three sets A, B, and C; i.e., $A \cup B \cup C = N$.[52]

Hence, each distinct set has the same cardinality and the union of all three has the same cardinality. The whole is not greater than its parts when it comes to infinite sets. The doctrine of the Trinity states that each person, though distinct, is equal to each other, but the whole is not greater than its parts. All three are fully God. Hence, in the mathematical context of infinite sets, we can catch a faint glimpse at the meaning of the eternal "One and the Many."

7.6.2 SOVEREIGN, ALMIGHTY, AND FAITHFUL

As already stated, the universe does not run by "natural law." God rules the universe directly. God, in His Son Jesus Christ, is currently "upholding all things by the word of His power" (Hebrews 1:3). God's word of sustaining power lies behind every human attempt to mathematically express the order, power, and beauty of the universe. In reference to the motion of the heavenly bodies, God's sustaining word is so precise that astronomical clocks are set to the movement of the stars.

7.6.3 WISE

The works of God display the wisdom of God (Psalm 104:24). There is an exactness, accuracy, wisdom, and logic in what God has made. All aspects of His creation interact with each other in a unified whole. Mathematics reflects the wisdom of God in His works. We have already made note of the efficient structure of the honeycomb. Internally, every aspect of the discipline of mathematics interacts with every other aspect creating a unified whole. If you want to get a correct answer to a mathematical problem, calculations need to be carried out orderly and accurately.

Christ is the wisdom of God and the *logos* of God through whom all things were made and in whom all things consist (John 1:1-4; Colossians 2:3; Proverbs 8:12-36; Colossians 1:17). As we understand, study, and use the tool of mathematics, we come to understand a small subset of the wisdom and *logos* of God. It is in Christ that our mathematical equations reflecting the order of the universe consist. It is in Christ that the marvelous unity and diversity of both the universe and the internal structure of mathematics cohere. It is in this way that Christ speaks both to the content and the philosophy of mathematics.

---

[52]The author credits this idea to Thomas E. Iverson, "God: All Sufficient or Infinite," *A Christian Perspective on the Foundations of Mathematics*, ed. Robert Brabenec (Wheaton: Wheaton College, 1977), pp. 121-125.

I. R. Shafarevitch, a Russian Orthodox Christian, mathematician and articulate critic of socialist economic policies, makes the following observations from the history of mathematics:

> A superficial glance at mathematics may give an impression that it is a result of separate individual efforts of many scientists scattered about in continents and in ages. However, the inner logic of its development reminds one much more of the work of a single intellect, developing its thought systematically and consistently using the variety of human individualities only as a means. It resembles an orchestra performing a symphony composed by someone.... One is overwhelmed by a curious feeling when one sees the same designs as if drawn by a single hand in the work done by four scientists quite independently of one another.[53]

Not only is it in Christ that mathematics consists and coheres, He is also King and ruler of the nations guiding every aspect of history (and the mathematics enterprise) toward His purposed ends (Revelation 1:5: Ephesians 1:11). The symphony of mathematics finds its origin in the only wise and providential God of Scripture.

---

[53]Cited in Philip J. Davis and Reuben Hersh, The *Mathematical Experience* (Boston: Houghton Mifflin, 1981), pp. 52-53. The following is a list of a few of the independent "discoveries" in mathematics made simultaneously by two or more persons working independently.

Law of inverse squares: In 1666 by Newton and in 1684 by Edmund Halley.

Logarithms: In 1620 by Joost Bürgi (1552-1632) and in 1614 by John Napier.

Calculus: In 1671 by Newton and in 1676 by Leibniz.

Non-Euclidean Geometry: From 1836-1840 by Lobachevsky, from 1826-1833 by Bolyai and in 1829 by Gauss.

Principle of least squares: In 1809 by Gauss and in 1806 by Adrien Marie Legendre (1752-1833).

Vector treatment without the use of co-ordinate systems: in 1843 by Sir William Rowan Hamilton (1805-1865) and in 1843 by Herman Günther Grassman (1809-1877).

Mathematics does not hold a monopoly on this type of occurrence. Here are a few examples from other fields:

Discovery of oxygen: in 1774 by Karl Wilhelm Scheele (1742-1786) and in 1774 by Joseph Priestley (1733-1804).

Liquefaction of oxygen: in 1877 by Louis Paul Cailletet (1832-1913) and in 1877 by Rasul Pictet.

Periodic law: in 1864 by Beguyer de Chancourtois (1820-1886), in 1864 by John Newlands (1837-1898), and in 1864 by Julius Lothar Meyer (1830-1895).

Law of periodicity of atomic elements: in 1869 by Julius Lothar Meyer and in 1869 by Dmitri Ivanovich Mendeléeff (1834-1907).

Law of conservation of energy: in 1843 by Julius Lothar Meyer, in 1847 by James Prescott Joule (1818-1889), in 1847 by Hermann Ludwig von Helmholz (1821-1894), in 1847 by Ludvig August Colding (1815-1888), and in 1847 by Sir William Thomson or Lord Kelvin (1824-1907).

Taken from William F. Ogburn, *Social Change* (New York, 1923), pp. 90-102 and cited in Leslie A. White, "The Locus of Mathematical Reality," *The World of Mathematics*, ed. James R. Newman (New York: Simon and Schuster, 1956), 4:2357-2358.

7.6.4 UNCHANGEABLE

God never differs from Himself. God cannot change, being unchanging and unchangeable (Malachi 3:6; James 1:17; Hebrews 13:8). If He did change, He would not be God. As an echo, the structure of mathematics voices this, for the structure of number systems reveals unchangeable laws. Also, in every mathematical procedure a fixed order of operations must be followed. For example, in a mathematical expression containing powers, expressions in parenthesis, multiplication and division, addition and subtraction, the law of operations requires doing the powers first, expressions in parenthesis second, multiplication and division third, and addition and subtraction last.

$$3^2 + 2(4 + 5^2) - 16 \cdot 8 + 24 \cdot 2^5 = 9 + 2(4 + 25) - 128 + 24 \cdot 32 =$$
$$9 + 2 \cdot 29 - 128 + 768 = 9 + 58 - 128 + 768 = 707$$

7.6.5 ALL-KNOWING

The knowledge of God covers all knowledge to infinity. Such knowledge takes in everything from the highest angel to the lowest amoeba.

God created, counts, and names the stars (Genesis 1:16; Psalm 147:4-5). The number of stars is estimated to be 10^{26}. This number is so large that we cannot begin to fathom its immensity.[54]

Figure 145: God counts the stars

From research into the structure of the DNA molecule, which is, as we have already noted, replete with mathematical relationships, the number of ge-

---

[54]The approximate number of stars is obtained by the following calculations. There are 10^{11} stars in every galaxy and 10^{11} galaxies within the range of the largest telescope. This means that, potentially, there are 10^{22} (10^{11} x 10^{11}) visible stars. Einstein estimated that the number of stars in total space is 10^4 times larger than the number of stars within telescopic range. Hence, 10^4 x $10^{22} = 10^{26}$. Counting one number every second, it would take three thousand trillion centuries to count to 10^{26}! But, radio telescopes can "hear" stars that give no visible light. It is estimated, in our galaxy alone, that there are 10^{11} such stars! Truly, the number of stars is *beyond comprehension*.

netic possibilities is estimated to be $3,000,000,000^{30}$.[55] The number of human beings that have ever lived, or ever will live, is a drop in the ocean compared to that number!

Not only does God know all things that were, are or shall be, He also knows all things that could have been, could be, and might be if He had so chosen. In this context, the work of the mathematician Kurt Gödel brings confirmation. In essence, he said that there is knowledge "out there" that man cannot know. Exhaustive knowledge belongs to God alone.

7.6.6 EVERYWHERE PRESENT

Man cannot flee from the presence of God (Psalm 139:7), for His immediate presence is everywhere. God fills the universe with Himself. He knows all things because He is immediately present to all things, the very foundation of creature existence.

Geometry, Euclidean and non-Euclidean, deals with the study of space. In Scripture, the spatial aspect of creation is used as a direction sign pointing to the immensity of God: "Do I not fill heaven and earth?" (Jeremiah 23:24).

7.6.7 GOOD

When God finished the creation, He made a pronouncement. He did not say that everything was almighty or wise. With infinite wisdom, God made all things by merely speaking all things into existence. Talk about power! Although made by His power and designed by His wisdom, the works of God reflected the goodness of God (Genesis 1:31).

Because God is good, His creation is good. As already mentioned (see section 6.8), the essence of goodness is to do good. To be good is to be generous, to give of oneself freely, extravagantly, and lavishly. God has showered His goodness upon what He has made.

Because God's creation is good, all of His works exist in order to serve. Every aspect of creation, inanimate and animate, serves and provides for other aspects of creation. All birds and animals exist for man in that they provide for his enjoyment, services, and food. The earth is filled with a wide variety of minerals and grows all kinds of vegetables – all for the use and delight of man. Water gives life, power, cleansing, and provides for transportation. Without oxygen, everything would die in a few minutes. The dry land is a suitable habitation for man and all other living creatures. The sun, moon, and the stars serve the earth; they are for our times and seasons (Genesis 1:14-18). A thorough study of Psalm 65, Psalm 104, Psalm 145, Psalm 148, and Job 38 would bring all of these perspectives to light.

Mathematics is a description of God's good creation. As we investigate the multitude of equations that reflect the workings of God's universe, let us not

---

[55]Ruth Haycock, *Bible Truth for Schools Subjects, Volume III: Science/Mathematics* (Whittier: Association of Christian Schools International, 1981), p. 101. This incomprehensible number is approximately equal to 2×10^{284}.

forget that, behind it all, is the providential touch of personality, a God who is kind, full of good will toward men, tenderhearted, sympathetic, open, and friendly.

7.6.8 ORDERLY, PRECISE, AND JUST

The finger of God penned a message one night for King Belshazzar bringing an end, not only to his gay and festive party, but to his kingdom, "MENE, MENE, TEKEL, PERES." Note the precision in God's judgments, "MENE: God has numbered your kingdom and finished it; TEKEL: You have been weighed on the balances, and found wanting; PERES: Your kingdom has been divided..." (Daniel 5:25-28). God is a righteous and good God and sin will be punished. Because God is, we are assured of the truth that there is right and there is wrong.

Noting the use of the mathematical terms "number", "weigh", and "divide" is not meant to illustrate a simplistic and naïve approach to "integrating mathematics with the Bible." This simplistic approach seeks to show that, for mathematical principles (e.g., addition, subtraction, division, etc.) to be true, they must be found first in the Bible.[56] You will not find the elements of differential and integral calculus or quaternions or matrix theory taught in Scripture. What you do find in Scripture is an overarching worldview perspective that gives meaning and context to every facet of the mathematics enterprise. In the words of Rousas J. Rushdoony, "... without the God of Bible and the revelation therein given no fact can be truly known, nor can its existence be even posited."[57]

MENE, MENE, TEKEL, PERES are not meant to illustrate mathematical operations, but God's inviolable justice.[58] In mathematics, we discover an echo of God's absolute justice, order, and precision (Matthew 5:18; Matthew 10:30). Solutions to mathematical problems are either right or wrong. Buildings must be constructed, in accordance with strict mathematical specifications, to withstand stresses and strains. If not, they will most assuredly collapse.

7.7 SUBDUE THE EARTH

Imagine trying to live without a working knowledge of numbers for a single day. You would survive, but without an understanding of the applications of mathematics, the ability to function adequately in this world would be severely limited.

Stanley L. Jaki comments, "A proposition of mathematical physics ... can even disclose previously unsuspected phenomena of nature, a prerequisite for

---

[56]For an example of this approach, see J. C. Keister, "Math and the Bible," *The Trinity Review* (September-October, 1982), no. 27 and Dave Gamble, "Mathematics in Christian Perspective," *On Teaching* (July/August, 1983), p. 1.

[57]Rushdoony, *By What Standard?*, p. 53.

[58]Biblical Christians thank God for the person of Christ through whom "justice and mercy" kissed on Calvary's Cross (Psalm 85:10).

achieving control in greater depth."[59] W. J. Harris, educator, said in 1898, "Mathematics ... furnishes the peculiar study that gives us ... the command of nature."[60]

According to Morris Kline:

Mathematics is the key to our understanding of the physical world; it has given us power over nature.... Mathematics has enabled painters to paint realistically, and has furnished not only an understanding of musical sounds but an analysis of such sounds that is indispensable in the design of the telephone, the phonograph, radio, and other sound recording and reproducing instruments. Mathematics is becoming increasingly valuable in biological and medical research ... mathematics is indispensable in our technology.[61]

7.7.1 NUMBER

Around the home, a mother is continuously working with numbers measuring ingredients for cooking and medicine for Johnny's little cold. Suppose you had a recipe that serves four people. How could you convert it into a meal for fifty? Simple mathematical ratios would provide the answer.

The business and economic world depends on numbers almost exclusively. Basic arithmetic is needed in order to write checks, balance bank statements, and prepare tax returns.

Why is the purchasing power of the dollar decreasing? Fifty years ago, five cents could buy what one-dollar buys today. Why? An understanding of economic practices plus basic mathematics provides the answer. Suppose that there are x number of dollars existing at any given moment in a nation's money supply. When banks loan money, they do not deduct the amount of the loan from the balances of their depositors. Instead, they simply create the money with a bookkeeping entry and check. The person who receives the loan then spends it into the economy. If the loan is $10,000, then the total number of dollars in the money supply increases to $x + 10,000$. As a result, the real value of each previously existing dollar has gone down, ever so slightly. Each dollar has decreased in real value by $x/(x + 10,000)$. For example, if the money supply is doubled, then the value of the money held before the doubling is cut in half (assuming the all-important factor that the production of consumer goods remains constant). One dollar now buys what fifty cents used to. Given the existence of "fiat" money, over a period of time, the real purchasing power of money will decrease significantly.

When dealing with the economic system of today, the Christian must not only be conversant in basic mathematics, but he must also know what the Bible says about the ethical dimensions of inflation (e.g., Leviticus 19:35-37; Deuter-

[59]Jaki, *Cosmos and Creator*, p. 28.
[60]W. J. Harris, *Psychologic Foundations of Education* (New York, 1898), p. 325.
[61]Morris Kline, *Why Johnny Can't Add: The Failure of the New Mathematics* (New York: Vintage, 1974), p. 174.

onomy 25:13-16; Isaiah 1:22; Micah 6:11; Proverbs 11:1; Proverbs 20:10; Proverbs 20:23) and how the doctrine of creation and man provides a proper foundation for a biblical Christian perspective of work and economics (Colossians 3:17; Colossians 3:22-4:1; I Corinthians 10:31).[62]

7.7.2 LOGIC

There are two aspects to logic. First, symbolic logic is foundational, not only to the development of electronic computer circuitry, but also to the logical flow of computer programs.[63] Second, the methods of mathematics train the mind in systematically approaching and solving, not only mathematical problems, but other problems as well; especially those germane to conflict resolution or ethical infractions. This does *not* mean that mathematical training guarantees ethical purity or that the ability to reason guarantees liberty (personally or culturally). Loving God in Christ (the *logos*) does not consist of ratiocination only; it also includes loving Him with all your heart, all your soul, and all your strength – and, loving your neighbor as yourself (Mark 12:29-31). Godly reason should produce loving obedience to God's ethical standards (John 15:9-17). Note that the leaders of the French Revolution thought that the ability to reason independently was the sole guarantor of liberty. They forgot that reason is merely a *tool* of liberty, not the *basis* of it. In their case, the deification of reason resulted, not in liberty, but in heads rolling from the guillotine.

The study of logic will also lesson the chances of either using or falling prey to common informal fallacies such as false dilemma, begging the question, guilt by association, and equivocation.[64] Also, the study of deductive geometry and Boolean truth tables will restrain the student from engaging in the formal logical fallacies of asserting the consequent or denying the antecedent.[65] The

---

[62]See John Jefferson Davis, *Your Wealth in God's World* (Phillipsburg: Presbyterian and Reformed, 1984), Tom Rose, *Economics: Principles and Policy from a Christian Perspective* (Mercer, PA: American Enterprise Publications, 1986), George Grant, *Bringing in the Sheaves: Transforming Poverty into Production* (Atlanta: American Vision Press, 1985), and David Chilton, *Productive Christians in an Age of Guilt Manipulators* (Tyler, TX: Institute of Christian Economics, [1981] 1985).

[63]See Kline, ed., *Mathematics: An Introduction to Its Spirit and Use*, pp. 166-197. George Boole (1815-1864) developed the study of symbolic logic. These rules also govern the construction and use of electric circuits (or any system whose components constitute a flow pattern); e.g., series switch-circuit systems, parallel switch-circuit systems, or complex switch-circuit systems.

[64]See Greg Bahnsen, *Always Ready: Directions for Defending the Faith* (Atlanta: American Vision and Texarkana: Covenant Media Foundation, 1996), pp. 133-149, Morris S. Engel, *With Good Reason*, 6th ed. (New York: Bedford Books/St. Martin's Press, 2000), and Irving M. Copi, *Introduction to Logic*, 10th ed. (Englewood Cliffs, NJ: Prentice-Hall, 1998).

[65]The fallacy of asserting the consequent occurs when one asserts a conditional premise (If P, then Q) and then affirms that the converse (If Q, then P) is true. Given: If $x < 5$, then $x < 10$. False conclusion: If $x < 10$, then $x < 5$ (x could be 7). The fallacy of denying the antecedent occurs when, given the same conditional premise (If P, then Q), one then affirms that the inverse (If not P, then not Q) is true. Given: If $x < 5$, then $x < 10$. False conclusion: If $x \geq 5$, then $x \geq 10$ (x could be 8).

student will also make productive use of the disjunctive syllogisms and the *reductio ad absurdum* argument.[66]

7.7.3 EXPONENTS AND LOGARITHMS

Logarithms are to exponents what subtraction is to addition. Both procedures "reverse" themselves. For example:
$$10^3 = 1,000 \text{ and } \log_{10}1,000 = 3$$

An exponential curve is of the form $y = a^x$ where $x > 0$. Musical instruments that generate sound from strings (e.g., the grand piano) or from columns of air (e.g., pipe organ) reflect the shape of the exponential curve in their structure.

Exponents and logarithms are used to describe population growth, depreciation, the intensity of a beam of light, the measurement of the magnitude of earthquakes in the Richter scale, the chemical

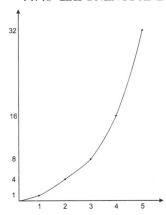

Figure 146: Exponential curve, $y = 2^x$

pH of a solution, the measurement of the loudness of sound in the Decibel scale, compound interest, and the observed brightness of stars. To illustrate some of these relationships, we will consider first the brightness of the stars and then compound interest.

Scientists use a measuring system called magnitude to express the observed (or apparent) brightness of stars using a system first introduced by Hipparchus (ca. 180-125 BC) and Ptolemy (ca. 85-ca. 165). They noted the twenty brightest stars in the sky and grouped them together as stars of the "first magnitude." Stars about 2.5 times fainter were

Astronomical Object	Apparent Magnitude
Sun	-26.5
Moon (full)	-12.5
Venus (brightest planet)	-4.0
Sirius (brightest star)	-1.42
Capella	0.06
Rigel	0.14
Beta Centauri	0.66
Acrux	0.87
Beta Crucis	1.28
Uranus	6.0
Neptune	9.0
Pluto	15.0

---

[66]A disjunctive syllogism begins with the premise that at least one of two propositions is true; i.e., P or Q. If you can prove that one of the propositions is not true; i.e., not Q, then you can validly infer that P is true. The *reductio ad absurdum* argument employs the following argument: Given the assertion of the conditional premise (If P, then Q) and given "not-Q," then the contrapositive "not-P" logically follows. Given: If $x < 5$, then $x < 10$. Valid conclusion: If $x \geq 10$, then $x \geq 5$.

classified as stars of the "second magnitude." Those 2.5 times fainter than second magnitude stars were classified as stars of the "third magnitude," and so on. Stars of the sixth magnitude were at the limit of naked-eye visibility. This system, virtually unaltered, is still in use today. Astronomers have set 2.512 as the exact ratio between magnitudes. When the difference in magnitude between two stars is five, it means one star is one hundred times brighter than the other; i.e., a star of magnitude one is one hundred times brighter than a star of magnitude six. Mathematically, the magnitude scale is logarithmic (2.512 is the fifth root of one hundred). The present limit for any existing terrestrial telescope is magnitude 24.5. With this telescope, you can see stars about *16 million* (2.512^{18}) times fainter than the faintest stars visible to the unaided eye in open country!

Instrument	Faintest Detectable Magnitude
Human eye (city)	3.0
Human eye (country)	6.5
Binoculars[67]	10.0
4 inch (10 cm) telescope	12.0
40 inch (1 m) telescope	17.0
Largest 100 inch (5 m) reflector telescope	24.5

The formula for calculating the transcendental number *e* is:

$$e = \lim_{n \to \infty} \left(1 + \frac{1}{n}\right)^n \approx 2.7182818284\ 59045 \ldots$$

Compare this formula with the formula for calculating earnings on a given principal: $A = P\left(1 + \frac{r}{n}\right)^{nt}$ where A = accumulated earnings, P = principal (original investment), r = interest rate, n = number of times the rate is compounded annually, and t = the number of years the money is in the savings account. This formula can be transformed into the formula for the transcendental number *e*

---

[67]Binoculars are good for viewing wide areas of the sky and the moon. Mathematically, these optical instruments are usually classified using a pair of numbers, like 7x50. The first number, in this case 7, indicates the magnifying power of the instrument. The letter "x" means "times." The number after the letter "x" represents the diameter of the objective lens (in millimeters). In general, lower powered binoculars have wider fields, higher-powered have narrower fields and are heavier and thus harder to hold still. Binoculars are often rated in terms of Relative Light Efficiency (RLE). A lower RLE means that the image is less bright. It is calculated by dividing the diameter of the objective lens in millimeters by the square of the power. In a 7x50 pair of binoculars the RLE is 50/7x7 = 1.02. If the RLE is less than 1, the instrument is only good for viewing the moon. A pair with an RLE of 1 will give you the ability to see objects about 20 times fainter than the unaided eye can detect.

when you let P = $10,000, r = 100%, and t = 1 (compounded indefinitely; i.e., let n → ∞). Given this scenario, after one year, a $10,000 investment will be worth $27,182.82 and *no more*.

Almost every investor is familiar with the Rule of 72; a formula for determining how many years it will take for an investment to double. The formula is 72=RT where R = rate of growth (or return) and T = number of years to double. This formula is a "rule of thumb" for the precise formula derived as follows:

$$2P = P\left(1+\frac{r}{n}\right)^{nt} \Rightarrow 2 = \left(1+\frac{r}{n}\right)^{nt} \Rightarrow \ln 2 = nt \ln\left(1+\frac{r}{n}\right) \Rightarrow t = \frac{\ln 2}{n \ln\left(1+\frac{r}{n}\right)}$$

$$Let\ n = 1,\ then: t = \frac{\ln 2}{\ln(1+r)} \ or\ t \approx \frac{.6931}{\ln(1+r)}$$

The following table compares the independent value *r* (rate of return or growth) with the dependent variable *t* (time in years) along with the approximation given by the Rule of 72:

r	t	Rule of 72
3%	23.4	24.0
4%	17.7	18.0
5%	14.2	14.4
6%	11.9	12.0
7%	10.2	10.3
8%	9.0	9.0
9%	8.0	8.0
10%	7.3	7.2
11%	6.6	6.5
12%	6.1	6.0
15%	5.0	4.8
20%	3.8	3.6
30%	2.6	2.4

7.7.4 MATHEMATICAL MODELING

The method of modeling will be discussed in section 8.4.1. It is the essential method of mathematics and is basic to the understanding and mastery of the real world. It is the foundation for atomic theory, especially quantum mechanics, and for the structuring of the periodic table of the chemical elements. It was this table, and the atomic numbers contained therein, that resuscitated the Pythagorean idea that the basic structure of the world can be understood as a pattern of numbers, specifically integral numbers.[68] Nuclear physicist Niels Bohr (1885-1962) considers the constitution of atomic theory as "an important step towards the solution of a problem which for a long time has been one of the boldest dreams of natural science, namely, to build up an understanding of the regularities of nature upon the consideration of pure numbers."[69]

7.7.5 CONIC SECTIONS

The parabolic curve describes cables on suspension bridges and reflecting surfaces where the focus[70] either collects what is received or reinforces what is transmitted. Surfaces like these are found in telescopes, radar transmitters, solar furnaces, sound reflectors, automobile headlights, searchlights, and satellite dishes. When a projectile is launched from the earth with a velocity less than the orbital velocity of the earth (about 17,000 miles per hour), its path will resemble a parabola. This is the shape of the arch followed by a thrown rock or a bullet or the path of any projectile.

Figure 147: Parabolic reflectors

The ceiling in Statuary Hall in the United States Capitol is parabolic. Until 1857, the House of Representatives met in this hall. John Quincy Adams (1767-1848), while a member of the House, was the first to notice its acoustical phenomenon. He noted that, from where he sat, he could clearly hear the conversations of other members of the House at the other side of the room! And, he could hear these conversations in spite of the other conversations taking place in the hall. Adams had located himself at a focal point of one of the parabolic reflecting ceilings and he could "eavesdrop" on other House members if they were located at the other focal point. In the architecture of Statuary Hall (see Figure 147), sound from focal point A bounces off the ceiling's dome (a para-

---

[68] See the chapter entitled "The World as a Pattern of Numbers" in Jaki, *The Relevance of Physics*, pp. 95-137. For the story of how scientists filled in the elements of the table developed by Dmitri Mendeléeff, see Isaac Asimov, *The Intelligent Man's Guide to Science: The Physical Sciences* (New York: Basic Books, 1960), 1:167-221. The story of Mendeléeff is described in Bernard Jaffe, *Crucibles: The Story of Chemistry* (New York: Dover Publications, 1976), pp. 150-163. For Mendeléeff's own personal testimony, see Dmitri Mendeléef, "Periodic Law of the Chemical Elements," *The World of Mathematics*, ed. James R. Newman (New York: Simon and Schuster, 1956), 2:913-918.
[69] Niels Bohr, *Atomic Theory and the Description of Nature* (Cambridge: Cambridge University Press, 1934), pp. 103-104.
[70] Focus is Latin for fireplace.

bolic reflector) and travels parallel to the opposite reflector, where it bounces to focal point B.

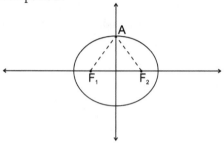

Figure 148: Focal points of an ellipse

The shape of the ellipse is used in the creation of whispering chambers (see Figure 148,[71] gears and springs, and the distribution of stress in machinery. Those who appreciate the mathematics of billiards would love to cue up on an elliptical pool table. A ball shot through one focus will always make it into the pocket at the other focus. When a projectile is launched from the earth with a velocity between the orbital velocity of the earth and escape velocity (about 25,000 miles per hour), its path will be either a circle or an ellipse. This is the orbital path of geosynchronous satellites circling the earth.[72]

As we saw in section 2.16, when the sound wave from a sonic boom intersects the earth, its path is a hyperbola. The hyperbola is essential to understanding LORAN (stands for LOng RAnge Navigation), a navigational technique used in locating ships at sea. The Cassegrain telescope (see Figure 149) makes special use of this shape. When a projectile is launched from the earth with a velocity greater than the escape speed, its path will be a hyperbola. This is the path traced out by spacecraft traveling to the moon or to other planets.

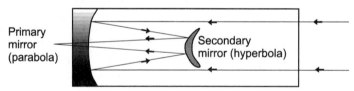

Figure 149: Cassegrain telescope

7.7.6 STATISTICS

Mathematical statistics is the study of ways to analyze data and the inferences drawn from this examination. Although ancient civilizations did gather and tabulate data, Carl Friedrich Gauss (1777-1855) was the first to develop the methods of statistical inference. These tools enable the mathematician to inves-

---

[71]Elliptical whispering galleries are like Statuary Hall in that a person at one focal point can hear the whispers of a person at the other focal point.
[72]Geosynchronous satellites travel above earth's equator from west to east at an altitude of approximately 22,300 miles and at a speed matching that of earth's rotation. Thus they remain stationary in the sky in relationship to earth.

tigate data sets that are too large or too complex to understand comprehensively. Random samples are taken of the given data set and inferences are derived from these samples. These conclusions are qualified by margins of error or confidence intervals.

These methods can be misused in order to infer virtually anything (based upon the biases of the one making the study). You can make statistics *say anything you want it to say*.[73] In order to evaluate the reliability of any statistical study, one must ask two questions:

1. Was the sample taken both random and representative of the data set?
2. Are the conclusions qualified by margins of error or confidence intervals?

This branch of mathematics must be braced by a biblical ethic. This ethic furnishes a standard for honesty in: (1) revealing all pertinent assumptions regarding the data to be sampled, (2) measuring the data, (3) reporting the data, and (4) following the basic mathematical principles of statistical inference when stating the conclusions drawn from the data. Without these ethical standards, this discipline will run amuck. Scripture warns:

> Diverse weights and diverse measures, they are both alike, an abomination to the Lord (Proverbs 20:10).

> You shall do no injustice in judgment, in measurement of length, weight, or volume. You shall have honest scales, honest weights, ... (Leviticus 19:35-36).

7.7.7 PROBABILITY

Closely related to statistics, probability is the study of chance occurrences and probability theory seeks to formulate rules that describe random variations. A subset discipline under this study is the theory of combinations and permutations.[74] Some of the seminal historical thinkers in this realm were the Frenchmen Pierre de Fermat (1601-1665), Blaise Pascal (1623-1662),[75] and Pierre-Simon de Laplace (1749-1827).

As in statistics, understanding and using probability engages considerations of an ethical nature. Most people link probability to gambling and, historically, probability theories were originally developed in the context of games of "chance" (e.g., dice games). The gambling conglomerate (either private or State)

---

[73]See Darrell Huff, *How to Lie with Statistics* (New York: W. W. Norton, [1954, 1982] 1993).

[74]The theory of combinations refers to the number of possible ways of arranging objects chosen from a sample size of n if you do not care about the order in which the objects are arranged. Mathematically expressed, the number of n things taken j at a time is $n!/[(n - j)!j!]$ where ! stands for factorial (e.g., $5! = 5 \cdot 4 \cdot 3 \cdot 2 \cdot 1$). Permutations are like combinations except that order is important. In this case, the formula is $n!/(n - j)!$.

[75]It is significant to note that Pascal used a probability wager as an "apologetic" for the existence of God. Pascal's "wager" is as such: If you cast your lot on the side of God, then you have nothing to lose in this life and everything to gain in the life to come. But, if you deny God's existence, then you jeopardize yourself for all eternity should God actually exist. In essence, Pascal was challenging men to "gamble" their lives on a fifty-fifty chance that Christianity might be true (not a very satisfactory argument for belief in God).

makes use of probability theory, playing on the "get rich quick" paradigm embraced by the fools who inhabit their establishments, in order to ensure that the "house" always wins. The "get rich quick" worldview blinds the participant to the laws of probability (the odds are overwhelmingly against one obtaining this chimera). There are many biblical proverbs that speak directly or indirectly to this flight of fancy and its underlying rejection of the biblical work ethic:

> The soul of a lazy man desires, and has nothing; but the soul of the diligent shall be made rich (Proverbs 13:4).

> The desire of the lazy man kills him, for his hands refuse to labor. He covets greedily all day long, but the righteous gives and does not spare (Proverbs 21:25-26).

> Will you set your eyes on that which is not? For riches certainly make themselves wings; they fly away like an eagle toward heaven (Proverb 23:5).

> A man with an evil eye hastens after riches, and does not consider that poverty will come upon him (Proverbs 28:22).

While some people succumb to the fallacy of disregarding low probabilities in order to obtain beneficial outcomes, others foolishly discount the low probabilities of inexpedient events. Many people constantly disregard planning for such seeming contingencies (e.g., earthquakes, floods, tornadoes, accidents, health problems, death, etc.). When such events happen to the unprepared, the results are often tragic. Proverbs 22:3 (NIV) warns, "A prudent man sees danger and takes refuge, but the simple keep going and suffer for it." In this context, we must bear in mind that the future is *not* in the hands of random chance happenings. Although the theory of probability enables one to prepare (with the help of insurance companies) for uncertainties that are outside of our finite and incomplete predictive capabilities, we must know with certainty that the future, and whatever it may bring, is in the hands of God and His providence. The God who "numbers the hairs on our head and the stars in the sky" knows the future exhaustively (Luke 12:7; Psalm 147:4). Those who love Him can trust in His beneficial providence (Romans 8:28). God does not act according to the laws of probability for He governs even the casting of lots:

> The lot is cast into the lap, but its every decision is from the Lord (Proverbs 16:33).

This field is essential to business and science. For example, insurance companies make special use of these disciplines in the preparation of rate schedules (actuarial science) helping the provident to prepare for the uncertain future. In the logic of computer programs, an understanding of combinatorics is critical if one wants to code for all logical possibilities of given conditions.[76] In the biological and chemical fields, an understanding of these basic principles is foundational to research in genetics.

7.7.8 TRIGONOMETRY

The principle of triangulation is used to find the heights of buildings, trees, and mountains. It is essential to the professional surveyor. It is also used to compute astronomical distances and to construct maps.

Trigonometry is essential to the study and application of all types of oscillatory motion. We have already looked at its application to musical sounds. Other similar motions include the ebb and flow of alternating electric current, the amplitude and frequency modulation of radio waves, and spring and bob motion.

7.7.8.1 THE MOTION OF A BOB ON A SPRING

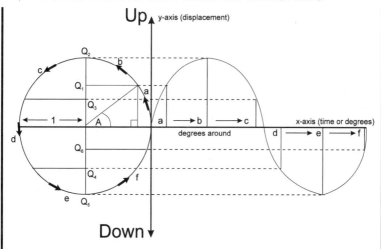

Figure 150: Spring and bob motion

Given: $\angle A$ in degrees. Then, $y = \sin A$.

To find a function in terms of time:

Suppose a point revolves around the unit circle (radius of 1) f times in one second. Then, for each revolution, $\angle A$ increases by 2π radians (the circumference of the circle when the radius equals 1) and the size of the angle which describes the amount of revolution of the point in one second is $2\pi f$.

---

[76]From the author's personal experience as a computer programmer and analyst, these conditions can be both numerous and complex. An understanding of combinatorics is a programmer's "lifesaver."

If the point revolves for t seconds and makes f revolutions per second, then it will make ft revolutions in t seconds. The angle generated during these ft revolutions with be $2\pi ft$. Hence, the value of $\angle$A in t seconds will be $2\pi ft$. Thus, $y = sinA$ or $y = sin(2\pi)ft$.

The maximum y-value is called the amplitude, D ($D = 1$ in this example). Hence, a complete mathematical description of the motion of a bob on a spring over time is: $y = D \, sin(2\pi)ft$.

Making deductions from Hooke's law, $F = kL$, where $F =$ force exerted by the spring, $k =$ constant for given spring representing its stiffness, and $L =$ stretching or compression length of the spring (amplitude), Newton's second law of motion, $F = ma$ where $F =$ force, $m =$ mass, and $a =$ acceleration ($32 \, ft/sec^2$), and $y = D \, sin(2\pi)ft$, we can conclude that the period, the time required for the bob to make one oscillation, is *independent* of the amplitude of the motion; i.e., whether one pulls the bob down a great distance or a short distance and then releases it, the time for the bob to go through each complete oscillation will be the same.

The motion will eventually die down due to air resistance and internal energy losses in the spring. Giving the bob a little push (adding energy to compensate this loss) will alter the amplitude, but *not* the period. Hence, these oscillations can be used to measure time or to regulate the motion of mechanical hands on a dial that would show time elapsed (i.e., a clock).

Robert Hooke (1635-1703), British scientist, developed a formula that described the quantitative relationship between a spring and an attached bob. The deductions based upon Hooke's law, basic trigonometric functions, and Newton's second law of motion, combined with the spiraled spring regulator patented by Christian Huygens (1629-1695) in 1675, laid the foundation for the solution of the greatest scientific problem of the 18th century. With these theoretical and mechanical tools, John Harrison (1693-1776), British carpenter and horologist, invented the first practical marine chronometer (the famous H4) that enabled navigators to compute accurately their longitude at sea.[77]

Harrison, whose singleness of purpose had made it possible for him to achieve what was impossible to other scientists of his era (including Newton), wrote:

I think I may make bold to say, that there is neither any other Mechanical or Mathematical thing in the World that is more beautiful or curious in texture than this my watch or Time-keeper for the Longitude ... and I heartily thank Almighty God that I have lived so long, as in some measure to complete it.[78]

---

[77]For the intriguing and fascinating story of how Harrison conquered the longitude problem, see Lloyd A. Brown, *The Story of Maps* (New York: Dover Publications, [1949] 1979), pp. 208-240 and Dava Sobel's *Longitude* (New York: Walker and Company, 1995).

[78]Cited in Rupert T. Gould, *The Marine Chronometer: Its History and Development* (London: J. D. Potter, 1923), p. 63.

With this tool, the seas became much safer to navigate. Not only did the marine chronometer aid European nations commercially, it also helped to expedite the burgeoning worldwide missionary thrust and, according to some modern horologists, this mechanical timepiece "facilitated England's mastery over the oceans, and thereby led to the creation of the British Empire – for it was by dint of the chronometer that Britannia ruled the waves."[79]

7.7.9 ALGEBRAIC EQUATIONS AND FUNCTIONS

Algebra is the essential language describing functional relationships. The real world is replete with such associations. For example, functional relationships are found in the mechanics of the pendulum, the pulley, and the lever. Driving, sailing, or flying distance is a function of speed and time. The length of an iron bar is a function of its temperature. Density is a function of mass and volume. Free fall motion is a function of time and distance (see Figure 151). Barometric pressure is a function of latitude and longitude at a certain location. Postage is a function of the parcel's weight. Water pressure is a function of depth. The speed of a satellite is a function of the diameter of its orbit. Air pressure changes in relationship to distance above sea level. Functional relationships describe the growth of bacteria, the cooling of liquids, the intensity and direction of light passing through a medium, the decay of sound, and the electrical conductivity of glass at various temperatures.

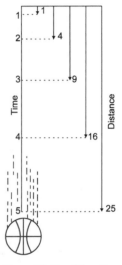

Figure 151: Free fall motion

For a parenthetical respite on the "lighter" side, one's waistline is a function of caloric intake minus vigorous exercise. The number of lies told by a politician is a function of the amount of time his lips are moving.

7.7.10 COMPLEX AND IMAGINARY NUMBERS

Complex numbers (of the form a + bi where the imaginary number i = $\sqrt{-1}$) were first identified by the Italian mathematician Girolamo Cardano (1501-1576) in his *Ars magna* (1545) and further developed by Carl Friedrich Gauss (1777-1855). Cardano referred to these numbers as "fictitious mental tortures" and they *appear* to be useless at first sight. Yet, they provide the mathematician with a powerful tool. These numbers describe the behavior of forces, velocities, and accelerations that act in definite directions. They are especially useful in the algebra of vectors, the study of how two forces, velocities or accelerations, are exerted from different directions on the same mass producing a third force in a new direction. For example, anyone with experience in shooting knows that on

---

[79]Sobel, p. 153.

a windy day a rifleman must compensate for the force of the wind by aiming the gun slightly "into" the direction of the wind. The vector analysis of complex numbers perfectly describes this situation.

These numbers also appear, and are indispensable, in certain map making procedures, electrical alternating circuit theory, and, as already mentioned, quantum mechanics.

7.7.11 GEOMETRY

Geometrical understanding illumines the methods of optics, cartography, and building construction. Solid geometry is used extensively by mechanical, electrical, civil, and architectural engineers. Simple machines reflect geometric principles in the design of gears, cams, cranks, screws, and pulleys. The fundamentals of elementary geometry are prerequisites not only for the practical purposes of art and industry, but also for the study of other branches of the mathematics tree. The deductive edifice developed by Euclid provides excellent training in refining and developing both the intuitive and logical thought processes. Why? A geometric argument requires concentration and comprehension. In the education of modernity, this Euclidean discipline has been too often neglected. J. L. Heilbron notes, "For many centuries, a good education included the study of Euclidean geometry. Most of the so-called geometry books now used in schools are not agents of culture, wisdom, or even education, however, but opportunities to exercise students in the use of pocket calculators."[80]

7.7.12 THE DIVINE PROPORTION

The number φ has been deliberately used in art, architecture, and even in the musical works of greats like Bach and Beethoven.[81]

7.7.13 SPIRALS

Man has made use of many different types of spiral helices. They are found in the spiral circular staircase, cables, screws, bolts, thermostat springs, nuts, ropes, and the familiar barbershop pole cane.

7.7.14 CRYSTALLOGRAPHY

Quartz crystals are used in watches, videocassette recorders, washing machines, pilotless stoves, microwave ovens, dishwashers, and automobile engines. They are used in sonar to convert sound waves into electrical signals. In the future, these geometric gems might be used as warning devices for impending earthquakes, as detectors of structural imbalances in a building or bridge, as

---

[80]J. L. Heilbron, *Geometry Civilized: History, Culture, and Technique* (New York: Oxford University Press, [1998] 2000), p. v.

[81]For more detail about the Divine Proportion, see Huntley, Matila Ghyka, *The Geometry of Art and Life* (New York: Dover Publications, [1946] 1977), and Roger Herz-Fischler, *A Mathematical History of the Golden Number* (New York: Dover Publications, [1987] 1998).

mechanisms for converting ocean waves to electrical energy, and as voice and pen recognition devices for personal identification.

7.7.15 LINEAR PROGRAMMING

This is a technique that uses algebra and geometry in solving problems that deal with efficient use of resources. A business can use this method to find the minimal cost for producing a given commodity or the revenue-maximizing output mix for a production facility with several capacity limitations.

7.7.16 THE CALCULUS

This method is applicable to many construction problems. How much metal would be needed to build a spherical tank? How much sheet metal is needed in the body of an automobile? How do you construct the wings of a jet? To what degree of inclination should you make a road when it passes through a mountain range? Calculus provides the answers.

The method can also be used to find the volume of an irregular container, to measure structural stress, to measure the ebb and flow of current in an electrical line, and to determine the build up of water force behind a dam. It is essential to the study of projectile motion, especially in finding the velocity necessary for a projectile to escape the earth's gravitational pull.

7.8 DISCIPLE ALL NATIONS

Is there any connection between mathematics and evangelism? In our historical survey, we have seen that the answer to this question is a resounding yes!

For example, at the time of Christ, the mathematicians in Alexandria, Egypt had developed quite a storehouse of useful navigational knowledge. Available in Egypt were the technological tools for reaching the ends of the earth with the Gospel. It was only as a result of the Reformation that the people of God began to use these navigational tools to do just that.

Today, missiology is combining statistical research with high-speed computer technology to map out the unreached peoples of the earth. Much of this work is being done by Dr. Ralph Winter and his associates at the United States Center for World Mission in Pasadena, California.

In 1976, the author discussed the issue of world missions with a missionary from Wycliffe Bible Translators. An interesting statement surfaced in the conversation: "The best training for linguistics is a mind trained in mathematics."

Language, like mathematics, is a structure. Laws govern its use. A person with a mathematical bent is best able to decipher the grammatical structure of an unwritten tribal language. The structure of language cannot be measured with a stopwatch or ruler. Hence, numerical methods would not apply. It has been shown that non-numerical mathematical methods such as set theory, modern abstract algebra, topology,[82] and mathematical logic can be applied to

---

[82]Topology is the study of the properties of geometric figures or solids that are not normally affected by changes in size or shape.

the task of Bible translation. These mathematical formulations are powerful enough to deal with the structured relationships found in the complexity of linguistic structures.

Ivan Lowe, a member of Wycliffe, comments:

> Our first results two years ago on group theory applied to pronouns and discourse looked to be very much "up in the clouds." They seemed to perfectly fit the pure mathematician's toast, "Here's to pure mathematics, may it never be of any use to anybody!" And yet within two years the very same theory is being applied to languages as far apart from each other as Bolivia, Peru and West Africa, and it is helping us understand better than before how stories are put together to sound meaningful in these languages.[83]

Wycliffe missionaries have also developed a full range of computer software that enables missionaries to take laptop computers directly to the field to develop dictionaries and grammars. This software is efficiently and dramatically shortening the time it takes to translate an unwritten tongue into the words of Scripture.

The Far East Broadcasting Company (F.E.B.C) has used linear programming techniques to help them to maximize their financial resources, personnel, strategy, program effectiveness, and use of broadcasting resources. Variables that come into play for efficient broadcasting include strength of available signal, size and language of target people groups, and hours available in the day.

Finally, many people live outside the circumference of classical missionary outreach. The university campuses of the world (including America!) are key mission fields where idea confronts idea. Evangelical groups like Campus Crusade, the Navigators, and InterVarsity are effective only to a degree and they are essentially peripheral in their impact. In addition, they are not able to penetrate countries officially closed to classical missionaries. Mathematical skills in research and teaching can be a passport enabling biblical Christians to penetrate these closed areas in any country, home or abroad, and to confront false philosophies head to head.

And *with what* do we challenge these false philosophies? Obviously, the truncated, impotent "gospel" of modern evangelicalism will not suffice.[84] False philosophies must be confronted with a robust, comprehensive, and systematic faith, a faith wide enough to speak to all of life. In the words of Rousas J. Rushdoony:

> But the gospel is for all of life: the good news is precisely that the whole of life is restored and fulfilled through Jesus Christ, that, in the counsel of God, the kingdom is destined to triumph in every sphere of life. This gos-

---

[83]Ivan Lowe, "Christian Mathematician, Where Are You?" *Translation* (January-March 1971), 7.
[84]To grasp how weak modern evangelicalism has become, we must understand its roots. See Douglas W. Frank, *Less Than Conquerors: How Evangelicals Entered the Twentieth Century* (Grand Rapids: Eerdmans, 1986) and Iain Murray, *Revival & Revivalism: The Making and Marring of American Evangelicalism 1750-1858* (Edinburgh: The Banner of Truth Trust, 1994).

pel cannot be proclaimed and the dominion of the kingdom extended except on Christian presuppositions. The answer to the question, how wide a gospel do we have, is simply this: as wide as life and creation, as wide as time and eternity. It rests in the decree of the self-contained and autonomous God; it is a faith grounded on a truly systematic theology.[85]

QUESTIONS FOR REVIEW, FURTHER RESEARCH, AND DISCUSSION

1. Thoroughly explain the biblical worldview of mathematics.
2. Explain three Christian approaches to mathematics education.
3. Develop the four objectives of a biblical Christian mathematics curriculum.
4. Explain how the revelation of the triune God answers the problem of the "one and the many."
5. Explain how the revelation of the triune God forms the foundation for understanding the true nature of "unity in diversity."
6. Explain the following statement, "No fact can be known truly unless it is interpreted in the context of and in complete dependence upon the ontological trinity."
7. Give as many examples as possible (which are *not* enumerated in the text) of the "unity in diversity" principle found in God's creation and in the structure of mathematics.
8. Explain how Christ speaks to both the content and philosophy of mathematics.
9. Present a biblical Christian perspective on the "laws of nature."
10. Explain how to relate the dominion mandate, in the ethical and cultural context, to the "healing of the nations."
11. Explain the ethical underpinnings that one must embrace in the study of statistics and probability.
12. Present an apologetic and evangelistic methodology whereby, beginning with mathematics, you can defend Christianity as a system of philosophy and proclaim the Gospel to a non-Christian.

---

[85]Rushdoony, *By What Standard?*, p. 176.

8: PEDAGOGY AND RESOURCES

The over-arching message of our mathematics courses should proclaim that mathematics is all about exploring and forming God's creation.[1]

Don Van Der Klok

Alfred North Whitehead was not only an erudite mathematician, but, as an educator, he was keenly interested in pedagogical principles. He observed:

> There is a widely-spread sense of boredom with the very idea of learning. I attribute this to the fact that they [students - J.N.] have been taught too many things merely in the air ... The whole apparatus of learning appears to them as nonsense.[2]

Making direct application to mathematics, Morris Kline has come to grips with some vital issues:

> I believe that the chief problem in teaching mathematics is motivation. It is difficult for beginners in mathematics to see just why one should study the subject. Vague answers, such as that scientists and engineers use it, or that it is needed to get a college degree, don't help at all. The motivation was sadly deficient in the traditional curriculum, and is entirely absent in the modern mathematics curriculum. By utilizing real problems chosen from

---

[1]Don Van Der Klok, "A Christian Mathematics Education?" *The Christian Educators Journal* (February 1983), 27.
[2]Alfred North Whitehead, *Essays in Science and Philosophy* (London: Rider and Company, 1948), p. 133.

the world in which the student lives and involving phenomena which he himself experiences, we may be able to motivate the study of mathematics. We would all agree that we want our subject to make sense to the students. The road to making sense is through the senses, and this means physical problems and situations.[3]

It is essential that mathematics appeal to the student *at the time he takes the course*. In Christian education, there is a true and therefore broader context for this appeal.

Larry Zimmerman warns:

A Christian school that is content only with the teaching of manipulatory skills of arithmetic, algebra, and geometry blinds the student's perception to all but a fraction of the glory of God reflected in the unique mirror of mathematics.[4]

Jay Adams, in his book *Back to the Blackboard*, queries:

What teacher in a thousand teaches mathematics for use in actual situations? Who demonstrates its use in real life while inviting his students to observe? Who uses it for life and ministry, demonstrating how practical for the kingdom of God knowledge of the multiplication table can be?[5]

A true answer to Adams' interrogations would be, "Hardly any." Why?

8.1 WEAKNESS IN MODERN MATHEMATICS INSTRUCTION

All of us have been reared on a diet of humanistic education. In the Christian classroom, it is extremely hard to break free from this habitual mindset. According to humanistic presuppositions, mathematics is just a mental construct or an intellectual game, an attempt by man to create order out of a multiverse of chaos. The essential emphasis is upon the autonomy of man's mind. The fruits of this type of thinking can be picked in most mathematics classrooms where mathematics is generally isolated and taught for the sake of mathematics alone. Mathematical thought is divorced from the real world except for occasional non-sequitur applications in exercises. The student who is taught mathematics in this context ends up sighing in disgust, "This is boring."

Ask any student why he has to take mathematics, and he will usually reply somberly:

"Because I have to," or "It is required for university."

Ask any teacher why he teaches mathematics, and he will usually reply, and in a very defensive tone:

---

[3]Morris Kline, "A Proposal for the High School Mathematics Curriculum," *The Mathematics Teacher* (April 1966), 329-330.
[4]Cited in Haycock, p. 101.
[5]Jay Adams, *Back to the Blackboard* (Phillipsburg: Presbyterian and Reformed, 1982), p. 97.

"Mathematics teaches logic," or "Mathematics is historically significant," or "Mathematics is practical."

Apparently, the student is not getting the message. Why? The technique employed by most mathematics teachers is not getting the point across.

8.2 BIBLICAL MOTIVATION

Mathematics does teach logic, mathematics is historically significant, and mathematics is practical. Yet, one vital ingredient is missing. The proper context for true motivation is the context of God's creation. Joseph Fourier remarked, "The profound study of nature is the most fertile source of mathematical discoveries."[6] Not only does God's creation provide the soil for mathematical discoveries, it also, in the words of Richard Courant, supplies the student "powerful motivation and inspiration."[7]

A subject that inspires a student will motivate him. The mathematician John von Neumann said, "Some of the best inspirations of modern mathematics (I believe, the best ones) clearly originated in the natural sciences."[8] W. E. Chancellor, educator, said, "The motive for the study of mathematics is insight into the nature of the universe. Stars and strata, heat and electricity, ... incorporate mathematical truths."[9]

To affirm a relationship to the physical reality of creation, the very life source of mathematics, is to ultimately affirm accountability to the Creator of that reality. This is the underlying reason why most mathematical pedagogy divorces itself from the context of creation. And, tellingly, when stress is laid upon the study of nature, it is the study of *nature* that is emphasized, not the study of *creation*. A profound and deep study of *creation* is a pathway that points the teacher and student directly to the Creator.

According to Scripture, the unbeliever in his unrighteousness is in an active state of demolishing this pathway linking the creation to God (Romans 1:18-21). Because of this condition, the road of science does not and cannot automatically lead the unbeliever to God contra to what Stanley L. Jaki affirms in the title of one of his truly profound works – *The Road of Science and the Ways to God*. This title reflects Jaki's uncritical respect and dependence upon the thinking of Thomas Aquinas (I understand [the road of science] in order to believe [the ways to God]). Biblically, the only way for the unbeliever to see and understand the true nature of creation is for his blinded eyes to be opened. Blind eyes are opened only through the regenerating power of the Holy Spirit applying the finished work of Christ's atonement to the heart of man as a free gift of grace.

The biblical Christian, being saved by God's grace through faith in Christ – the whole package a gift from God so that no man can boast (Ephesians 2:8-9),

---

[6]Fourier, p. 7.
[7]Courant, "Mathematics in the Modern World," 48.
[8]Neumann, 4:2054.
[9]W. E. Chancellor, *A Theory of Motives, Ideals and Values in Education* (Boston and New York, 1907), p. 406.

will walk on those scientific roads by the same faith (Colossians 2:6). The journey of transforming mind renewal (Romans 12:1-2; Ephesians 4:17-24) will always lead to the revelation of the marvelous and wonderful ways of God; i.e., the wisdom (*logos*) of Christ. For the biblical Christian, the gift of faith saves his reason (he believes so that he can truly, but not exhaustively, understand). As Cornelius Van Til states:

> Here is thinking done on the basis of the self-authenticating revelation of God. Here is a theology in which the primacy of faith over reason means that reason or intellect is saved from the self-frustration involved in the denial, virtual or open, of such a God and of such a Christ. Only those who know that they are not infallible, but are, by virtue of ever present sin within them in spite of their regeneration by the Holy Spirit, inclined to suppress this revelation, also know that they need such a God, such a Christ and his infallible word to tell them the truth which alone can set them free. For theirs is the knowledge that only by having such a God as their personal God does their search for knowledge have any meaning.[10]

Biblical faith provides the *only* rational basis for thinking; it is the *only* fuel that powers the scientific and mathematical reasoning engine. It is the biblical faith that enables man to discern in truth the patterns (the *logos* and wisdom of Christ) and map the road of science in a universe of God's making.

The unbeliever travels the same road of science. Violating God's copyright laws, he fills his tank of reason with the fuel of Christian theism; i.e., using the gift of Christ's enlightenment (John 1:9), he discerns the patterns of creation without honoring Christ as the source of those patterns. His journey along the road of science is thereby interrupted by thousands of self-placed detour signs (marked "fools ➡"), each of which leads the unsuspecting traveler into a blind alley of his own making (Romans 1:21-28). These signs are so numerous that the entire horizon is eclipsed. The traveler is incapable of seeing the *suntéleia* (συντελεια), the consummated goal, of the trip and can only envision and map the pseudo-goal of pure pragmatism.[11] This is why when the unbeliever emphasizes the need to put mathematics in the context of the study of nature he is *always* either silent or obscurant about the Creator of that nature. In the consequent construction of the pedagogical superstructure, nature and reason become the center of truth instead of the edge. In the words of Glenn R. Martin, "If we do not love God, we shall forever be at the edge of Truth, and to crown our folly, we shall view the edge of Truth as the center."[12]

---

[10]Cornelius Van Til, *The Case for Calvinism* (Nutley, NJ: The Craig Press, 1964), p. 24.

[11]Descartes equated empirical science with truth. For the pragmatic post-Kantian world of modern education, truth is what works.

[12]Glenn R. Martin, "Biblical Christian Education: Liberation for Leadership – An Address Given to the Marion College Faculty," *Marion College* (August 30, 1983), 16.

Hence, through the deliberate omission of the creation context, most mathematics teaching today denies the reality of the Creator God.[13] Mathematics is *not* neutral. Christ said, "He who is not with Me is against Me...." (Luke 11:23). The only way for mathematics to be taught *truly* is for mathematics to be taught *deliberately* out from the foundation of the Creator God and His revelation in His word and His works.

True motivation and inspiration for mathematics lies in the observance of God's created order. The wonder of creation, as we have seen, reveals a God *who is not boring*. Hence, true mathematics teaching should be taught in a room, not with mirrors on the walls, but with windows wide open to the outside world.

To teach mathematics just for the sake of mathematics isolates the subject from much of its inspiration, versatility, and power. The relationship of mathematics to other disciplines like science, history, art, and economics is not one of isolation, but one of dynamic interaction.

8.3 TEACHER PREPARATION

The saying goes that "teachers are a dime a dozen, but a good teacher is worth his weight in gold." A good teacher is a teacher who is excited about what he teaches. If what you teach thrills your heart, then your students cannot help but catch the same "germ."

The prophet Isaiah, before the throne of God, heard the seraphim crying, "Holy, holy, holy is the Lord of hosts; the whole earth is full of His glory" (Isaiah 6:3). The apostle John, in the spirit on the Lord's day, heard the twenty-four elders fall before Him who sits on the throne, saying:

> You are worthy, O Lord, to receive glory and honor and power; for You created all things, and by Your will they existed and were created (Revelation 4:11).

True motivation, either in teaching or in learning, is generated by worship. As teachers, we need to open up the windows of our spirit and mind to behold the glory of God in creation. There is a treasure to be found in the mathematical structure of God's creation. We only need to learn to how to find it and worshipfully appreciate it.

As an aid to help you find this treasure, subscribe to mathematics or science journals. Study physics, chemistry, and biology textbooks. Become knowledgeable of science. In 1961, Morris Kline observed:

> The primary value of mathematics is that it is the language and essential instrument of science. High school mathematics teachers will have to become somewhat better informed in science. Only about 5%, roughly, of

---

[13]This omission is self-consciously deliberate in a minority of cases (by unbelievers who know what they believe and why) and unconscious (unthinkingly blind) in the majority of cases.

the present high school teachers have any idea of how mathematics is used in science. Most haven't any notion of why anyone wants mathematics.[14]

The percentage of mathematics teachers who know how mathematics is used in science may be a little higher today, but not very much.

Keep your eyes open for mathematical insights in newspapers and magazines. Get out into God's creation and investigate. Take pictures and collect flowers, pine cones, shells, etc.

8.4 TEACHING PRINCIPLES

Before delving into teaching techniques, it must be noted that some secular educators are also striving to reform the mathematical curricula along the same lines delineated by the author (minus the biblical foundation). The author heartily approves of such renovation and encourages the reader to plunder such material where appropriate (especially where the material interacts with science and history) recognizing the gift of God's common grace to the unbeliever. Because of this gift, even in an unbeliever's rebellious rejection of God as the foundation for all knowledge, much of their material is valid and true, albeit upside down. As believers, we only need to turn their work right side up, making God instead of man the center of it all. Then, and only then, will we be able to see the wonderful and marvelous facts of creation as God intended us to see them; that everywhere and in every fact we see His sign posted saying, "It is Mine."

8.4.1 MODELING

A mathematical model is an abstract construction derived from a concrete, physical situation. The ability to think abstractly is a gift of God. Without this gift, doing mathematics would be impossible.

A cursory glance at a dictionary reveals that abstract words, the names of things that cannot be seen, come from words for actual objects or actions. For example, the word "understand" in German and English comes from the idea of "standing under." In French, it comes from the idea: "Can you take hold of that?"

In mathematics and science, abstraction is the process of stripping away nonessential and unimportant details and considering only those properties necessary to the solution of a problem. The process of abstraction always starts with a concrete, physical problem. From a concrete foundation, mathematicians and scientists either diagram the problem or restate it in mathematical equations. In doing so, they leave out all details except those absolutely necessary to solve the problem.

---

[14]Morris Kline, "Math Teaching Reforms Assailed as Peril to U.S. Scientific Progress," NYU *Alumni News* (October 1961).

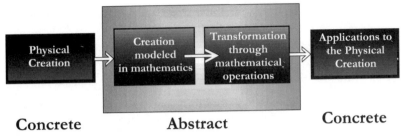

| Physical Creation | Creation modeled in mathematics | Transformation through mathematical operations | Applications to the Physical Creation |

Concrete   **Abstract**   **Concrete**

Figure 152: The abstract/concrete connection

Abstract thought has concrete foundations. The problem in most of today's mathematics textbooks and instruction is that both tend to absolutize the abstract construction, the axiomatic-deductive approach, and ignore the concrete foundations. Without this foundation, all our students see and hear is an endless and "logical" procession of meaningless signs, words, and rules; mere chicken scratches, for want of a better word. No wonder children are bored with the whole process! Richard Courant and Harold Robbins made these remarks in 1941:

> There seems to be a great danger in the prevailing over-emphasis on the deductive-postulational character of mathematics. True, the element of constructive invention, of directing and motivating intuition, is apt to elude a simple philosophical formulation; but it remains the core of mathematical achievement, even in the most abstract fields. If the crystallized deductive form is the goal, intuition and construction are at least the driving forces. A serious threat to the very life of science is implied in the assertion that mathematics is nothing but a system of conclusions drawn from definitions and postulates that must be consistent but otherwise may be created by the free will of the mathematician. If this description were accurate, mathematics could not attract any intelligent person. It would be a game with definitions, rules, and syllogisms, without motive or goal. The notion that the intellect can create meaningful postulational systems at its whim is a deceptive half-truth. Only under the discipline of responsibility to the organic whole ... can the free mind achieve results of scientific value.[15]

In essence, these eminent mathematicians are stating that mathematics must be placed under a discipline of responsibility to its concrete foundations; i.e., the physical world. Morris Kline confirms this:

> Nature is the matrix from which ideas are born. The ideas must then be studied for themselves. Then, paradoxically, a new insight into nature, a

---

[15]Courant and Robbins, p. xvii.

richer, broader, more powerful understanding, is achieved, which in turn generates deeper mathematical activities.[16]

Speaking in the context of the trigonometric ratios being placed under the influence of the functional concept made potent by development of symbolic algebra, Alfred North Whitehead compares abstractions with concrete fact:

> The science of trigonometry arose from that of the relations of the angles of a right-angled triangle, to the ratios between the sides and the hypotenuse of the triangle. Then, under the influence of the newly discovered mathematical science of the analysis of functions, it broadened out into the study of the simple abstract periodic functions which these ratios exemplify. Thus trigonometry became completely abstract; and in thus becoming abstract, it became useful. It illuminated the underlying analogy between sets of utterly diverse physical phenomena; and at the same time it supplied the weapons by which any one such set could have its various features analysed and related to each other. Nothing is more impressive than the fact that as mathematics withdrew increasingly into the upper regions of abstract thought, it returned back to earth with a corresponding growth of importance for the analysis of concrete fact.... The paradox is now fully established that the utmost abstractions are the true weapons with which to control our thought of concrete fact.[17]

For some specific, historical examples of the power of abstraction, Kline continues:

> Mathematics, by abstracting concepts and properties from the physical objects, is able to fly on wings of thought beyond the sensible world of sight, sound and touch. Thus mathematics can "handle" such "things" as bundles of energy, which perhaps can never be qualitatively described because they are apparently beyond the realm of sensation. Mathematics can "explain" gravitation, for example, as a property of a space too vast to visualize. In like manner mathematics can treat and "know" such mysterious phenomena as electricity, radio waves, and light for which any physical picture is mainly speculative and always inadequate. The abstractions, that is the mathematical formulas, are the most significant and the most useful facts we have about these phenomena.[18]

To teach effectively, the coherence between abstract mathematical models and concrete reality must be emphasized. The parables of Jesus illustrate this coherence beautifully. Jesus used parables as concrete illustrations taken from daily life or nature in order to explain abstract truth. The abstract truth is contained, like a gold nugget, in the concrete illustration.

---

[16]Kline, *Mathematical Thought from Ancient to Modern Times*, p. 205.
[17]Whitehead, *Science and the Modern World*, p. 41.
[18]Kline, *Mathematics in Western Culture*, p. 465.

This methodology can be readily applied to mathematics instruction. Seek to present the concrete as a foundation for the exposition of the abstract. The abstract principle is in the concrete contained; the concrete is in the abstract explained. Abstract mathematical concepts should always be presented in the context of the study of real, and largely physical problems.

8.4.2 INTEGRATION AND HISTORY

Early in the 20th century, Alfred North Whitehead identified one of the chief causes of the weakness of what is today known as the traditional mathematics curriculum:

> It is entirely out of relation to the real exhibition of the mathematical spirit in modern thought, with the result that it remained satisfied with examples which were both silly and unsystematic. Now the effect which we want to produce on our pupils is to generate a capacity to apply ideas to the concrete universe.[19]

He continues:

> There can be nothing more destructive of true education than to spend long hours in the acquirement of ideas and methods which lead nowhere.[20]

Anyone who has seen, let alone afford, a college mathematics textbook comes away with the impression that mathematics consists simply of a mass of dry definitions, theorems, and formulae. Philip J. Davis and Reuben Hersh, two seasoned mathematics professors, comment with sagacity on this state of affairs:

> The presentation in textbooks is often "backward." The discovery process is eliminated from the description and is not documented. After the theorem and its proof have been worked out, by whatever path and by whatever means, the whole verbal and symbolic presentation is rearranged, polished and reorganized according to the canons of the logico-deductive method. The aesthetics of the craft demands this. Historical precedents – the Greek tradition – demand it. It is also true that the economics of the book business demands maximum information in minimum space. Mathematics tends to achieve this with a vengeance. Brevity is the soul of mathematical brilliance or wit. Fuller explanations are regarded as tedious.[21]

Whitehead gives directives for reform:

> … eradicate the fatal disconnection of subjects which kills the vitality of our modern curriculum. There is only one subject-matter for education,

---

[19]Whitehead, *Essays in Science*, p. 135.
[20]*Ibid.*, p. 133.
[21]Davis and Hersh, *The Mathematical Experience*, p. 282.

and that is Life in all its manifestations. Instead of this single unity, we offer children Algebra, from which nothing follows; Geometry, from which nothing follows....[22]

The best way to foster integration is to teach concepts in their historical context. Walter W. Sawyer (1911-), professor in mathematics, comments, "Mathematics is like a chest of tools: before studying the tools in detail, a good workman should know the object of each, when it is used, how it is used, what it is used for."[23] History gives us this knowledge. In the words of mathematics historian Carl Boyer:

> No scholar familiar with the historical background of his specialty is likely to succumb to that specious sense of finality which the novitiate all too frequently experiences. For this reason, if for no other, it would be wise for every prospective teacher to know not only the material of his field but also the story of its development.[24]

To divorce mathematics from history is to divorce mathematics from life. The mathematics teacher must not only know science, but also the historical flow of mathematical thought. As an instructor in the history of mathematics, Eli Maor confirms the importance of using history in mathematical pedagogy:

> As one who has taught at all levels of university instruction, I am well aware of the negative attitude of so many students toward the subject. There are many reasons for this, one of them no doubt being the esoteric, dry way in which we teach the subject. We tend to overwhelm our students with formulas, definitions, theorems, and proofs, but we seldom mention the historical evolution of these facts, leaving the impression that these facts were handed to us, like the Ten Commandments, by some divine authority. The history of mathematics is a good way to correct these impressions. In my classes I always try to interject some morsels of mathematical history or vignettes of the persons whose names are associated with the formulas and theorems.[25]

History teaches us that the first concepts to be accepted and utilized by mathematicians were those concepts that were most appealing to the intuition; concepts like counting numbers, fractions, and plane and solid geometry. Those concepts that defied the intuition were developed and used later; concepts like irrational numbers, negative numbers, complex numbers, algebra, and the calcu-

---

[22]Whitehead, *The Aims of Education and Other Essays*. p. 10.
[23]Walter W. Sawyer, *Mathematician's Delight* (Harmondsworth Middlesex: Penguin, 1943), p. 10.
[24]Boyer, *The History of the Calculus and Its Conceptual Development*, preface to the second printing.
[25]Maor, *e: The Story of a Number*, pp. xiii-xiv. Concerning Maor's "historical evolution" concept, it is valid in that it reflects man's historical growth in the knowledge of the ordered patterns of God's universe. It is invalid in its omission of the Creator God as the author and the ultimate teacher of man in this quest for the categorization of such patterns.

lus. These concepts were only accepted because they explained some aspect of the physical world.

8.4.3 INDUCTION AND DEDUCTION

Inductive thinking is a method of reasoning in which the study of particular facts leads to general conclusions. It can also be denoted as empirical thinking and corresponds to *a posteriori* or "look and see" epistemology. Deductive thinking starts with basic axioms, or the general, and reasons to particular conclusions. This way of thinking corresponds to *a priori* or "stop and think" epistemology. As we have seen, man-centered approaches to the theory of knowledge see either induction or deduction as the *sole* pathway to truth. For the biblical Christian, the fear of the Lord, not induction or deduction, is the substructure of knowledge (Proverbs 1:7). Absolutizing either induction or deduction as the sole substructure of knowledge is erroneous because both procedures would endeavor to understand the facts of God's universe *without God*. A biblical Christian epistemology will use induction and deduction as *tools*, not divining rods, to aid in the understanding of the God-created, God-dependent, and God-sustained facts of the universe.

Today, mathematics is seen as a deductive science. Historically, the application of deductive logic has always followed the trail of inductive observation. Deductive thinking deals with the abstract while inductive thinking deals with the concrete.

In mathematics, induction and deduction complement each other. Neither approach should be absolutized. Unfortunately, the deductive approach reigns supreme today. Morris Kline comments:

> The formal logical style is one of the most devitalizing influences in the teaching of school mathematics. An ordered logical presentation of mathematics may have aesthetic appeal to the mathematician but serves as an anaesthetic to the student.[26]

The basic approach should be induction first, deduction last. In teaching, kindle motivation in the mind of the student by presenting mathematical principles in the form of pictures, analogies, physical, and heuristic arguments. Act upon the fact that God has gifted the human mind with the capacity to observe, explore, formulate, explain, and predict. Deductive logic should be used as the final step in the verification of your results; it should act as a control and guide. Then, it can be used as a launching pad to further applications. Albert Einstein's comments about the discovery of physical laws is apropos at this point:

> The supreme task of the physicist is to arrive at those universal elementary laws from which the cosmos can be built up by pure deduction. There is

---

[26]Kline, *Why Johnny Can't Add: The Failure of the New Mathematics*, p. 61.

no logical path to these laws; only intuition, resting on sympathetic under-standing of experience, can reach them.[27]

To this conclusion add the thoughts of Johann Friedrich Herbart (1776-1841), German philosopher and educator:

The great science [mathematics – J.N.] occupies itself at least just as much with the power of imagination as with the power of logical conclusion.[28]

8.4.4 KNOWLEDGE, UNDERSTANDING, AND WISDOM

Scripture speaks amply of the interplay between knowledge, understanding, and wisdom. Here are a few examples:

And Moses said to the children of Israel, "See, the Lord has called by name Bezalel the son of Uri, the son of Hur, of the tribe of Judah; and He has filled him with the Spirit of God in wisdom, and understanding, in knowl-edge and all manner of workmanship, to design artistic works, to work in gold and silver and bronze, in cutting jewels for setting, in carving wood, and to work in all manner of artistic workmanship" (Exodus 35:30-33).

The fear of the Lord is the beginning of knowledge, but fools despise wis-dom and instruction (Proverbs 1:7).

For the Lord gives wisdom; from His mouth come knowledge and under-standing; He stores up sound wisdom for the upright; He is a shield to those who walk uprightly (Proverbs 2:6-7).

The fear of the Lord is the beginning of wisdom, and the knowledge of the Holy One is understanding (Proverbs 9:10).

Wise people store up knowledge (Proverbs 10:14).

Blessed be the name of God forever and ever, for wisdom and might are His. And He changes the times and the seasons; He removes kings and raises up kings; He gives wisdom to the wise and knowledge to those who have understanding (Daniel 2:20-21).

The path to knowledge, understanding, and wisdom commences with the fear of the Lord. The foundation for a genuine knowledge of anything is a humble and worshipful acknowledgment of God as the creative and sustaining bedrock of every aspect of each particular fact of the universe (inanimate and animate). The learning process builds upon this bedrock presupposition. First, knowledge is learning the rudiments or particulars of a subject (e.g., multiplica-tion table, calculation rules, etc.). Second, understanding is the ability (1) to ap-ply the biblical worldview to unerringly "account for counting" thereby discern-ing truth from falsehood and (2) to be able to relate one aspect of knowledge to

---

[27]Einsten, *Essays in Science*, p. 4.
[28]Cited in Moritz, p. 31.

another and to grasp how the particulars relate to universals (seeing unity in the diversity of things). Third, wisdom is the ability (1) to effectively arrange and articulate knowledge and understanding, (2) to apply knowledge and understanding to a new situation, and (3) to reposit more knowledge and understanding based upon the previous two steps.

Applying the above discussion generically to mathematics education, a mathematics teacher should not teach just mere facts or skills (the particulars of the subject). Unthinking, robotic computation does not lay the foundation for mature understanding and wisdom. Knowledge by itself is worthless unless it is linked with understanding.

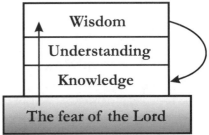

Figure 153: Building blocks to wisdom

The teacher should provide a foundation not only for understanding, but also for wisdom. The student should learn how to do mathematics (knowledge), why the methods of mathematics work (understanding), and how to apply these methods to new situations (wisdom).

We can apply this learning process to specific disciplines in mathematics; e.g., algebra, geometry, trigonometry, and elementary calculus. The first step is to learn the rudiments of the discipline. The second step is to learn why these rudiments work, how they have worked in history and science, and how they relate to other mathematical disciplines (e.g., the interplay between algebra and geometry in the form of coordinate geometry). The third step is to apply these rudiments to new areas of knowledge (using history and science as a guide). On the secondary level (year levels 7-12), students are ready for this type of study, analysis, and application and this is why, traditionally, these four subjects are taught at these age levels. Year levels K-6 are preparatory years; years in which the student should first focus on learning the rudiments (knowledge) of arithmetic (year levels K-3) with a gradual and incremental emphasis on understanding and wisdom (year levels 4-6) by showing how algebra, geometry, and trigonometry build upon arithmetical foundations.

8.5 ANNOTATED RESOURCES

Concerning resources, the author is tempted to say, "Just refer to the bibliography." Yes, a rich gold mine is located there, but, for the sake of service to the reader, the author will extract and polish a few of the nuggets contained therein.

Before we unearth some of these treasures, we must recognize that mathematics, like every other academic discipline, needs to be thoroughly rewritten within a biblical framework. Christian scholars should not be content to merely "baptize" secular textbooks; they should be the leaders in educational textbook writing. Although biblical Christian scholars have done some work in mathematics, 99% of all histories, textbooks, and other resources in mathematics are by authors who write from a humanistic presuppositional base. The author has carefully screened the following list of books and considers them useful in the Christian school context. But, *it must be emphasized that virtually no popular work in existence gives honor to Whom honor is due in the realm of mathematics.* A few good books are out of print and are so indicated. Many times, these books are reprinted. It would be well worth the effort to keep an eye out for them. All other books are known to be in print.

8.5.1 WORLDVIEW ANALYSIS

The epistemological premise of a self-consciously biblical Christian must be the Living God as revealed by Holy Scripture. A biblical Christian believes that knowledge is possible *only* because God created all things. God created a real and objective world. He also created our minds in His image and so established them so that we can know and verify knowledge. Because of the infectious nature of sin, our minds are hampered in this knowledge quest. Redemption of the heart by grace through Christ and renewal of the mind in God's word is the *only valid starting point* for epistemological certainty and confidence.

In amplifying and applying these concepts, the works of Cornelius Van Til are indispensable. Van Til, a prolific writer, taught apologetics at Westminster Theological Seminary for over half a century. His biblical fervor and intellectual rigor are in a class of its own. Because of his somewhat obscurant writing style, his thought may be confusing in particular areas. Two works have helped clarify some of this confusion. First, John Frame's (1939-) book entitled *Cornelius Van Til: An Analysis of His Thought* (Presbyterian and Reformed) is an excellent and comprehensive exposition. Second, before his untimely death, Greg L. Bahnsen (1948-1995) compiled a masterful 764-page compendium of Van Til's thought. It is entitled *Van Til's Apologetic: Readings and Analysis* (Presbyterian and Reformed). Both books are *must reads.* Bahnsen's much shorter *Always Ready: Directions for Defending the Faith* (Covenant Media Foundation/American Vision) contains many gems of erudite epistemological insights. Rousas J. Rushdoony's *By What Standard?* (Thoburn Press) represents his own analysis of the philosophy of Van Til. Less technical works reflecting Van Til's approach are *Persuasions: A Dream of Reason Meeting Unbelief* (Canon Press) by Douglas Wilson and *Every Thought Captive* (Presbyterian and Reformed) by Richard L. Pratt, Jr. Gary DeMar's *War of the WorldViews: A Christian Defense Manual* (American Vision) is also quite useful. Chris Schlect, instructor at Logos School in Moscow, Idaho, has developed an excellent high school syllabus entitled *Christian Worldview and Apologetics.* It reflects a decade or so of Schlect's thinking and teaching in the Van Til perspective.[29]

Once a student of Van Til, Francis Schaeffer's writings are worthy of recommendation. See the five-volume set entitled *The Completed Works of Francis A.*

---

[29]Obtainable by writing to Logos School, 110 Baker Street, Moscow, ID 83843.

Schaeffer: A Christian Worldview (Crossway Books). David Noebel, founder of Summit Ministries (a worldview training institute), has written a benchmark guide for worldview investigation entitled *Understanding the Times* (Harvest House). Also helpful is *Turning Point: A Christian Worldview Declaration* (Crossway Books) by Herbert Schlossberg and Marvin Olasky. Finally, mentored by C. S. Lewis, Harry Blamires, in *The Christian Mind* (Servant), examines how Christians have surrendered the realm of the mind to the enemies of God and how to re-conquer this lost domain by thinking self-consciously as a Christian.

8.5.2 CHRISTIAN RESOURCES

An excellent little resource has been collected by Professors Gene B. Chase and Calvin Jongsma and is entitled *Bibliography of Christianity and Mathematics (1910-1983)*.[30] Materials are listed if they argue for or against some relationship holding between Christianity and mathematics or if they discuss mathematics or Christianity in terms of or in connection with the other one. Works include titles from Roman Catholic, Evangelical, and Reformed perspectives. Topics include integration concerns, Christian influences on philosophy of mathematics, mathematics to defend, clarify, or influence Christianity, specific mathematics topics, specific philosophical topics, vocational concerns, and humor. This book is a small gold mine.

One noteworthy article included in the above list is by Vern S. Poythress and is entitled "A Biblical View of Mathematics" It is one of many articles in *Foundations of Christian Scholarship* (Ross House Books), edited by Gary North. It is a pioneering study aimed directly at university professors of mathematics and is therefore replete with appropriate terminology. See also his less technical essay entitled "Creation and Mathematics; or What does God have to do with Numbers?" found in *The Journal of Christian Reconstruction*, 1:1 (1974).

Two books, one by Reijer Hooykaas entitled *Religion and the Rise of Modern Science* (Eerdmans – out of print) and the other by Christopher B. Kaiser entitled *Creation and the History of Science* (Eerdmans) attempt to interface Christian perspectives with the history of science and mathematics.

The Soul of Science: Christian Faith and Natural Philosophy (Crossway Books), by Nancy Pearcey and Charles Thaxton, is an excellent analysis of the historical interaction of the Christian faith with physics (Newtonian and Einsteinian), biology, chemistry, and mathematics.

Concerning Christian perspectives on the history of science, the erudite and extraordinary works by Stanley L. Jaki are indispensable.[31] Jaki, a Benedictine priest, holds doctorates in both systematic theology and nuclear physics. Some of his best works are *The Relevance of Physics* (Scottish Academic Press), *Science and Creation: From Eternal Cycles to an Oscillating Universe* (Scottish Academic Press), *The Road of Science and the Ways to God* (Scottish Academic Press), *The Ori-*

---

[30]Obtainable from Dordt College Press, Sioux Center, Iowa 51250.
[31]If the reader cannot find his works, almost all of them may be ordered by writing directly to the author: Rev. Stanley L. Jaki, P. O. Box 167, Princeton, NJ 08542-0167.

gin of Science and the Science of Its Origin (Scottish Academic Press), *The Milky Way: An Elusive Road for Science* (David & Charles), *Cosmos and Creator* (Scottish Academic Press), *God and the Cosmologists* (Real View Books), and *The Savior of Science* (Eerdmans). Although Jaki emphasizes the doctrine of creation as foundational for understanding issues in the history of science; i.e., this doctrine provides the epistemological and metaphysical presupposition for the understanding and doing of science, he does unfortunately capitulate to modern scientific and liberal theological views regarding the age of the universe. He also dismisses the validity of biblical chronology[32] and discredits any attempt to take Genesis 1 as a recounting of actual events of history. To him, Genesis 1 says that God created the world but this description is *just a picture or symbol* – it is not a record of actual history. Thus, Jaki says that Genesis 1 has nothing to do with historical events or science, except as a symbolic and philosophical description of the creation of the universe by God and the absolute dependence of that creation upon God. For amplification of these themes, see Jaki's *Genesis 1 Through the Ages* (Real View Books). For a critique, see James Jordan's essay entitled "Stanley Jaki on Genesis One" (Institute for Christian Economics). For an exegetical defense of the traditional reading of the first chapter of Genesis, see Jordan's *Creation in Six Days* (Canon Press). For further exegetical analysis in which both biblical details and scientific data are compared, see Douglas F. Kelly's *Creation and Change: Genesis 1.1-2.4 in the Light of Changing Scientific Paradigms* (Christian Focus Publications).

Concerning the history of science, especially the relationship between medieval science and modern science, there are differing scholarly opinions, all of which are undergirded by presuppositions. The two schools of thought are: (1) there is continuity between medieval science and modern science and (2) there is discontinuity between the two. Some of the proponents of the first school of thought are science historians Pierre Duhem (1861-1916), Charles Homer Haskins (1870-1937), Lynn Thorndike (1882-1965), Marshall Clagett (1916-), Anneliese Maier (1905-1971), Alistair C. Crombie (1915-1996), Reijer Hooykaas (1906-1994), and Stanley L. Jaki (1924-). Some of the proponents of the second school are science historians Alexandre Koyré (1892-1964),[33] A. Rupert Hall (1920-), and the "paradigm-shift" paradigm of Thomas Kuhn (1922-1996).[34] In reality, there are realms of continuity (e.g., theory of impetus) and discontinuity (e.g., cosmology) between these two important periods of science history and one cannot and should not radically divorce the two epochs.[35] If we are to ever hope to appreciate and understand the modern scientific world, then we cannot afford to be ignorant of the road that brought it to us.

---

[32]See Philip Mauro, *The Wonders of Bible Chronology* (Swengel, PA: Reiner Publications, n.d.).

[33]Koyré is the leader of this school of thought. For an analysis, see Jaki, *The Road of Science and the Ways to God*, pp. 230-235.

[34]For an analysis of Kuhn's thought, see *Ibid.*, pp. 236-245.

[35]For an analysis of these ideas, see Lindberg, *The Beginnings of Western Science*, pp. 355-368. Lindberg admits that he belongs to the discontinuity school of thought although he does see value in the continuity school. Significantly, he is silent on the work of Stanley L. Jaki and Reijer Hooykaas.

8.5.3 HISTORY

Men of Mathematics (Simon and Schuster), by Eric Temple Bell, is an excellent resource although he sometimes sacrifices truth for excitement. Christians like Isaac Newton and Leonhard Euler are discussed with a relative degree of objectivity. *Of Men and Numbers: The Story of the Great Mathematicians* (Dover Publications), by Jane Muir, offers a fascinating read. Herbert Westren Turnbull's excellent book, *The Great Mathematicians* (Barnes & Noble), makes accessible to the common reader an instructive survey of men and ideas. As a further resource in this area, a comprehensive index of mathematicians, called "The MacTutor History of Mathematics Archive," has been developed by the School of Mathematics and Statistics at the University of St. Andrews, Scotland.[36]

It is helpful to read about great mathematicians, but nothing is better than reading their writings. Thomas L. Heath's reproduction of Euclid's *Elements* entitled *Euclid: The Thirteen Books of The Elements* (Dover Publications) is available in three volumes. David E. Smith has collected 125 classic mathematical essays in *A Source Book in Mathematics* (Dover Publications). See also Dirk Struik's excellent compilation entitled *A Source Book in Mathematics 1200-1800* (Princeton University Press). Dover Publications has also reprinted Galileo's *Dialogues Concerning Two New Sciences* and Isaac Newton's *Opticks*. Prometheus Books has also made available what is considered to be among the finest scientific works ever published – Newton's *The Principia*. You can access the best of Kepler and Copernicus in volume 16 of the *Great Books of the Western World* (Encyclopaedia Britannica) series or through the reprints done by Prometheus Books.

Mathematics (Time-Life), by David Bergamini, is a very readable and enlightening documentation. Full color photographs highlight the text. It is out of print, but should be available in libraries and used bookshops.

The Mathematical Experience (Houghton and Mifflin), by Philip J. Davis and Reuben Hersh, is a combination history and philosophy of mathematics in which the authors make mathematics accessible to the educated layman. The mystery of mathematical effectiveness is described, but not conclusively explained, and one chapter presents the arguments of I. R. Shafarevitch, a Russian mathematician who presents a biblical perspective of the subject.

An Introduction to the History of Mathematics (Holt, Rhinehart and Winston), by Howard Eves, introduces a wealth of historical problems that complement the text. Bright students at the senior high school level can benefit from this work.

Mathematics in the Time of the Pharaohs (Dover Publications), by Richard J. Gillings, is a detailed account of Egyptian mathematics.

Art and Geometry: A Study in Space Intuitions (Dover Publications), by William M. Ivins, Jr., is a critical appraisal of Greek mathematics, its strengths and

---

[36]Their World Wide Web site is <www-history.mcs.st-and.ac.uk/history/index.htm>. The postal address of the school (founded in 1411) is University of St. Andrews, Fife KY16 9AJ, Scotland.

weaknesses, along with a look at some of the accomplishments of the mathematicians of the Scientific Revolution.

The History of the Calculus and its Conceptual Development (Dover Publications), by Carl Boyer, provides an excellent background for understanding the intellectual and theological impetuses that resulted in the discovery of the calculus.

Mathematics in Civilization (Dover Publications), by H. L. Resnikoff and R. O. Wells, Jr., is a liberal arts university text that presents a very good survey of the historical flow of mathematical method. The chapters on navigation and cartography are excellent.

Number Words and Number Symbols: A Cultural History of Numbers (Dover Publications), by Karl Menninger, is a detailed and scholarly study of how numbers have developed in a wide variety of cultures. The author's superlative narrative ability makes the presentation a joy to read.

Dover Publications has recently bound Florian Cajoiri's classic and comprehensive two-volume work entitled *A History of Mathematical Notations* in one volume. It is *the* definitive source book on the origin and development of mathematical notation.

Specific historical texts by Morris Kline include *Mathematical Thought from Ancient to Modern Times* (Oxford), a monumental (1,200 pages) and highly acclaimed work. Its scope is broad and its documentation detailed. In *Mathematics in Western Culture* (Oxford), he offers a less technical and a more culturally related overview. It is good reading for beginners. *Mathematics: The Loss of Certainty* (Oxford) is his own humanistic appraisal of modern mathematics. It reveals Kline's somewhat schizophrenic attitude toward the milieu called mathematics. First, Kline approves of the view that sees mathematics as the study of the structure of nature (i.e., nature understood without biblical presuppositions). Second, because of the 19th and early 20th century foundation debacles, Kline sees this view as suspect. He then embraces an epistemological admixture of (1) empiricism in which truth in mathematics is dependent upon its applicability to the physical world and (2) neo-Kantianism in which man imposes structure on the world through the matrix of his mind. Because of this conundrum, Kline's thought reflects a continuous tension between two extremes: (1) in order to be valid, mathematics must be closely anchored to science and (2) there really is no connection between mathematics and the physical world and there can be no explanation for its apparent correlation (in essence, mathematics is *purely and only* man-made). His presuppositions drive him and his analysis of the status of mathematics into the "twilight zone" of uncertainty.

A number of excellent topical mathematical histories are available. Some of the better ones are *A History of π (pi)* (St. Martin's Press) by Petr Beckman, *An Imaginary Tale: The Story of √-1* (Princeton) by Paul J. Nahin, *The Nothing That Is: A Natural History of Zero* (Oxford) by Robert Kaplan, *A Mathematical History of the Golden Number* (Dover Publications) by Roger Herz-Fischler, *A History of the Circle: Mathematical Reasoning and the Physical Universe* (Rutgers) by Ernest Zebrowski, Jr., *The Advent of the Algorithm* (Harcourt) by David Berlinski, and three rich studies by Eli Maor (all published by Princeton): *To Infinity and Beyond*, *Trigonometric Delights*, and *e: The Story of a Number*.

The History of Science: From Augustine to Galileo (Dover Publications), by Alistair C. Crombie, is a wonderful and readable analysis of the ideas and concepts that engendered the Scientific Revolution. Stanley L. Jaki highly recommends this quintessential study in the history of science.

The Measure of Reality: Quantification and Western Society, 1250-1600 (Cambridge), by Alfred W. Crosby, is a fascinating and provocative historical tour. The author explores with precision the epochal shift from qualitative to quantitative perception in Western Europe during the High Middles Ages. Because of this mental shift, Western Europeans became world leaders in science, technology, armaments, navigation, business practice (with its unfortunate concomitant bureaucracy), and created many of the greatest masterpieces of music and painting. This work is an excellent supplement to the historical analysis of Stanley L. Jaki.

Geometry Civilized: History, Culture, and Technique (Oxford), by J. L. Heilbron, adroitly connects history and geometry. It contains a complete introduction to the techniques of plane geometry and trigonometry within an engaging historical backdrop.

Sanderson Smith's *Agnesi to Zeno: Over 100 Vignettes from the History of Math* (Key Curriculum Press) contains a series of blackline activity masters that provide historical augmentation to a wide variety of mathematical topics. When using this resource, bear in mind that some of the vignettes reflect the author's moderated capitulation to the multiculturalism and feminism of modernity.

8.5.4 CREATIONAL STUDIES AND DOMINION MANDATE APPLICATIONS

Theodore Andrea Cook's *The Curves of Life* (Dover Publications) and Matila Ghyka's *The Geometry of Art and Life* (Dover Publications) both explore the mathematical curves of creation and then show how man uses them in art and life. Both works are profusely illustrated.

The Divine Proportion: A Study in Mathematical Beauty (Dover Publications), by H. E. Huntley, is a delightful book to read. The author traces the appearance of the Golden Ratio in creation.

Universal Patterns, by Rochelle Newman and Martha Boles, is the first in a series of books published by Pythagorean Press that explores the multifaceted applications of the Golden Ratio. The book is unique in its interdisciplinary approach to the study of art and mathematics as they relate to the geometrical patterns found in creation. Unfortunately, it is out of print.

Science and Music (Dover Publications), by Sir James Jeans, is a rare book in which the author combines his very distinguished scientific abilities with musical knowledge and simple explanations. Note carefully his evolutionary introduction.

Algebra in the Real World (Dale Seymour – out of print), by Leroy C. Dalton, is a superbly organized enrichment textbook. He explores the world of computer logic, packing problems, the number *e*, algebraic functions, the Golden Ratio, the Fibonacci Sequence, and music. It is an able launching pad for real world applications. Elementary Algebra is an assumed prerequisite.

Mathematics in Action (Dover Publications), by Sir Oliver Graham Sutton, is a middle-level exposition of the application of advanced mathematics to the study of creation. The author demonstrates how mathematics is applied in ballistics, theory of computers, waves and wavelike phenomena, the theory of fluid flow, meteorological problems, statistics, flight, and similar phenomena. Advanced topics such as differential equations, Fourier series, group theory, eigen values,[37] Planck's constant, and airfoil theory are explained so clearly in everyday language that almost anyone can derive benefit from reading this book.

Two books by Morris Kline, *Mathematics and the Physical World* (Dover Publications) and *Mathematics and the Search for Knowledge* (Oxford), reveal the author's unique ability to explain mathematical concepts clearly and simply. Both are classical historical and physical commentaries. The second book takes a special look at the power of mathematics in making "visible" the imperceptible world of gravitation, electromagnetism, and quantum mechanics.

William C. Vergara's books are delightful and informative. Two valuable books, although out of print, are *Mathematics in Everyday Things* (Harper and Brothers) and *Science, the Never-Ending Quest* (Harper and Row). The first is written in a question and answer format and is geared for junior high and high school students. It is a collection of mathematical concepts and their relationship to the real world. The second is a good historical survey that focuses on how scientists have used mathematics in taking control over the physical world.

For a wide-ranging survey of the relationship of the real numbers with the real world see David Berlinski's *A Tour of the Calculus* (Random House). According to a review by the San Francisco Chronicle, the author "is both poet and genius. And he's funny…. The writing is clean and powerful."

Do not forget that mathematics interacts, not just with science, but with other sectors of life. We need to use it everyday in the areas of earnings and taxes, banking, investments, loans, eating at restaurants, foreign travel, sports, hobbies, cooking, home improvement, etc. A book that answers all your practical day-to-day math needs is *The Only Math Book You'll Ever Need* (Dell Publishing) by Stanley Kogelman, Ph.D. and Barbara R. Heller.

8.5.5 MENTAL MATHEMATICS

In the author's formative years of education, the slide rule was considered an essential tool of the mathematician or physicist. Today, a $10 calculator not only does everything that a slide rule used to do, but does it with much greater precision. The unfortunate downside of this electronic revolution is disappear-

---

[37]Eigen values and eigen functions are components of the mathematics of quantum matrix theory.

ance from the curriculum of an emphasis on training the mind of the student to do simple calculations mentally.[38]

For example, consider the problem of multiplying 80 by 45. Students should be taught to estimate the answer first (between 3000 and 4000) or apply the distributive rule to get the answer without using paper or punching keys: $80 \times 45 = (80 \times 40) + (80 \times 5)$; $80 \times 40 = 3200$, $80 \times 5 = 400$; hence, the answer is 3600. It is not uncommon to observe a student mindlessly punching keys on the calculator (or computer keyboard) when faced with a problem of this sort, miss a key (e.g., 0 in this case), and confidently record the answer to 80×45 as 360!

The distributive rule also comes in handy when mentally figuring the 15% tip for excellent service on the before-tax restaurant bill. For example, if the bill is $26.45, round off to $26. 10% of $26 is $2.60. Take half of $2.60 (the other 5%) and get $1.30. Add $2.60 to $1.30 to get the tip of $3.90.

There are many books available that expand the mind's capability to do mental mathematical calculations. A few of the better ones are *How to Calculate Quickly* (Dover Publications) by Henry Sticker, *Short-Cut Math* (Dover Publications) by Gerard W. Kelly, *Rapid Math Without a Calculator* (Carol Publishing Group) by A. Frederick Collins, *The Trachtenburg Speed System of Basic Mathematics* (Greenwood Press) by Jakow Trachtenburg, *Becoming a Mental Math Wizard* (Shoe Tree Press) by Jerry Lucas, and the runaway bestseller *Math Magic* (HarperCollins) by Scott Flansburg.

8.5.6 GENERAL INTRODUCTION

The monumental *Mathematics: From the Birth of Numbers* (W. W. Norton), written by physician Jan Gullberg, is a gently guided, profusely illustrated grand tour of the world of mathematics. Heed the endorsement by long-time mathematics teacher Harold Jacobs (about whom we will hear more shortly):

> Having spent the last 35 years teaching mathematics and having acquired a large library of reference books on the subject during that time, I only wish Jan Gullberg had written *Mathematics: From the Birth of Numbers* long ago. His writing style makes you feel as if you are a fortunate student spending some time with a gifted teacher. The abundance of appropriate examples, historical illustrations, and quotations also help to enlighten and entertain the reader. The scope of this book is astonishing and I am convinced that it will serve as a valuable reference work for teachers and students for years to come.[39]

---

[38]The author views the calculator and personal computer as indispensable tools in the educational process. These tools enable one to model equations and functions with ease and quickness. The tool should be employed, however, *only after the student has mastered computational fundamentals* (the "grammar" of mathematics). Too many students use these tools simply to get fast answers (in reality, electronic cheating). To take a calculator away from them is like taking a crutch from a handicapped person; they are incapacitated without it.

[39]Harold Jacobs, in Gullberg, *Mathematics: From the Birth of Numbers*, back jacket cover.

Martin Gardner is best known as the author of Scientific American's "Mathematical Games" column. *Aha! Gotcha* and *Aha! Insight* (W. H. Freeman) are a compilation of some of his works. The first book covers topics in logic, probability, numbers, geometry, time, and statistics dealing with paradoxes that defy logic and reason. The second is a confounding collection of brain-twisters that challenge the reader's reasoning power and intuition.

Walter W. Sawyer has no peer as a popularizer of mathematics. *A Prelude to Modern Mathematics* (Dover Publications) and *A Concrete Approach to Abstract Algebra* (Dover Publications), both dealing with advanced topics, are in print. Other out of print books, all published by Penguin, are *A Search for Pattern, Vision in Elementary Mathematics, Mathematician's Delight,* and *A Path to Modern Mathematics.*

Mathematics: An Introduction to its Spirit and Use (W. H. Freeman), edited by Morris Kline, is an excellent introduction to the manifold functions of mathematics. It contains a wealth of highly original articles by highly acclaimed mathematicians that deal with topics in history, number, algebra, geometry, statistics, probability, symbolic logic, computers, and real world applications.

In 1956, James R. Newman completed a masterful compilation entitled *The World of Mathematics* (Simon and Schuster). Subtitled "A small library of the literature of mathematics from A'h-mose the scribe to Albert Einstein presented with commentaries and notes," this massive 4-volume anthology contains much material that is almost impossible to obtain elsewhere, including original papers and large sections of out of print books. The volume titles are: (1) Men and numbers, (2) World of laws and the world of chance, (3) Mathematical way of thinking, and (4) Machines, music and puzzles. No mathematics teacher should be without this monumental work. The Simon and Schuster edition is out of print. In 1988, Microsoft Press reprinted this choice work but this edition has also gone out of print. Fortunately, in 2000, Dover Publications has brought it back into print.

Mathematical Snapshots (Oxford), by Hugo Steinhaus, is an excellent introduction to mathematical concepts. Using striking photographs and diagrams, the author explains with erudition mathematical concepts, from simple to complex.

Dan Pedoe's *The Gentle Art of Mathematics* (Dover Publications) is a good introduction to more advanced topics of mathematics such as infinity, the laws of probability, and the algebra of classes. Knowledge of algebra is necessary to follow the author's presentation.

The best seller *Math Without Tears* (G. K. Hall & Company), by Roy Hartkopf, contains excellent insights for mathematics teachers. He covers a wide range of mathematical topics, from basic concepts of arithmetic through logarithms, graphs, trigonometry, series, and the calculus. His approach is refreshing and clear.

In Rózsa Péter's book, *Playing with Infinity: Mathematical Explorations and Excursions* (Dover Publications), the infinite collection of counting numbers; i.e., 1,

2, 3, etc., is the starting point. With complete clarity and much originality, and without being technical or superficial, the author proceeds to grow every branch of the mathematics tree from that seed foundation.

Mathematics for the General Reader (Dover Publications), by E. C. Titchmarsh, is a truly marvelous introduction to basic mathematics, through to the calculus, by a top-flight mathematician who writes clearly, simply, and beautifully.

Mathematics teacher and consultant Theoni Pappas is committed to demystifying mathematics and to eliminating the elitism and fear often associated with it. Some of her many books (all published by Wide World Publishing/Tetra) are *The Joy of Mathematics, More Joy of Mathematics, The Magic of Mathematics, Mathematics Appreciation,* and *Mathematical Footprints.*

Keith Devlin's *The Language of Mathematics: Making the Invisible Visible* (W. H. Freeman) is a marvelous tour of the numeric underpinnings of everyday life. Some of the chapter titles are What is Mathematics?, Why Numbers Count, Patterns of the Mind, Mathematics in Motion, Mathematics Gets into Shape, The Mathematics of Beauty, and Uncovering the Hidden Patterns of the Universe.

Finally, for a delightful and entertaining introduction to many mathematical principles, try to obtain and view the video entitled *Donald in Mathmagic Land* (The Walt Disney Company).

8.5.7 TEXTBOOKS

Most people have been trained, under the auspices of pedagogical modernity, to think that there is a widespread agreement across the line with regard to the mathematical corpus. How has this been accomplished?

First, the mathematical professional community – especially the bureaucratic university complex – generally holds rank; i.e., they tend to not allow disagreements with the *status quo.* Those with radically differing outlooks, either pedagogically or philosophically, tend to be ignored; they are "outsiders" and not invited to the "party." For example, the late Dr. Morris Kline was generally ostracized by the mathematical community because of his viewpoints expressed in *Why Johnny Can't Add: The Failure of the New Mathematics, Why the Professor Can't Teach: The Dilemma of University Education,* and *Mathematics: The Loss of Certainty.*

Second, most textbooks are standardized so that they can be understood by all.[40] In this standardization, mathematics has lost most, if not all, of its life. Pick up almost any university text on mathematics and notice the "dry" formalism or logical "chicken scratches" contained therein.[41] There is little or no ac-

---

[40]Standardization, in educational circles, is a politically correct word that stands for, in the author's opinion, "sanitization."

[41]In his university training, this dry formalism was taught to the author semester after semester. As the course material advanced beyond differential and integral calculus, it seemed like he had entered into an n-dimensional domain of transcendent abstract analysis, aimed not at the Elysian fields of delight, but at the specter of the null and the void. Numbed and dizzied by such spiraling integrations into the void, he still managed to receive his undergraduate degree in mathematics in 1973 (graduating *summa cum laude* – which proves that getting good grades does not mean that you have

knowledgment of the history behind the formulae and theorems – the personalities, the mistakes and misconceptions, the back alleys, and the dead ends. The link between science and mathematics is virtually non-existent. There is almost no interaction with worldview (philosophy) and if so, Christianity usually gets bad press. A student therefore gets the impression of a uniform and decontaminated outlook with regard to the milieu called mathematics. This dry standardization caused Walter W. Sawyer to react, "Most textbooks contain vast amasses of information, the object of which is not always obvious."[42] The following books are a few of the exceptions to that rule.[43]

Master teacher Harold Jacobs has written three high school texts, all published by W. H. Freeman, which makes mathematics interesting to the student. They are *Elementary Algebra, Geometry*, and *Mathematics: A Human Endeavor*. All are excellently written and attempt to build mathematical principles on real world situations teaching from concrete foundations to abstract principles. The Geometry text includes an illustrative unit on non-Euclidean geometry plus a survey of solid and coordinate geometry. The supportive teacher's guides contain a wealth of teaching insights and resources.

In contrast to the traditional deductive approach to geometry, Michael Serra has produced an inductive geometry high school text that is growing in popularity. It is entitled *Discovering Geometry: An Inductive Approach* (Key Curriculum Press). Key Curriculum Press has also produced a wealth of ancillary and innovative mathematics materials worthy of investigation and use.[44]

James Edgar Thompson has ably written a series of self-teaching texts (published by Van Nostrand and unfortunately out of print). They are *Arithmetic for the Practical Worker, Algebra for the Practical Worker, Geometry for the Practical Worker, Trigonometry for the Practical Worker*, and *Calculus for the Practical Worker*. Thompson was a master at interweaving skill instruction with historical context and methods that enable the reader to *apply* the *principles* taught in any way. American physicist Richard P. Feynman cut his mathematical teeth on some of Thompson's texts in his early teenage years.

Mathematics for the Non-mathematician (Dover Publications), by Morris Kline, is probably the *best mathematics curriculum on the market*. Suited for high school level work, it is a compilation of all of his works in a textbook format. The instruction is entirely historical and every mathematical concept is therefore seen in its proper context. Exercises augment the text in a real life manner. Topics

---

been properly educated). On graduation day the author vowed, "I will never open another math book again as long as I live." Five years later, he annulled this vow and, after studying Morris Kline's *Mathematics and the Physical World*, his eyes were opened (so, "this" is what "that" meant!) and he entered the mathematical Elysium; a region furnished with rich fields, groves, shades, streams – a domain of delight consisting of the pleasurable beholding of God's great and marvelous works (Psalm 111:2).

[42]Sawyer, p. 10.

[43]Note again, the recommendation of these books does not imply that the author commends them in their every detail.

[44]Write to Key Curriculum Press, 1150 65th Street, Emeryville, CA 94608-1109. Their World Wide Web site is <www.keypress.com>.

covered are: Why Mathematics?, Historical Orientation, Logic and Mathematics, Number: The Fundamental Concept, Algebra, the Higher Arithmetic, The Nature and Uses of Euclidean Geometry, Charting the Heavens and the Earth, The Mathematical Order of Nature, The Awakening of Europe, Mathematics and Painting in the Renaissance, Projective Geometry, Coordinate Geometry, The Simplest Formulas in Action, Parametric Equations and Curvilinear Motion, The Application of Formulas to Gravitation, The Differential Calculus, The Integral Calculus, Trigonometric Functions and Oscillatory Motion, The Trigonometric Analysis of Musical Sounds, Non-Euclidean Geometries and Their Significance, Other Arithmetics and Their Algebras, The Statistical Approach to the Social and Biological Sciences, The Theory of Probability, and the Nature and Values of Mathematics. Care must be taken to avoid being trapped by his rationalistic approach.

The AIMS Education Foundation <www.aimsedu.org>, located in Fresno, California, has launched a major program to develop a pattern-based math/science curriculum and has produced a superb ancillary series of biographical mathematics supplements. Authored by Luetta and Wilbert Reimer,[45] the two-volume set entitled *Mathematicians are People, Too* (Dale Seymour) is ideally suitable for Grades 3-7. The three-volume set by the same authors entitled *Historical Connections in Mathematics: Resources for Using History of Mathematics in the Classroom* (AIMS Education Foundation) can be used in both elementary and high school contexts.

John Saxon (1923-1996) has authored some commendable textbooks, all published by Saxon Publishers, which aid the student in both comprehension and retention of basic mathematical skills. He facilitates this goal by using the "incremental approach," a methodology of gentle repetition, extended over a considerable period of time. Some of the junior high and high school titles available are *Math 76*, *Math 87*, *Algebra 1/2*, *Algebra I*, *Algebra II*, *Advanced Mathematics*, *Calculus*, and *Physics*. These texts have proven themselves to be successful and the demand for them is quite high (especially in the home schooling market). Unfortunately, the author fails to incorporate the context of history, philosophy, and the physical world although his goal is to prepare the student to use mathematics in the sciences. Also, no complete course in deductive geometry can be found in the Saxon series. This omission is symptomatic of a pathological problem with the Saxon approach to learning. In the absolutization of the "manipulatory skill set" or "incremental" approach to learning the student becomes conditioned, in the Pavlovian context, to do math problems as a "reflex reaction."[46] A complete deductive approach to geometry – consisting of arguments that require concentration and comprehension – is missing from this curriculum because you cannot apply "rote learning" to it.

---

[45]Wilbert Reimer was the author's high school mathematics instructor in the mid-1960s at Immanuel Academy High School, Reedley, California.
[46]Ivan Petrovich Pavlov (1849-1936) was a Russian psychologist who analyzed conditional or reflex responses.

A Beka Books and Bob Jones University Press are two publishers that have tried to produce a Christian program of mathematics textbooks ranging from kindergarten to the upper high school levels. Note that A Beka's approach to mathematics is more traditional (and anti-set theory) having reprinted and upgraded many textbooks from the pre-set theory era of the 1940s and 50s. A Beka, because of its radicalization of certain fundamentalist doctrinal positions (which creates several serious blind spots or misjudgments, especially in their history curriculum), is also weak on worldview analysis and there is a tendency to "baptize" some of the mathematics texts with sporadic Bible verses. The range of books published by Bob Jones University Press attempts to incorporate concepts, skills, speed, accuracy, and practicality. This publisher recognizes that mathematics is not a neutral subject and therefore attempts to place mathematics in its creational context.

Need a thorough treatment of the calculus? As an introductory text, *What is Calculus About?* (Random House), by Walter W. Sawyer, is a clear and informative introduction to the method of the calculus. The language is non-technical and the writing style is easy and interesting. Unfortunately, it is out of print. Silvanus P. Thompson's *Calculus Made Easy* (St. Martin's Press) was originally written in 1910; its continued popularity confirms the author's unique ability to present the rudiments of the subject in an understandable way. A more advanced and complete text is authored by Morris Kline titled *Calculus: An Intuitive and Physical Approach* (the two John Wiley and Sons editions are out of print, but Dover Publications has reprinted the first edition). Kline uses history and the physical world as starting points and discusses the logical basis for the calculus in the last chapter. Professor Alexander J. Hahn of the University of Notre Dame has developed a unique, historical textbook entitled *Basic Calculus: From Archimedes to Newton to its Role in Science* (Springer-Verlag). The only perspective that this book lacks is a thorough analysis of the interaction between philosophy and the calculus.

Why Johnny Can't Add: The Failure of the New Mathematics (Vintage Books – out of print), by Morris Kline, is not a textbook, but it evaluates the theory, the methods, and the writers of a great majority of mathematics curricula. He is highly critical of the set theoretical approach used in the 1960s and presents direction for reforming it. Because of the 1960s context, portions of this book may be somewhat dated.

As of the date of the second edition of this book, the author has developed a textbook statement (including tentative titles and approximate year levels) summarizing the topics that should be included in a biblical Christian curriculum (from junior high to the first year of college).[47] In order to present a thorough biblical view of mathematics, each topic *must interact* with philosophical issues, history, and science. Not only must we teach the skills of the subject, we

---

[47]This program will be developed and marketed through Christian Heritage Academy, International, a division of Patria Institute, LLC. See the *About the Author* section (page 409) for contact information.

must also teach the student how the biblical worldview "accounts for counting" (philosophy), God's providence in the historical development of mathematics, and how mathematics serves as a dominion tool of science.

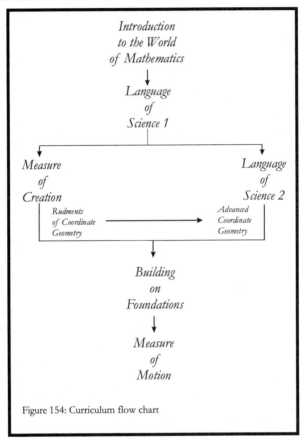

Introduction to the World of Mathematics

Language of Science 1

Measure of Creation

Rudments of Coordinate Geometry

Language of Science 2

Advanced Coordinate Geometry

Building on Foundations

Measure of Motion

Figure 154: Curriculum flow chart

1. *Introduction to the World of Mathematics* (pre-algebra, general introduction): broadens the student's view of mathematics in terms of revealing some of its profound, yet simple, concepts; sparks student interest in the beauty and wonder of mathematics encouraging motivated learning of the subject (Year 7-8). Prerequisite: mastery of the "grammar" of mathematics (i.e., the principles of elementary arithmetic).

2. *Language of Science I* (Algebra I): builds the basic principles of algebra upon its historical interaction with science, specifically physics (Year 8-9). Prerequisite: *Introduction to the World of Mathematics*.

3. *Measure of Creation* (Geometry, Trigonometry): includes plane, non-Euclidean geometry, and rudimentary Cartesian geometry incorporating a

balanced inductive and deductive approach (Year 9-10). Prerequisite: *Language of Science* I.

4. *Language of Science II* (Algebra II/Advanced Mathematics): same pedagogical procedure as *Language of Science I*; includes extension of Cartesian Geometry with its applications to conics and trigonometric functions (Year 10-11). Prerequisite: *Language of Science I* and *Measure of Creation.*

5. *Building on Foundations* (pre-calculus): includes the nature of the internal structure of number systems, the different branches of the mathematics tree, and the review and extension of Geometry, Algebra, and Trigonometry (Year 11-12). Prerequisite: *Language of Science I*, *Language of Science II*, and *Measure of Creation.*

6. *Measure of Motion* (a liberal arts approach to the calculus): basic introduction to differential and integral calculus incorporating a physical and historical approach (Year 12 or first year of college). Prerequisite: *Building on Foundations.*

As we look at the place of mathematics in the future of biblical Christian education, let us envision an army of effective and motivated teachers using a whole series of refined textbooks that thoroughly and completely expose the student to the wonders of God's creation and the enormous power of mathematics; books that truly give honor to Whom honor is due in *every* facet of the mathematics enterprise. Without this sort of vision, Christian education will perish. It is the hope of the author that this work will act as part of a catalyst to make such a vision a reality. As we work to implement this vision let us always be encouraged by the certainty that our toil is not in vain (I Corinthians 15:58). In the words of Cornelius Van Til, we labor "in the dawn of everlasting results."[48]

Soli Deo Gloria

QUESTIONS FOR REVIEW, FURTHER RESEARCH, AND DISCUSSION

1. Explain the weaknesses of modern mathematics curricula and instruction.
2. Explain the power and truth perspective that biblical revelation and motivation bring to mathematics instruction and study.
3. Carefully critique the following statement, "The road of science will lead any reasoning person to the ways of God."
4. Carefully explain the following statement, "Abstract or pure mathematics is the only way to control our thought of concrete fact."
5. By what guidelines can the biblical Christian "plunder" the mathematical reform work of some secular educators?
6. Describe the process of abstraction.
7. Explain the danger of absolutizing abstraction in mathematics pedagogy.
8. Why is mathematics essential to science?

---

[48]Cornelius Van Til, *Essays on Christian Education* (Phillipsburg: Presbyterian and Reformed, [1971] 1979), p. 207.

9. Prove the following statement, "With mathematics, man can grasp and utilize what is beyond the realm of sensation."
10. Why is knowledge of history essential to the teaching of mathematics?
11. Present biblical, historical, and pedagogical perspectives on the relationship between induction and deduction.
12. Explain the pedagogical interplay between knowledge, understanding, and wisdom.
13. What is the danger in absolutizing the use of calculators or personal computers in mathematics pedagogy?
14. What is the danger in absolutizing the "skill set" approach to mathematics pedagogy?
15. Explain the rationale and methodology used in teaching the following from a biblical Christian perspective.
    a. Arithmetic.
    b. Number theory.
    c. Algebra.
    d. Geometry (Euclidean, non-Euclidean, and coordinate).
    e. Trigonometry.
    f. The calculus.

TIMELINES

FROM ADAM TO CHRIST

Adam (ca. 4000 BC)
Worldwide Flood (ca. 2400 BC)
Abraham (ca. 2000 BC)
Israel in Egypt (ca. 1700-ca. 1500 BC)
The Rhind Papyrus (1650 BC): numerical problems
Moses and Exodus from Egypt (ca. 1450 BC)
Reign of King David and King Solomon (ca. 1055-ca. 975 BC)
Assyrian Empire (800-600 BC)
Rome founded (753 BC)
Northern Kingdom, Israel, fell to Assyrians (ca. 721 BC)
Jeremiah, the prophet (ca. 650-ca. 586 BC)
Thales of Miletus (636-546 BC): everything is "full of the gods"; everything can be
    reduced to water; beginnings of deductive geometry
Anaximander (611-547 BC): boundless reality
Babylonian Empire (600-540 BC)
Classical Greek period (600-300 BC)
Epimenides (ca. 600 BC): poet
Anaximenes (588-526 BC): everything can be reduced to air
Southern Kingdom, Judah, fell to Babylonians (ca. 586 BC)
The prophets Zechariah and Haggai (ca. 575-ca. 515 BC)
Pythagoras (572-492 BC): arithmetic, geometry, and music
Siddhartha Gautama or Buddha (ca. 563-ca. 483 BC): Indian mystic
Confucius (ca. 551-479 BC): Chinese philosopher
Heraclitus (544-484 BC): everything is the "fire" logos; reality is constantly chang-
    ing
Medes and Persian Empire (540-430 BC)
Parmenides (515-450 BC): everything is unchangeable; sphericity of the earth
Anaxagoras (ca. 500-428 BC): elementary atomism; order (*nous*) out of chaos
Zeno of Elea (495-430 BC): paradoxes of motion
Empedocles (ca. 490-430 BC): everything is made up of a plurality of things –
    earth, air, fire, and water.
Phidias (ca. 490-430 BC): Greek sculptor
Battle of Thermopylae (480 BC)
Malachi, the prophet (ca. 480-ca. 430 BC)
Socrates (ca. 470- ca. 399 BC): philosopher; mentor of Plato
Democritus (ca. 460-ca. 370 BC): a priori atomism
Hippocrates (ca. 460-ca. 377 BC): physician

Peloponnesian War between Athens and Sparta (431-404 BC)

Plato (427-347 BC): philosopher; idealism/dualism, mathematics in the training of the mind

Aristotle (384-322 BC): philosopher; emphasis on particulars, systematizer of deductive logic

Pyrrho (ca. 360-ca. 272 BC): Skeptic philosopher

Alexander the Great (356-323 BC): King of Macedonia; world conqueror

Menaechmus (ca. 350 BC): Tutor of Alexander the Great; conic sections

Epicurus (341-270 BC): philosopher; founder of Epicureanism

Alexandria, Egypt founded (336 BC)

Zeno of Citium (ca. 335-ca. 263 BC): Stoic philosopher

Hellenistic Greek period (330-63 BC)

Euclid (ca. 330-ca. 275 BC): deductive systematization of plane geometry

Aristarchus (ca. 310-230 BC): heliocentricism

Euclid's *Stoicheia* or *Elements* (ca. 300 BC)

Archimedes (ca. 287-212 BC): circle and sphere, mechanics, hydrostatics, infinite series

Eratosthenes (ca. 275-194 BC): geography

First Punic War (264-241 BC)

Apollonius of Perga (ca. 262-ca. 190 BC): conic sections

Hannibal (237-ca. 183 BC): Carthaginian general; second Punic war

Second Punic War (218-201 BC)

Hipparchus (ca. 180-125 BC): trigonometry and astronomical tables

Greece fell to Rome (146 BC)

Marcus Tullius Cicero (106-43 BC): Roman statesman, orator, and philosopher

Posidonius (ca. 100 BC): Greek philosopher; circumference of the earth

Julius Caesar (100-44 BC): Roman general and statesman

Lucretius (94-55 BC): Roman philosopher and poet

Judea annexed by Rome (63 BC)

Strabo (ca. 63 BC-ca. 24 AD): Greek geographer and historian

Julius Caesar's calendar reform (45 BC)

Herod the Great (37-ca. 4 BC)

Reign of Caesar Augustus (27 BC-14 AD)

John the Baptist (ca. 6 BC-28 AD)

Life of Christ (ca. 4 BC-ca. 30 AD)

FROM CHRIST TO WYCLIFFE

Heron or Hero (ca. 100 BC or ca. 100 AD): Alexandrian scientist; mechanics

Paul, the Apostle (ca. 2–65)

John, the Apostle (ca. 6-ca. 98)

Pliny (ca. 23-79): Roman encyclopedist

Stephen martyred and Apostle Paul converted (35)

Plutarch (ca. 46-ca. 120): Greek biographer and essayist

Council of Jerusalem (48)

Reign of Roman Emperor Nero (54-68)
Apostle Paul's letter to the Romans (57)
Fire of Rome; Nero launched persecution of Christians (64)
Apostle Peter and Apostle Paul executed in Rome (65)
Polycarp (ca. 69-155): Christian prelate, bishop of Smyrna
Jerusalem destroyed by Roman legion led by Titus (70)
Last books of the New Testament completed (ca. 70)
Claudius Ptolemy (ca. 85-ca. 165): trigonometry and geocentrism
Menelaus (ca. 100): Alexandrian mathematician; trigonometry
Justin the Martyr (100-165): Christian apologist
Marcus Aurelius (121-180): Roman Emperor and philosopher
Claudius Ptolemy's *Syntaxis mathematica* (ca. 150)
Clement of Alexandria (ca. 150-ca. 220): Christian apologist
Tertullian of Carthage (150-230): Christian theologian and apologist
Polycarp martyred (155)
Origen (185-254): Christian theologian
Plotinus (204-270): neo-platonism
Mani (ca. 216-ca. 276): Persian prophet; radical dualism
Diophantus (ca. 250): Alexandrian mathematician; theory of numbers and syncopated algebra
Arius (ca. 256-336): Greek theologian; defective view of the person of Christ
Eusebius of Caesarea (ca. 260-ca. 340): Palestinian theologian; church historian
Athanasius (ca. 293-373): Greek bishop of Alexandria; defender of Christianity against Arianism
Pappus (ca. 300): Alexandrian mathematician; collections and commentaries
Great persecution of Christians began by Emperor Diocletian (303)
Reign of Constantine (306-337)
Conversion of Constantine (312)
Edict of Milan or the toleration of Christians (313)
Eusebius' *Ecclesiastical History* (324)
Council of Nicea (325): condemnation of Arius
Ambrose (ca. 340-397): mentor of Augustine
Theodosius (ca. 346-395): Roman Emperor; banned Greek works
Aurelius Augustinus (354-430): theologian and writer
Hypatia (370-415): Alexandrian mathematician; commentaries of the works of Apollonius and Diophantus
Christianity as state religion of the Roman Empire (381)
Augustine converted to Christianity (386)
Patrick (ca. 389-ca. 461): Christian prelate; Apostle of Ireland
Theon of Alexandria (ca. 390): Alexandrian mathematician; commentaries on Euclid and Ptolemy
Split of Empire between East and West (395)
Waterwheel technology began in Europe (5th century)
European cold front commenced (5th century)

Rome sacked by Visigoths (410)

Patrick began mission to Ireland (432)

Council of Chalcedon (451): established that Christ has both a divine and a human nature, which exist inseparably within Him

Anicius Manlius Severinus Boethius (ca. 475-524): Roman philosopher; trivium and quadrivium

Fall of Roman Empire (ca. 476)

Aryabhata the Elder (476-ca. 550): Arabic astronomer; trigonometry and astronomical tables

St. Benedict (480-543): founder of Benedictine monastic order

Justinian (483-565): Emperor of Rome and Constantinople; Justinian law code

Invention of the heavy plough (6th century)

Continuation of Barbarian chaos (6th century)

Revival of learning in Ireland (6th and 7th centuries)

John Philoponus (ca. 500 AD): Alexandrian Monophysite theologian; critique on Aristotle, impetus theory of motion

Codex of Justinian published (529)

St. Benedict's monastic Rule (540)

Pope Gregory I (590-604)

Hindus began working with the number zero, negative numbers, and the idea of positional notation (7th century)

Muslim invasions (7th century)

Birth of Islam (622)

Muslim armies ransacked Egypt (640)

Last library at Alexandria burned (641)

Bede (ca. 673-735): Anglo-Saxon theologian; calendar

Boniface (ca. 675-754): English Benedictine missionary; Apostle of Germany

European cold front reversed (8th century)

Introduction of the three-field system in Europe (8th century)

Organization of the self-contained tenant-farmed estate in Europe (8th century)

Boniface began mission to the Germans (716)

Bede's *Ecclesiastical History* (731)

Alcuin of York (ca. 735-804): Anglo-Saxon prelate and scholar; leader of revival in learning

Charlemagne the Great (768-814): Holy Roman Emperor; leader of cultural rebirth known as the Carolingian Renaissance

Mohammed ibn Musa al-Khowarizmi (ca. 780-ca. 850): Arabic mathematician; syncopated algebra

Alcuin of York became royal advisor to Charlemagne (781)

Viking invasions of Europe (9th century)

Invention of the iron horse shoe in Europe (9th century)

Charlemagne the Great crowned Holy Roman Emperor (800)

Claudius Ptolemy's *Almagest* (827)

Mohammed ibn Musa al-Khowarizmi's *Hisab al-jabr w'al-muqabalah* (ca. 830)

Alfred the Great (849-899): King of West Saxons; scholar and lawmaker

Al-Battani (877-918): Arab mathematician; trigonometry

Abu-al-Wafa (940-ca. 997): Arab mathematician; trigonometry

Avicenna (980-1037): Persian physician and philosopher; ophthalmology

Beginning of the Cistercian order (11th century): agriculture

Leif Ericsson, Norwegian navigator, discovered North American mainland (1000)

Gregory VI or Hildebrand (ca. 1023-1085): Roman Catholic pontiff; lay investiture controversy with Henry IV

William the Conqueror (ca. 1027-1087): king of England

Anselm of Cantebury (1033-1109): Italian-born Augustinian theologian and author

Henry IV (1050-1106): Holy Roman Emperor; lay investiture

Omar Khayyam (ca. 1050-1122): Persian astronomer, mathematician, and poet; rudiments of coordinate geometry

Split between Eastern and Western Church (1054)

Crab Nebula, the result of a supernova, recorded in China (1054)

Norman Conquest of England (1066)

Henry IV submitted to Gregory VI over investiture controversy (1077)

Peter Abelard (1079-1142): French philosopher and theologian; Aristotelian logic, challenged indirectly the traditions of the church

Domesday Book (1086): waterwheel documentation

Bernard of Clairvaux (1090-1153): French monastic reformer

The Crusades (1095-1272)

Anselm of Cantebury's *Cur Deus Homo* (1099)

Rise of the University (12th century)

Gerard of Cremona (ca. 1114-1187): trigonometry

Bernand of Clairvaux founded monastery at Clairvaux (1115)

John of Salisbury (ca. 1115-1180): English bishop and scholar; diplomacy, logic, Aristotelian philosophy

Adelard of Bath (ca. 1120): English scholar; first to translate Euclid into Latin

Peter Abelard's *Sic et Non* (1122)

Averroës (1126-1198): Spanish-Arab physician and philosopher; ophthalmology

Gothic architecture (1140-1250)

Pope Gregory IX (ca. 1147-1241)

Universities of Paris and Oxford founded (1150)

Genghis Khan (1162-1227): Mongol ruler

Leonardo of Pisa, Fibonacci (ca. 1170-1250): Italian mathematician; arithmetic, algebra, geometry

Robert Grossteste (1170-1253): English philosopher and mathematician; empirical method

Waldensian reform movement began (1173)

Saint Francis of Assisi (1182-1226): Italian mystic; founder of Franciscan order

Leonardo of Pisa's (Fibonacci) *Liber Abaci* (1202)

Peak of the Cistercian order (12th and 13th centuries)

Roger Bacon (1214-1291): English friar, scientist, and philosopher; empirical method
Magna Carta (1215)
Thomas Aquinas (1226-1274): Italian Dominican monk, philosopher and theologian; nature-grace dichotomy
Marco Polo (1254-1324): Italian explorer
Dante Alighieri (1265-1321): Italian poet
Pope Urban IV died (1264)
Thomas Aquinas' *Summa Theologica* (1266-1273)
Roger Bacon's *Opus Majus* (1267)
Marco Polo began his travels (1271)
Condemnation of Aristotle by Bishop Etienne Tempier (1277)
Paper mills reported in Italy (1280)
William of Occam (ca. 1285-ca. 1349): English scholastic philosopher; nominalism
Thomas Bradwardine (1290-1349): English theologian; forerunner of symbolic algebra
Richard of Wallingford (1292-1336): English mathematician; trigonometry
Jean Buridan (ca. 1295-1358): French scholastic philosopher; impetus theory of motion (law of inertial motion)
First recorded examples of the weight-driven mechanical clock (14th century)
Discussions of the nature of the continuum and of maxima and minima problems (14th century)
John Manduith (ca. 1320): English mathematician; trigonometry
John Wycliffe (ca. 1320-1384): English reformer
Dante Alighieri's *Divine Comedy* (1321)
Nicole Oresme (ca. 1323-1382): French theologian, forerunner of coordinate geometry, function concept, and dynamics (law of inertial motion)
William Langland (ca. 1332-ca. 1400): English poet and societal critic
Geoffrey Chaucer (ca. 1343-1400): English poet
Black Plague began (1347)
Jan Hus (1369-1415): early Bohemian reformer
John Wycliffe's translation of the Bible into English (1381)
Geoffrey Chaucer's *Canterbury Tales* (1390)

FROM WYCLIFFE TO EULER

Johann Gutenberg (ca. 1400-ca. 1468): German printer
Nicholas Cusa (1401-1464): German cardinal, scholar, mathematician, scientist, and philosopher; treatises on quantitativism
Joan of Arc burned at the stake (1431)
Johann Müller or Regiomontanus (1436-1476): German mathematician and astronomer; trigonometry
Luca Pacioli (ca. 1445-ca. 1509): Italian mathematician; divine proportion, double-entry bookkeeping.
Rise in wealth and productivity in Europe (1450)

Isabella I (1451-1504): Queen of Castille
Christopher Columbus (1451-1506): Italian explorer; discoverer of the New World
Ferdinand V (1452-1516): King of Castille
Leonardo da Vinci (1452-1519): Florentine artist and innovator
Fall of Constantinople to the Turks; end of Byzantine Empire (1453)
Amerigo Vespucci (1454-1512): Italian navigator
Symbols for plus (+) and minus (-) first appear in print (1456)
Vasco da Gama (ca. 1460-1524): Portuguese explorer
Desiderius Erasmus (ca. 1466-1536): Dutch moral reformer
Niccolò Machiavelli (1469-1527): Italian politician and author of *The Prince*
Mathematical and scientific works began to be printed with new printing press
     (1470s)
Albrecht Dürer (1471-1528): German artist
Nicholaus Copernicus (1473-1543): Polish astronomer; heliocentrism
Pope Leo X (1475-1521)
Michelangelo (1475-1564): Italian artist
Spanish Inquisition inaugurated (1479)
Ferdinand Magellan (ca. 1480-1521): Portuguese explorer
First printed edition of Euclid's *Elements* (1482)
Martin Luther (1483-1546): German reformer
Michael Stifel (1487-1567): German mathematician; law of exponents
Henry VIII (1491-1547): King of England; founder of the Church of England
Christopher Columbus discovered the New World (1492)
William Tyndale (ca. 1492-1536): English reformer and Bible translator
Luca Pacioli's *Summa de arithmetica, geometrica, proportioni et proportionalita* (1494)
Philipp Melanchthon (1497-1560): German reformer
Girolamo Cardano (1501-1576): Italian mathematician; negative and complex
     numbers, cubic equations and quartic equations
Francis Xavier (1506-1552): Spanish Jesuit missionary; Apostle of the Indies
Luca Pacioli's *De Divina proportione* (1509)
Reign of Henry VIII (1509-1547)
John Calvin (1509-1564): French/Swiss reformer
Robert Recorde (ca. 1510-1558): English mathematician
John Knox (ca. 1513-1572): Scottish reformer; founder of Presbyterianism in Scot-
     land
Most Greek works translated and published (1515)
Desiderius Erasmus' Greek New Testament (1516)
John Foxe (1516-1587): English clergyman
Martin Luther's 95 Theses posted on the church door in Wittenberg (1517)
Martin Luther excommunicated (1521)
The Peasants' Revolt erupted (1524)
William Tyndale's New Testament published (1525)
Beginning of the Anabaptist movement (1525)
Johann Müller's *De triangulis omnimodis libri quinque* printed (1533)

John Calvin's *Institutes of the Christian Religion* (1536)
François Viète (1540-1603): French mathematician; symbolic algebra
Nicholaus Copernicus' *De Revolutionibus* (1543)
Michael Stifel's *Arithmetica integra* (1544)
Girolamo Cardano's *Ars magna* (1545)
Council of Trent (1545-1563)
Tycho Brahe (1546-1601): Danish astronomer
Simon Stevin (1548-1620): Dutch mathematician; symbolic algebra
Book of Common Prayer (1549)
Francis Xavier began mission to Japan (1549)
John Napier (1550-1617): Scottish mathematician; logarithms
Matteo Ricci (1552-1610): Italian Jesuit missionary, mathematician, and scientist
Joost Bürgi (1552-1632): Swiss clockmaker; logarithms
Robert Recorde's *The Whetstone of Witte* introduced two long parallel lines (=) as a
     symbol for equality (1557)
Elizabeth became Queen of England (1558)
Bartholemäus Pitiscus (1561-1613): German clergyman; trigonometry
Francis Bacon (1561-1626): English philosopher; radical empiricism
Henry Briggs (1561-1631): English mathematician; logarithms and long division
John Foxe's *Book of Martyrs* published (1563)
William Shakespeare (1564-1616): English author and playwright
Galileo Galilei (1564-1642): Italian mathematician; laws of falling bodies, parabolic
     path of projectiles, motions of pendulums, mechanics and the strength of ma-
     terials
King James I (1566-1625): first king of the United Kingdom; King James Bible
     named after him
Jan Lippershey (ca. 1570-ca. 1619): Dutch spectacle maker; telescope
Johannes Kepler (1571-1630): German astronomer; laws of planetary motion
Saint Bartholomew's day massacre of the Protestant Huguenots in France (1572)
Supernova discovered in the Constellation Cassiopeia by Tycho Brahe (1572)
William Oughtred (1574-1660): English mathematician; slide rule
Edmund Gunter (1581-1626): English astronomer and mathematician; primitive
     slide rule
Matteo Ricci began mission to China (1582)
Calendar reform of Pope Gregory XIII – loss of ten days to correct drift in calen-
     dar (1582)
Thomas Hobbes (1588-1679): English mechanistic cosmology philosopher and
     political theorist
John Napier's *A Plaine Discovery of the whole Revelation of Saint John* (1593)
First printing of *Trigonometriae sive de dimensione triangulorum libri quinque* by Bar-
     tholemäus Pitiscus (1595)
Johannes Kepler's first edition of *Mysterium cosmographicum* (1596)
René Descartes' (1596-1650): French mathematician and philosopher, coordinate
     geometry

Oliver Cromwell (1599-1658): Lord Protector of England
The "century of genius" (17th century)
Pierre de Fermat (1601-1665): French mathematician; number theory, probability, coordinate geometry
First telescope patented by Dutch spectacle maker Jan Lippershey (1608)
John Milton (1608-1674): English poet
First Baptist church in England (1609)
King James Bible (1611)
John Napier's *Mirifici logarithmorum canonis descriptio* (1614)
John Owen (1616-1683): Puritan clergyman and author
John Wallis (1616-1703): English mathematician; arithmetical theory of limits, binomial formula, and calculus
Jeremiah Horrocks (1618-1641), English popularizer of Kepler's three laws of planetary motion
Johannes Kepler's *Epitome astronomiae Copernicanae* (1618-1621)
Thirty Years' War (1618-1648)
Johannes Kepler's *Harmonice mundi* (1619)
Savilian professorships at Oxford established (1619)
Mayflower at Plymouth Rock (1620)
First slide rules invented (1620-1630)
Johann Heinrich Rahn (1622-1676): German mathematician
Blaise Pascal (1623-1662): French mathematician and philosopher; conics and probability theory
Nicholaus Bernoulli (1623-1708): Swiss patriarch of the Bernoulli family
Giovanni Domenico Cassini (1625-1712): Italian-born French astronomer
Robert Boyle (1627-1691): English chemist who divorced the discipline from its alchemaic roots and known for developing gas laws
John Ray (1627-1705): English naturalist, established "species" as the basic classification of living things
John Bunyan (1628-1688): English pastor and author
Christian Huygens (1629-1695): Dutch physicist and astronomer; discovered Saturn's rings, spring regulator for clocks
William Oughtred introduces the symbol "x" to stand for multiplication (1631)
Galileo Galilei condemned by the Inquisition (1632)
Baruch Spinoza (1632-1677): Dutch philosopher and pantheist
John Locke (1632-1704): English philosopher
Christopher Wren (1632-1723): English architect, scientist, and mathematician; designed St. Paul's Cathedral in London
Robert Hooke (1635-1703): English astronomer, physicist and inventor; developed the law governing the motion of the spring and bob
Arcturus, Alpha Boötes, seen in broad daylight with a telescope (1635)
Harvard College founded (1636)
René Descartes publishes *Discourse on Method* in which "*La Géométrie*" appears as an appendix (1637)

Blaise Pascal's computing machine (1642)
Isaac Newton (1642-1727): English mathematician; calculus, theory of equations, law of gravitation, planetary motion, infinite series, hydrostatics and dynamics
Louis XIV crowned (1643)
Rene Descartes' *Principia* anticipated Isaac Newton's laws of motion (1644)
Blaise Pascal's triangle (1644)
Olaus Roemer (1644-1710): Danish astronomer; speed of light
Gottfried Wilhelm Leibniz (1646-1716): German mathematician; calculus, mechanical calculator
Westminster Confession of Faith (1646)
Charles I executed (1649)
Oliver Cromwell named Lord Protector of England (1653)
Blaise Pascal converted to Christianity (1654)
Jakob Bernoulli (1654-1705): first great Swiss mathematician of the Bernoulli family
Edmund Halley (1656-1742): English astronomer
Johann Heinrich Rahn introduced the symbol "÷" to stand for division (1659)
Royal Society of London founded (1661)
Lucasian professorships at Cambridge established (1663)
Queen Anne of England (1665-1714)
French Academy of Paris founded (1666)
Girolamo Saccheri (1667-1733): Italian Jesuit priest; rudiments of non-Euclidean geometry
John Milton's *Paradise Lost* (1667)
Mechanical multiplier invented by Gottfried Wilhelm Leibniz (1671)
Greenwich observatory founded (1675)
First calculation of the speed of light by Danish astronomer Olaus Roemer (1676)
John Bunyan's *Pilgrims Progress* (1678)
Pierre de Fermat's *Introduction to Plane and Solid Loci* appears in print (1679)
Johann Sebastian Bach (1685-1750): German composer
George Berkeley (1685-1753): Irish prelate and philosopher
George Frederick Handel (1685-1759): German-born composer
Isaac Newton's *The Principia* (1687)
England's Bloodless Revolution (1688)
Alexander Pope (1688-1744): English poet
Jakob Bernoulli's *Acta eruditorum* (1690)
John Harrison (1693-1776): English clockmaker; marine chronometer for determining the longitude at sea
Voltaire (1694-1778): French philosopher and Enlightenment leader
Nicholas Bernoulli III (1695-1726): Swiss mathematician; geometry of curves
Daniel Bernoulli I (1700-1782): Swiss mathematician; mathematical physics
John Wesley (1703-1791): British evangelist
Isaac Newton's *Opticks* (1704)
Steam engine designed by English inventor Thomas Newcomen (1705)

Johann Sebastian Bach's first musical work (1707)

Carolus Linnaeus (1707-1778): Swedish naturalist; zoological classification

Leonhard Euler (1707-1783): Swiss mathematician; calculus, complex variables, applied mathematics

FROM EULER TO GÖDEL

Popularity of Deism (17th and 18th centuries)

Samuel Johnson (1709-1784): British writer and lexicographer

David Hume (1711-1776): Scottish skeptical philosopher, advocate of radical sensorial empiricism

Jean Jacques Rousseau (1712-1778): French romantic philosopher

Denis Diderot (1713-1784): French philosopher and encyclopedist

George Whitefield (1714-1770): British evangelist and leader of the American Great Awakening

Jean Baptiste le rond d'Alembert (1717-1783): French mathematician

Adam Smith (1723-1790): Scottish political economist and philosopher

Immanuel Kant (1724-1804): German idealist philosopher

James Cook (1728-1779): British navigator and explorer

The Great Awakening in England and America (1730-1770)

Sextant independently invented by English mathematician John Hadley and American inventor Thomas Godfrey (1731)

First Moravian missionaries (1732)

George Washington (1732-1799): first president of the United States

Girolamo Saccheri's *Euclides ab omni naevo vindicatus* (1733)

Joseph Priestley (1733-1804): British chemist and discoverer of oxygen

George Berkeley's *The Analyst* (1734)

George Whitefield converted to Christianity (1735)

Joseph-Louis Lagrange (1736-1813): French mathematician and philosopher; differential equations, calculus of variations

James Watt (1736-1819): Scottish inventor and mechanical engineer; steam engine

Jean Sylvain Bailly (1736-1793): French historian of astronomy and politician

Edward Gibbon (1737-1794): British historian

John Wesley converted to Christianity (1738)

Sir William Herschel (1738-1822): German-born British astronomer

Peak of the Great Awakening (1740)

Marquis de Sade (1740-1814): French novelist and philosopher

David Hume's *A Treatise of Human Nature* (1740)

George Frederick Handel's *Messiah* (1742)

Karl Wilhelm Scheele (1742-1786): German-born Swedish chemist and discoverer of oxygen

Leonhard Euler's *Introductio in analysin infinitorum* (1748)

John Playfair (1748-1819): Scottish physicist and mathematician; articulates modern rendering of Euclid's parallel postulate.

Pierre-Simon de Laplace (1749-1827): French astronomer and mathematician; differential equations, planetary motion, probability

England's Industrial Revolution (1750-1850)

Adrien Marie Legendre (1752-1833): French mathematician; number theory and elliptic integrals

Great Britain and American colonies accepted the Gregorian calendar – eleven days dropped from the calendar (1752)

Samuel Johnson's *A Dictionary of the English Language* (1755)

Wolfgang Amadeus Mozart (1756-1791): Austrian composer

Halley's Comet reappeared as per Edmund Halley's prediction (1758)

Noah Webster (1758-1843): American lexicographer

William Wilberforce (1759-1833): British statesman and reformer

William Carey (1761-1834): British cobbler; founder of Protestant missionary movement

John Quincy Adams (1767-1848): 6th President of the United States; noted parabolic reflectors in the Statuary Hall of the Capitol Building

James Cook discovered Australia (1768)

Joseph Fourier (1768-1830): French mathematician and physicist; periodic functions and conduction of heat

James Watt made Thomas Newcomen's steam engine more efficient (1769)

Napoleon Bonaparte (1769-1821): French emperor

Ludwig van Beethoven (1770-1827): German composer

Georg Wilhelm Friedrich Hegel (1770-1831): German dialectic philosopher

William Wordsworth (1770-1850): British poet

Samuel Taylor Coleridge (1772-1834): British poet and philosopher

Nathaniel Bowditch (1773-1838): early American astronomer and mathematician

Declaration of Independence (1776)

Adam Smith's *Wealth of Nations* (1776)

Edward Gibbon's *Decline and Fall of the Roman Empire* (1776)

James Cook's Pacific voyages (1776-1778)

Johann Friedrich Herbart (1776-1841): German philosopher and educator

Carl Friedrich Gauss (1777-1855): German mathematician; number theory, differential geometry, algebra, astronomy

Immanuel Kant's *Critique of Pure Reason* (1781)

Uranus discovered by Sir William Herschel (1781)

Friedrich Wilhelm Bessel (1784-1846): Prussian astronomer

United States Constitution (1787)

First British colony of convicts arrived in Botany Bay, Australia (1788)

Adoniram Judson (1788-1850): American missionary

United States Bill of Rights (1789)

French Revolution (1789-1799)

Louis Jacques Mandé Daguerre (1789-1851): French artist and inventor; daguerreotype process for obtaining photographic prints

Augustin-Louis Cauchy (1789-1857): French mathematician; theory of functions, calculus

Michael Faraday (1791-1867): British physicist and chemist; electromagnetic induction and field theory

Samuel F. B. Morse (1791-1872): American artist and inventor

Charles Babbage (1792-1871): British mathematician and inventor; anticipated the modern electronic computer

Charles Grandison Finney (1792-1875): American evangelist and educator; changed the concept of revival from God's sovereign and gracious act to a mere matter of "right methods"

William Carey sailed for India (1793)

Worship of God abolished in France (1793)

Nikolai Ivanovich Lobachevsky (1793-1856): Russian mathematician; non-Euclidean geometry

Reign of Terror began in France (1794)

Auguste Comte (1798-1857): French positivist philosopher

Janos Bolyai (1802-1860): Hungarian mathematician; non-Euclidean geometry

Napoleon Bonaparte made Emperor (1804)

British and Foreign Bible Society founded (1804)

Lewis and Clark expedition (1804-1806)

William Rowan Hamilton (1805-1865): Irish mathematician; founder of the algebra of quaternions

Gas light introduced in London (1807)

William Wilberforce took the lead in the abolition of the slave trade (1807)

Abraham Lincoln (1809-1865): 16th president of the United States

Herman Günther Grassman (1809-1877): German mathematician; co-founder of quaternions

Charles Darwin (1809-1882): British scientist; theory of evolution

Joseph Liouville (1809-1882): French mathematician; transcendental numbers

American Board of Commissions for Foreign Missions founded (1810)

Harriet Beecher Stowe (1811-1896): American author

Adoniram Judson began missionary trip to India (1812)

David Livingstone (1813-1873): Scottish missionary and African explorer

First locomotive steam engine constructed (1814)

George Boole (1815-1864): British mathematician and logician; symbolic logic

Ludvig August Colding (1815-1888): Danish physicist; law of conservation of energy

Karl Weierstrass (1815-1897): German mathematician; theory of complex functions

Karl Marx (1818-1883): German political philosopher; founder of communism

James Prescott Joule (1818-1889): British physicist; mechanical theory of heat and first law of thermodynamics

Beguyer de Chancourtois (1820-1886): French geologist; periodic law

Augustin-Louis Cauchy eliminated all logical inconsistencies from the limit concept of the calculus (1821)

Hermann Ludwig von Helmholz (1821-1894): German scientist; physiology, optics, acoustics, and electrodynamics

Arthur Cayley (1821-1895): British mathematician; inventor of matrices

Louis Pasteur (1822-1895): French chemist and biologist; microbiology, pasteurization

Charles Hermite (1822-1901): French mathematician; transcendence of e

Leopold Kronecker (1823-1891): German mathematician; theory of equations, elliptic functions

William Thomson or Lord Kelvin (1824-1907): Irish physicist; law of conservation of energy

First railway opened in England (1825)

William Spottiswoode (1825-1883): British mathematician

T. H. Huxley (1825-1895): British biologist; champion of the theory of evolution

George Friedrich Bernhard Riemann (1826-1866): German mathematician; integration theory, complex variables, non-Euclidean geometry

Noah Webster's *An American Dictionary of the English Language* (1828)

William Booth (1829-1912): English evangelist; Salvation Army

Julius Lothar Meyer (1830-1895): German chemist; periodic law

James Clerk Maxwell (1831-1879): British physicist; electromagnetic theory and the kinetic theory of gases

Louis Paul Cailletet (1832-1913): French metallurgist; liquefaction of oxygen

Andrew Dickson White (1832-1918): popularizer of warfare between science and Christianity paradigm

Alfred B. Nobel (1833-1896): Swedish chemist and engineer; bequeathed his fortune to institute the Nobel Prizes

Charles Spurgeon (1834-1892): British "prince of preachers"

Gottlieb Daimler (1834-1900): German engineer; automobile manufacturer, internal combustion engine

Dmitri Ivanovich Mendeléeff (1834-1907): Russian chemist; law of periodicity of atomic elements

Eugenio Beltrami (1835-1900): Italian mathematician

Samuel Langhorne Clemens (1835-1910): American author and humorist who wrote under the pen name of Mark Twain

First electric telegraph invented by Samuel F. B Morse (1837)

John Newlands (1837-1898): British chemist; periodic law

Reign of British Queen Victoria (1837-1901)

First star distances measured by Friedrich Wilhelm Bessel (1838)

Ernst Mach (1838-1916): Austrian physicist

First photographic likeness by daguerreotype (1839)

Charles Sanders Peirce (1839-1914): American pragmatic philosopher and mathematician

David Livingstone sailed for Africa (1840)

First penny postage stamp, the penny post, introduced in England (1840)

Edouard Lucas (1842-1891): French mathematician; Lucas numbers, prime numbers

Existence of transcendental numbers proved by Joseph Liouville (1844)

Georg Cantor (1845-1918): Russian-born German mathematician; theory of infinite sets

William Conrad Roentgen (1845-1923): German physicist; X rays

Neptune discovered by German astronomer Johann Gottfried Galle (1846)

First American postage stamps printed (1847)

Alexander Graham Bell (1847-1922): Scottish-born American inventor; telephone

Thomas Edison (1847-1931): American inventor

Karl Marx's *The Communist Manifesto* (1848)

George Romanes (1848-1894): Canadian naturalist and psychologist

Felix Klein (1849-1925): German mathematician; geometry, group theory

Ivan Petrovich Pavlov (1849-1936): Russian psychologist; reflex responses

First star, Vega or Alpha Lyrae, photographed by daguerreotype (1850)

Harriet Beecher Stowe's *Uncle Tom's Cabin* (1851)

Henri Poincaré (1854-1912): French mathematician and physicist; topology, differential equations

Sigmund Freud (1856-1939): Austrian physician; psychoanalysis

Heinrich Hertz (1857-1894): German physicist; radio waves

First Atlantic cable laid (1858)

Max Planck (1858-1947): German physicist; quantum mechanics

Charles Darwin's *The Origin of the Species* (1859)

American War between the States (1860-1865)

Herman Hollerith (1860-1929): American inventor; founder of IBM

David E. Smith (1860-1944): American mathematics historian

Pierre Duhem (1861-1916): French scientist and mathematician; dynamics and seminal science historian

Alfred North Whitehead (1861-1947): British philosopher and mathematician; logicism

David Hilbert (1862-1943): German mathematician; integral equations, formalist foundations in mathematics

Henry Ford (1863-1947): American automobile manufacturer; gasoline powdered automobile

London Mathematical Society founded (1865)

Pasteurization process developed by Louis Pasteur (1865)

Dynamite invented by Alfred B. Nobel (1866)

Karl Marx's *Das Kapital* (1867)

Wilbur Wright (1867-1912): American aviator

Charles Homer Haskins (1870-1937): American science historian

Orville Wright (1871-1948): American aviator

Bertrand Russell (1872-1970): British philosopher and mathematician (logicism)

Transcendence of e proved by Charles Hermite (1873)

Gilbert K. Chesterton (1874-1936): British writer and conservative critic
Winston Churchill (1874-1965): British Prime Minister; writer
Microphone invented by Alexander Graham Bell (1876)
Telephone patented by Alexander Graham Bell (1876)
Phonograph invented by Thomas Edison (1877)
James Jeans (1877-1946): British astronomer, mathematician, and physicist; kinetic theory of gases
Godfrey H. Hardy (1877-1947): British pure mathematician
Lighting by electricity inaugurated (1878)
William Booth started the Salvation Army (1878)
Incandescent lamp invented by Thomas Edison (1879)
Joseph Stalin (1879-1953): Communist dictator of Russia
Albert Einstein (1879-1955): German-born American nuclear physicist; general and special theory of relativity
Philip E. B. Jourdain (1879-1914): British mathematician
Oswald Spengler (1880-1936): German philosopher
Max Caspar (1880-1956): German historian; recognized dean of Kepler scholars
Sir Alexander Fleming (1881-1955): British bacteriologist; penicillin
William M. Ivins, Jr. (1881-1961): American art curator
Arthur Stanley Eddington (1882-1944): British astronomer, mathematician, and physicist
Franklin Delano Roosevelt (1882-1945): 32nd president of the United States
Percy W. Bridgman (1882-1961): American physicist, 1946 Nobel Prize winner in high-pressure physics
Lynn Thorndike (1882-1965): American science historian
Luitzen Brouwer (1882-1966): Dutch mathematician; topology, intuitionism (constructivism)
Mihály Babits (1883-1941): Hungarian poet
Richard E. Von Mises (1883-1953): Austrian mathematician and positivist philosopher
Tobias Dantzig (1884-1956): Russian-born American mathematician
Kenneth Scott Latourette (1884-1968): American church historian
Internal combustion engine invented by Gottlieb Daimler (1885)
Hermann Weyl (1885-1955): German mathematician
Niels Bohr (1885-1962): Danish nuclear physicist; quantum mechanics
John W. N. Sullivan (1886-1937): British scientific philosopher
Erwin Schrödinger (1887-1961): Austrian nuclear physicist
E. N. da Costa Andrade (1887-1971): British physicist
John Logie Baird (1888-1946): British electrical engineer; television, radar, fiber optics
Richard Courant (1888-1972): German-born American mathematician
Robin G. Collingwood (1889-1943): British philosopher
Adolf Hitler (1889-1945): German dictator
Edwin Hubble (1889-1953): American astronomer

Arnold Toynbee (1889-1975): British historian

Girolamo Saccheri's *Euclides ab omni naevo vindicatus* marketed by Eugenio Beltrami (1889)

Herman Hollerith used punched cards to facilitate the tabulation of the United States census (1890)

Alexandre Koyré (1892-1964): French science historian

Louis-Victor de Broglie (1892-1987): French physicist; quantum mechanics.

Gasoline powered automobile developed by Henry Ford (1893)

Clive Staples Lewis (1893-1963): British writer and critic; professor of medieval literature

Mao Tse-Tung (1893-1976): Chinese dictator

Omar Bradley (1893-1981): American World War II general

First theater showing of motion pictures (1895)

X rays discovered by William Conrad Roentgen (1895)

Cornelius Van Til (1895-1987): Dutch-born American pioneer in presuppositional worldview analysis

John L. Synge (1897-1995): Irish physicist

Arend Heyting (1898-1980): Dutch mathematician; intuitionism

Will Durant (1898-1981): American historian; wrote *The Story of Civilization* over a period of 40 years

Salomon Bochner (1899-1982): Austrian-born American mathematician

Jorge Luis Borges (1899-1986): Argentine short story writer

Otto Neugebauer (1899-1990): Austrian science historian

Joseph Needham (1900-1995): British science historian; specialized in the history of science in China

Boxer Rebellion in China (1900)

Werner Karl Heisenberg (1901-1976): German nuclear physicist

Ernest Nagel (1901-1985): American mathematician

Charles Lindbergh (1902-1974): American aviator

Paul A. M. Dirac (1902-1984): British mathematician, physicist, and 1933 Nobel Prize winner in atomic theory

Eugene Wigner (1902-1995): American nuclear physicist and 1963 Nobel Prize winner in the structure of the atom and its nucleus

Orville Wright's first successful flight of a manned, heavier-than-air, self-propelled aircraft (1903)

Ford Motor Company founded (1903)

John von Neumann (1903-1957): Hungarian-born American mathematician; quantum mechanics, game theory, computers

Louis Leakey (1903-1972): British paleoanthropologist

Marshall Stone (1903-1989): American mathematician

Albert Einstein's Special Theory of Relativity (1905)

Neon signs first displayed (1905)

Anneliese Maier (1905-1971): German historian specializing in medieval science

Carl G. Hempel (1905-1997): German-born American mathematician

San Francisco earthquake (1906)

Chester F. Carlson (1906-1968): American physicist and inventor; xerography

Carl Boyer (1906-1976): American mathematics historian

Kurt Gödel (1906-1978): Austria/Hungarian born American logician; mathematical systems and incompleteness theorems

Grace Murray Hopper (1906-1992): American navy officer, mathematician, and pioneer in data processing

Reijer Hooykaas (1906-1994): Dutch science historian

James R. Newman (1907-1966): American mathematics historian

Loren Eiseley (1907-1977): American evolutionary anthropologist

Lynn T. White (1907-1987): American professor of medieval history

Jacob Bronowski (1908-1974): Polish-born British mathematician/humanist

Morris Kline (1908-1992): American mathematician; electromagnetism and mathematics historian

Ford's Model T introduced (1909)

Edwin Land (1910-1991): American inventor; Polaroid camera

William Carroll Bark (1910-1997): American professor of medieval history

Robert K. Merton (1910-): American social historian

Howard Eves (1911-2004): American mathematics historian

Walter W. Sawyer (1911-): British mathematics professor and mathematics popularizer

R. M. S. Titanic sinks south of the Grand Banks of Newfoundland on its maiden voyage (1912)

Francis Schaeffer (1912-1984): American philosopher and worldview analyst

Verna Hall (1912-1987): American historian

Thomas F. Torrance (1913-): Chinese-born American theologian and science historian

World War I (1914-1918)

Sir Fred Hoyle (1915-): British astronomer and mathematician

Alistair C. Crombie (1915-1996): British science historian

Richard W. Hamming (1915-1998): American mathematician

Albert Einstein's General Theory of Relativity (1916)

Friedrich Heer (1916-1983): Austrian medieval historian

Francis Crick (1916-): British biophysicist; DNA

Rousas J. Rushdoony (1916-2001): American reformed theologian; Van Til popularizer and educator

Marshall Clagett (1916-): American science historian

Russian revolution (1917)

Alexandr Solzhenitsyn (1918-): Soviet dissident and writer

Henry Morris (1918-): American Creation Science pioneer

Oswald Spengler's *Decline of the West* (1918-1922)

Richard P. Feynman (1918-1988): American physicist; 1965 Nobel Prize winner in physics

Jean Gimpel (1918-1996): Medieval historian

Sir Edmond Hillary (1919-): New Zealand mountaineer and explorer
A. Rupert Hall (1920-): Science historian
John Glenn (1921-): American astronaut and politician
Thomas Kuhn (1922-1996): American historian and philosopher of science
John Saxon (1923-1996): Lt. Colonel, US Air Force; founder and author of Saxon
  math
Reuben Hersh (1923-): American mathematician
Chuck Yeager (1923-): American test pilot and United States Air Force officer
IBM founded (1924)
Stanley L. Jaki (1924-): Hungarian-born American nuclear physicist and science
  historian
Adolf Hitler's *Mein Kampf* (1925)
Television demonstrated by John Logie Baird (1926)
Werner Heisenberg's principle of indeterminacy (1927)
Solo transatlantic flight by Charles Lindbergh (1927)
First talking motion picture (1927)
Philip J. Davis (1927-): American mathematician
Penicillin discovered by Sir Alexander Fleming (1928)
James Watson (1928-): American biochemist; DNA
Crash of the Stock Market (1929)
Pluto discovered by American astronomer Clyde William Tombaugh (1930)
Neil Armstrong (1930-): American astronaut
Clive Staples Lewis converted to Christianity (1931)
Kurt Gödel's Incompleteness Theorems (1931)

AFTER GÖDEL

Franklin Delano Roosevelt's four presidential terms (1933-1945)
Fluorescent lamp invented (1934)
Wycliffe Bible Translators founded (1934)
Yuri Gagarin (1934-1968): Soviet cosmonaut
Paul J. Cohen (1934-): American mathematician; Cantor's continuum hypothesis
Douglas DC-3 propeller airliner service inaugurated (1935)
Radar invented by British physicist Sir Robert Watson-Watt (1935)
German air ship Hindenburg engulfed in flames over New Jersey (1937)
Xerography invented by American printer Chester F. Carlson (1937)
Flight of first turbojet aircraft in Germany (1939)
John Frame (1939-): American reformed theologian; apologetics
World War II (1939-1945): atomic research, mathematics and codebreaking
Pearl Harbor bombed (1941)
Stephen Jay Gould (1941-): American paleontologist; theory of evolution popular-
  izer
Stephen Hawking (1942-): British theoretical physicist and popularizer of Big Bang
  cosmogony

John Mauchley, an American physicist, and J. Presper Eckert, an American engineer, build the first successful general digital computer, the ENIAC (1942-1946)

Atomic bombs dropped on Hiroshima and Nagasaki (1945)

Dead Sea Scrolls discovered (1947)

Polaroid camera invented by Edwin Land (1947)

Transistor invented (1947)

Supersonic flight achieved by Chuck Yeager flying the X-1 rocket plane (1947)

Long playing records invented (1948)

Greg L. Bahnsen (1948-1995): American reformed theologian; renowned public defender of the Christian faith

Nicholas Bourbaki (ca. 1950): pseudonym for a group of French mathematicians

Christian missionaries forced to leave China (1950)

Korean War (1950-1953)

The summit of Mount Everest attained by Sir Edmond Hillary and Tenzing Norgay (1953)

Color television perfected (1953)

Andrew Wiles (1953-): British mathematician; solver of Fermat's "Last Theorem"

Vietnam War (1954-1975)

Bill Gates (1956-): American entrepreneur; founder of Microsoft

FORTRAN, a scientific programming language, developed by IBM (1956)

Russian Sputnik launched (1957)

Boeing 707 jet service inaugurated (1958)

COBOL, a business programming language, developed by Grace Hopper (1959-1961)

Laser invented by American research physicist Theodore Mairman (1960)

Yuri Gagarin, Soviet cosmonaut, became the first person to travel in space (1961)

Cuban Missile crisis (1962)

Discovery of DNA genetic code by Francis Crick and James Watson (1962)

Astronaut John Glenn became the first American to orbit the Earth (1962)

Paul J. Cohen showed that Cantor's continuum hypothesis is undecidable (1963)

John F. Kennedy, 35th President of the United States, assassinated in Dallas, Texas (1963)

The ARPAnet, a decentralized network and the genesis of the Internet, created by the Department of Defense to facilitate communications in the event of a nuclear attack (1969)

Apollo 11: first men, Neil Armstrong and Edwin Aldrin, landed on the moon (1969)

First Boeing 747 jumbo jet flight (1970)

Apollo 13: rupture of oxygen tank forced the cancellation of moon mission and heroic return to earth four days later (1970)

Apollo 17: last men on the moon (1972)

Richard Nixon, 37th President of the United States, forced to resign from office (1974)

Microsoft Corporation founded (1975)

The Altair 8800 is considered to be the first personal computer (1975)

Concorde supersonic air service inaugurated (1976)

Eruption of Mount St. Helens (1979)

Keuffel & Esser, a leading manufacturer of scientific instruments, ceased production of slide rules (1980)

First IBM personal computer with MS-DOS operating system (1981)

First US space shuttle launched (1981)

The Apple Macintosh personal computer featured icons and windows dialog (1984)

Challenger Space Shuttle disaster (1986)

The World Wide Web developed (1989)

Hubble Space Telescope deployed (1990)

Microsoft Corporation introduced its first version of the Windows operating system (1990)

Andrew Wiles of Princeton University solved Fermat's famous "Last Theorem" (1994)

BIBLIOGRAPHY

Abbott, Edwin A. *Flatland: A Romance of Many Dimensions.* New York: Dover Publications, [1884] 1992.

Aczel, Amir D. *Fermat's Last Theorem.* New York: Dell Publishing, 1996.

Adams, Jay. *Back to the Blackboard.* Phillipsburg: Presbyterian and Reformed, 1982.

Adler, Mortimer J. *How to Read a Book.* New York: Simon and Schuster, 1967.

Alic, Margaret. *Hypatia's Heritage: A History of Women in Science from Antiquity through the Nineteenth Century.* Boston: Beacon Press, 1986.

Aristotle. *The Metaphysics.* trans. John H. McMahon. Amherst: Prometheus Books, 1991.

Armitage, Angus. *John Kepler.* London: Faber, 1966.

Arnold, Sir Thomas and Alfred Guillaume, eds. *The Legacy of Islam.* London: Oxford University Press, 1931.

Artmann, Benno. *Euclid: The Creation of Mathematics.* New York: Springer-Verlag, 1999.

Asimov, Isaac. *The Intelligent Man's Guide to Science.* 2 vol. New York: Basic Books, 1960.

____. *Understanding Physics.* Dorset Press, [1966] 1988.

Atkins, P. W. *Periodic Kingdom.* New York: Basic Books, 1995.

Augustinus, Aurelius. *City of God.* trans. Marcus Dods. New York: Modern Library, 1950.

____. "On Christian Doctrine." *Great Books of the Western World: Augustine.* vol. 18. ed. R. M. Hutchins. Chicago: Encyclopaedia Britannica, 1952.

____. *The Confessions of St. Augustine.* Grand Rapids: Baker Book House, 1977.

Bacon, Francis. *The Works of Francis Bacon.* ed. J. Spedding, R. L. Ellis, D. D. Heath. 14 vol. Boston: Taggard and Thompson, 1857-1874.

Bahnsen, Greg L. *Always Ready: Directions for Defending the Faith.* Atlanta: American Vision and Texarkana: Covenant Media Foundation, 1996.

____. *Van Til's Apologetic: Readings and Analysis.* Phillipsburg: Presbyterian and Reformed, 1998.

Bailly, Jean Sylvain. *Historie de l'astronome ancienne depuis son origine jusqu'à l'établissment de l'Ecole d'Alexandrie.* Paris: chez les Frères Debure, 1775.

Ball, W. W. Rouse. *A Short Account of the History of Mathematics.* New York: Dover Publications, [1908] 1960.

Banks, Robert B. *Slicing Pizzas, Racing Turtles, and Further Adventures in Applied Mathematics*. Princeton: Princeton University Press, 1999.

Bark, William Carroll. *Origins of the Medieval World*. Stanford: Stanford University Press, 1958.

Barker, Stephen F. *Philosophy of Mathematics*. Englewood Cliffs, NJ: Prentice-Hall, 1964.

Barnes, Thomas G. *Physics of the Future: A Classical Unification of Physics*. El Cajon: Institute for Creation Research, 1983.

Barrow, John D. *Pi in the Sky: Counting, Thinking, and Being*. Oxford: Clarendon Press, 1992.

Baumgardt, Carola. *Johannes Kepler: Life and Letters*. New York: Philosophical Library, 1951.

Beckmann, Petr. *A History of π (pi)*. New York: St. Martin's Press, 1971.

Beechick, Ruth. *A Biblical Psychology of Learning*. Denver: Accent Books, 1982.

Behe, Michael J. *Darwin's Black Box: The Biochemical Challenge to Evolution*. New York: The Free Press, 1996.

Bell, A. E. *Christian Huygens and the Development of Science in the Seventeenth Century*. London: Eduard Arnold and Company, 1947.

Bell, Eric Temple. *Mathematics: Queen and Servant of Science*. New York: McGraw-Hill, 1951.

___. *Men of Mathematics*. New York: Simon and Schuster, [1937, 1965] 1986.

___. *The Development of Mathematics*. New York: Dover Publications, [1940, 1945] 1992.

___. *The Magic of Numbers*. New York: Dover Publications, [1946] 1991.

Benacerraf, Paul and Hilary Putnam, ed. *Philosophy of Mathematics: Selected Readings*. Cambridge: Cambridge University Press, [1964] 1983.

Bentley, W. A. and W. J. Humphreys. *Snow Crystals*. New York: Dover Publications, [1931] 1962.

Bergamini, David. *Mathematics*. Alexandria, VA: Time-Life Books, 1980.

Berkhof, Louis. *The History of Christian Doctrines*. Grand Rapids: Baker, [1937] 1975.

Berlinski, David. *A Tour of the Calculus*. New York: Vintage Books, 1995.

___. *The Advent of the Algorithm*. New York: Harcourt, 2000.

Berman, Bob. *Secrets of the Night Sky*. New York: HarperCollins, 1996.

Berry, Arthur. *A Short History of Astronomy from Earliest Times through the Nineteenth Century*. New York: Dover Publications, [1898] 1961.

Black, Jim Nelson. *When Nations Die*. Wheaton: Tyndale House Publishers, 1994.

Blackwell, William. *Geometry and Architecture*. Berkeley: Key Curriculum Press, 1994.

Blamires, Harry. *Repair the Ruins: Reflections on Educational Matters from the Christian Point of View*. London: Geoffrey Bles, 1950.

___. *The Christian Mind: How Should a Christian Think?* Ann Arbor: Servant Books, 1963.

Bloom, Allan. *The Closing of the American Mind*. New York: Simon and Schuster, 1987.

Boa, Kenneth D. and Robert M. Bowman, Jr. *An Unchanging Faith in a Changing World*. Nashville: Thomas Nelson, 1997.

Bochner, Salomon. *The Role of Mathematics in the Rise of Science*. Princeton: Princeton University Press, 1966.

Bohm, David. *Causality and Chance in Modern Physics*. London: Routledge & Kegan Paul, 1957.

Boice, James Montgomery. *Foundations of the Christian Faith*. Downers Grove: InterVarsity Press, 1986.

Bonola, Roberto. *Non-Euclidean Geometry*. New York: Dover Publications, [1912] 1955.

Boorstin, Daniel J. *The Discoverers*. New York: Vintage, [1983] 1985.

Born, Max. *Natural Philosophy of Cause and Chance*. New York: Oxford University Press, 1949.

Bourbaki, Nicholas. "The Architecture of Mathematics." *American Mathematical Monthly*, 57 (1950), 221-232.

Boyer, Carl. *A History of Mathematics*. rev. Uta C. Merzbach. New York: John Wiley & Sons, [1968] 1991.

____. *The History of the Calculus and Its Conceptual Development*. New York: Dover Publications, [1949] 1959.

____. *The Rainbow: From Myth to Mathematics*. New York: Thomas Yoseloff, 1959.

Brabenec, Robert, ed. *A Christian Perspective on the Foundations of Mathematics*. Wheaton: Wheaton College, 1977.

____. *A Fifth Conference on Mathematics from a Christian Perspective*. Wheaton: Wheaton College, 1985.

____. *A Fourth Conference on Mathematics from a Christian Perspective*. Wheaton: Wheaton College, 1983.

____. *A Sixth Conference on Mathematics from a Christian Perspective*. Wheaton: Wheaton College, 1987.

____. *A Third Conference on Mathematics from a Christian Perspective*. Wheaton: Wheaton College, 1981.

Bray, Gerald. *The Doctrine of God*. Downers Grove: InterVarsity Press, 1993.

Bridgman, Percy W. *The Logic of Modern Physics*. New York: Macmillan, 1927.

Broglie, Louis-Victor de. *Physics and Microphysics*. trans. M. Davidson. New York: Pantheon Books, 1955.

Bronowski, Jacob. *Science and Human Values*. Harmondsworth Middlesex: Pelican, 1964.

Brooke, John Hedley. *Science and Religion: Some Historical Perspectives*. Cambridge: Cambridge University Press, 1991.

Brown, Colin. *Philosophy and the Christian Faith*. Downers Grove: InterVarsity Press, 1968.

Brown, Lloyd A. *The Story of Maps*. New York: Dover Publications, [1949] 1979.

Brown, Vinson. *The Amateur Naturalist's Handbook*. Englewood Cliffs, NJ: Prentice-Hall, 1980.

Budiansky, Stephen. *Battle of Wits: The Complete Story of Codebreaking in World War II*. New York: The Free Press, 2000.

Bunim, Miriam. *Space in Medieval Painting and the Forerunners of Perspective*. New York: Columbia University Press, 1940.

Bunt, Lucas N. H., Phillip S. Jones, and Jack D. Bedient. *The Historical Roots of Elementary Mathematics*. New York: Dover Publications, [1976] 1988.

Burke, James. *Connections*. Boston: Little, Brown and Company, 1978.

____. *The Day the Universe Changed*. Boston: Little, Brown and Company, 1985.

____. *The Pinball Effect*. Boston: Little, Brown and Company, 1996.

Burnham, Robert, Jr. *Burnham's Celestial Handbook*. 3 vol. New York: Dover Publications, 1978.

Burtt, Edwin A. *Metaphysical Foundations of Modern Science*. London and New York: Routleluge & Kegal Paul, 1925.

Butler, Pierce. *The Origin of Printing in Europe*. Chicago: University of Chicago Press, [1940] 1966.

Butterfield, Herbert. *The Origins of Modern Science*. New York: The Macmillan Company, 1961.

Byrne, H. W. *A Christian Approach to Education*. Milford: Mott Media, 1977.

Cahill, Thomas. *How the Irish Saved Civilization: The Untold Story of Ireland's Heroic Role from the Fall of Rome to the Rise of Medieval Europe*. New York: Doubleday, 1995.

Cajori, Florian. *A History of Mathematical Notations*. New York: Dover Publications, [1928, 1929] 1993.

Calvin, John. *Institutes of the Christian Religion*. trans. Henry Beveridge. Grand Rapids: Eerdmans, [1536, 1559] 1989.

Campbell, Lewis and William Garnett. *The Life and Times of James Clerk Maxwell*. London: Macmillan and Company, 1882.

Campbell, Norman. *What is Science?* New York: Dover Publications, [1921] 1952.

Carson, Clarence. *Basic Economics*. Wadley, AL: American Textbook Committee, 1988.

Caspar, Max. *Kepler*. trans. C. Doris Hellman. New York: Dover Publications, [1959] 1993.

Casti, John L. and Werner DePauli. *Gödel: A Life of Logic*. New York: Perseus Publishing, 2000.

Chancellor, W. E. *A Theory of Motives, Ideals and Values in Education*. Boston and New York, 1907.

Chapman, Colin. *The Case for Christianity*. Icknield Way, England: Lion Publishing, 1981.

Chase, Gene, ed. *A Seventh Conference on Mathematics from a Christian Perspective*. Grantham, PA: Messiah College, 1989.

Chenu, M. D. *Nature, Man, and Society in the Twelfth Century*. Chicago: University of Chicago Press, [1957] 1968.

Chilton, David. *Productive Christians in an Age of Guilt Manipulators*. Tyler, TX: Institute of Christian Economics, [1981] 1985.

Cicero, *Tusculan Disputations*. trans. J. E. King. Cambridge, MA: Harvard University Press, 1951.

Clagett, Marshall. *Studies in Medieval Physics and Mathematics*. London: Variorum, 1979.

___. *The Science of Mechanics in the Middle Ages*. Madison: University of Wisconsin Press, 1959.

Clark, Gordon H. *A Christian Philosophy of Education*. Grand Rapids: Eerdmans, 1969.

___. *The Philosophy of Science and Belief in God*. Nutley, NJ: The Craig Press, 1964.

Clawson, Calvin C. *Mathematical Mysteries: The Beauty and Magic of Numbers*. Cambridge: Perseus, 1996.

Cochrane, Charles Norris. *Christianity and Classical Culture*. New York: Oxford University Press, 1957.

Cohen, Jack and Ian Stewart. *The Collapse of Chaos: Discovering Simplicity in a Complex World*. New York: Penguin, 1994.

Collingwood, Robin G. *An Essay on Metaphysics*. London: Oxford University Press, 1940.

Collins, A. Frederick. *Rapid Math Without a Calculator*. New York: Carol Publishing Group, [1956, 1987] 1989.

Colson, Charles and Nancy Pearcey. *How Now Shall We Live?* Wheaton: Tyndale House Publishers, 1999.

Comstock, Anna Botsford. *Handbook of Nature Study*. Ithaca and London: Cornell University Press, [1911, 1939, 1967, 1974] 1986.

Conner, Kevin J. *The Name of God*. Portland: Bible Press, 1975.

Cook, Theodore A. *The Curves of Life*. New York: Dover Publications, [1914] 1979.

Copernicus, Nicholaus. *On the Revolution of Heavenly Spheres*. trans. Charles Glenn Wallis. Amherst: Prometheus Books, [1543, 1939] 1995.

Copi, Irving M. *Introduction to Logic,* 10th ed. Englewood Cliffs, NJ: Prentice-Hall, 1998.

Courant, Richard. "Mathematics in the Modern World." *Scientific American*, 211 (1964), 48-49.

Courant, Richard and Harold Robbins. *What is Mathematics? An Elementary Approach to Ideas and Methods*. rev. Ian Stewart. New York: Oxford University Press, [1941] 1996.

Coxeter, H. S. M. *Introduction to Geometry*. New York: John Wiley & Sons, 1961.

Crombie, Alistair C. *Robert Grosseteste and the Origins of Experimental Science, 1100-1700*. Oxford: Clarendon Press, 1953.

___. *The History of Science from Augustine to Galileo*. New York: Dover Publications, [1959, 1970, 1979] 1995.

___., ed. *Scientific Change*. New York: Basic Books, 1963.

Crosby, Alfred W. *The Measure of Reality: Quantification and Western Society, 1250-1600*. Cambridge: Cambridge University Press, 1997.

Culver, Roger B. *Astronomy*. New York: Barnes and Noble, 1979.

Cummings, David B. *The Purpose of Christian Education.* Phillipsburg: Presbyterian and Reformed, 1979.

Curtius, E. R. *European Literature and the Latin Middle Ages.* trans. W. R. Trask. London: Routledge and Kegal Paul, 1953.

Dabney, R. L. *On Secular Education.* ed. Douglas Wilson. Moscow, ID: Canon Press, 1996.

Dalton, LeRoy C. *Algebra in the Real World.* Palo Alto: Dale Seymour Publications, 1983.

Dampier-Whetham, William C. D. *A History of Science and Its Relations with Philosophy and Religion.* Cambridge: Cambridge University Press, 1966.

Dantzig, Tobias. *Number: The Language of Science.* New York: Doubleday Anchor, [1930] 1954.

Dauben, Joseph W. *Georg Cantor: His Mathematics and Philosophy of the Infinite.* Cambridge, MA: Harvard University Press, 1979.

Davis, John Jefferson. *Your Wealth in God's World.* Phillipsburg: Presbyterian and Reformed, 1984.

Davis, Philip J. *The Lore of Large Numbers.* New York: Random House, 1961.

Davis, Philip J. and Reuben Hersh. *Descartes' Dream: The World According to Mathematics.* Boston: Houghton Mifflin, 1986.

___. *The Mathematical Experience.* Boston: Houghton Mifflin, 1981.

Dawson, Christopher. *Religion and the Rise of Western Culture.* New York: Doubleday, [1957], 1991.

Dawson, John W. Jr. *Logical Dilemmas: The Life and Work of Kurt Gödel.* Wellesley: A K Peters, 1997.

DeJong, James A. *As the Waters Cover the Sea: Millennial Expectations in the Rise of Anglo-American Missions 1640-1810.* Kampen: J. H. Kok N.V., 1970.

DeJong, Norman. *Education in the Truth.* Phillipsburg: Presbyterian and Reformed, 1969.

DeMar, Gary. *War of the Worldviews: A Christian Defense Manual.* Atlanta: American Vision, 1994.

Denton, Michael. *Evolution: A Theory in Crisis.* Bethesda, MD: Adler & Adler, 1985.

Devlin, Keith. *Mathematics: The New Golden Age.* London: Penguin Books, 1988.

___. *The Language of Mathematics: Making the Invisible Visible.* New York: W. H. Freeman, [1998] 2000.

Dimmick, Walter F. *The Christian History of Science.* Gilroy: Dimmick Research Labs, [1990] 1993.

Dirac, Paul A. M. "The Evolution of the Physicist's Picture of Nature." *Scientific American,* 208 (1963), 53.

Dixon, Robert. *Mathographics.* New York: Dover Publications, [1987] 1991.

Donald in Mathmagic Land. Burbank: The Walt Disney Company, 1959.

Downing, Douglas. *Dictionary of Mathematical Terms.* Hauppauge, NY: Barron's Educational Series, [1987] 1995.

Dreyer, John L. H. *A History of Astronomy from Thales to Kepler.* New York: Dover Publications, [1906] 1953.

___. *Tycho Brahe: A Picture of the Scientific Life and Work in the Sixteenth Century*. New York: Dover Publications, [1890] 1963.

Duhem, Pierre. *Le Systéme du monde: Histoire Des Doctrines Cosmologiques*. 10 vol. Paris: Librairie Scientifique A. Hermann et fils, 1913-1959.

___. *Medieval Cosmology*. Chicago: University of Chicago Press, 1985.

___. *The Aim and Structure of Physical Theory*. trans. Philip P. Weiner. Princeton: Princeton University Press, 1954.

___. *To Save the Phenomena: An Essay on the Idea of Physical Theory from Plato to Galileo*. Chicago: University of Chicago Press, [1908] 1969.

Duncan, David E. *Calendar: Humanity's Epic Struggle to Determine a True and Accurate Year*. New York: Avon, 1998.

Dye, David L. *Faith and the Physical World: A Comprehensive Review*. Grand Rapids: Eerdmans, 1966.

Dykes, Charles. "Medieval Speculation, Puritanism, and Modern Science." *The Journal of Christian Reconstruction: Symposium on Puritanism and Progress* (6:1), ed. Gary North. Vallecito, CA: Chalcedon, 1979.

Dzielska, Maria. *Hypatia of Alexandria*. New York: Harvard University Press, 1995.

Easton, Stewart C. *Roger Bacon and His Search for a Universal Science*. New York: Columbia University Press, 1952.

Eddington, Arthur Stanley. *Space, Time and Gravitation: An Outline of the General Theory of Relativity*. Cambridge: Cambridge University Press, 1920.

Ehrlich, Eugene. *Amo, Amas, Amat and More*. New York: Harper & Row, 1985.

Einstein, Albert. *Essays in Science*. New York: Philosophical Library, 1934.

___. *Lettres À Maurice Solovine*. Paris: Gauthier-Villars, 1956.

___. *Out of My Later Years*. New York: Citadel Press, [1950, 1956, 1984] 1991.

___. *What I Believe*. London: George Allen & Unwin, 1966.

Einstein, Albert and Leopold Infeld. *The Evolution of Physics: The Growth of Ideas from the Early Concepts to Relativity and Quanta*. Cambridge: Cambridge University Press, 1938.

Eiseley, Loren. *Darwin's Century*. Garden City: Doubleday and Company, 1958.

Elliot, Gil. *Twentieth Century Book of the Dead*. New York: Charles Scribner's Sons, 1972.

Engel, Morris S. *With Good Reason*, 6th ed. New York: Bedford Books/St. Martin's Press, 2000.

Euler, Leonhard. *Introduction to the Analysis of the Infinite, Book II*. trans. John D. Blanton. New York: Springer-Verlag, [1748] 1990.

Evans, Christopher. *The Making of the Micro: A History of the Computer*. Oxford: Oxford University Press, 1983.

Eves, Howard. *An Introduction to the History of Mathematics*. New York: Holt, Rhinehart and Winston, [1953, 1964, 1969] 1976.

Fahie, J. J. "The Scientific Works of Galileo." *Studies in the History and Method of Science*. ed. Charles Singer. New York: Oxford University Press, 1921.

Fakkema, Mark. *Christian Philosophy and Its Educational Implications*. Chicago: National Association of Christian Schools, 1952.

Farrington, Benjamin. *Greek Science*. Nottingham: Spokesman, 1980.

Feynman, Richard P. *Six Easy Pieces*. Reading: Helix Books, [1963] 1996.

___. *The Character of Physical Law*. Cambridge, MA: The MIT Press, 1967.

___. *The Pleasure of Finding Things Out*. Cambridge, MA: Perseus Books, 1999.

Field, J. V. *Kepler's Geometrical Cosmology*. Chicago: University of Chicago Press, 1988.

Fieldbook. Irving, TX: Boy Scouts of America, [1944] 1967.

Flansburg, Scott. *Math Magic*. New York: HarperCollins, [1993] 1994.

Fourier, Joseph. *The Analytical Theory of Heat*. trans. Alexander Freeman. Cambridge: Cambridge University Press, 1878.

Frame, John. *Apologetics to the Glory of God*. Phillipsburg: Presbyterian and Reformed, 1994.

___. *Cornelius Van Til: An Analysis of His Thought*. Phillipsburg: Presbyterian and Reformed, 1995.

France, R. T. *The Living God*. London: Inter-Varsity Press, 1972.

Frank, Douglas W. *Less Than Conquerors: How Evangelicals Entered the Twentieth Century*. Grand Rapids: Eerdmans, 1986.

Friedman, Milton & Rose. *Free to Choose: A Personal Statement*. New York: Harcourt Brace Jovanovich, [1979] 1980.

Funkenstein, Amos. *Theology and the Scientific Imagination from the Middles Ages to the Seventeenth Century*. Princeton: Princeton University Press, 1986.

Gaebelein, Frank E. *The Pattern of God's Truth*. Chicago: Moody Press, 1968.

Galilei, Galileo. *Dialogues Concerning the Two Chief World Systems: Ptolemaic and Copernican*. trans. Stillman Drake. Berkeley: University of California Press, 1962.

___. *Dialogues Concerning Two New Sciences*. trans. Henry Crew and Alfonso de Salvio. New York: Dover Publications, [1914] 1954.

___. *Discoveries and Opinions of Galileo*. trans. Stillman Drake. Garden City: Doubleday, 1957.

Gamow, George. *One, Two, Three – Infinity*. New York: The Viking Press, 1947.

___. *Thirty Years That Shook Physics*. New York: Dover Publications, [1966] 1985.

Gardner, Martin. *Aha! Gotcha: Paradoxes to Puzzle and Delight*. San Francisco: W. H. Freeman, 1982.

___. *Aha! Insight*. San Francisco: W. H. Freeman, 1978.

Gauss, Karl Friedrich. *Werke*. 12 vol. Göttingen: Königliche Akademie der Wissenschaftern, 1877.

Gay, Peter. *The Enlightenment: The Rise of Modern Paganism*. New York: W. W. Norton, [1966] 1977.

Geisler, Norman L. and Paul D. Feinberg. *Introduction to Philosophy*. Grand Rapids: Baker Book House, 1980.

Gentry, Kenneth L., Jr. *The Greatness of the Great Commission*. Tyler, TX: Institute for Christian Economics, 1990.

Ghyka, Matila. *The Geometry of Art and Life*. New York: Dover Publications, [1946] 1977.

Gibilisco, Stan and Norman Crowhurst. *Mastering Technical Mathematics.* New York: McGraw-Hill, [1961] 1999.

Gies, Frances & Joseph. *Cathedral, Forge, and Waterwheel: Technology and Invention in the Middle Ages.* New York: HarperCollins, 1994.

___. *Leonardo of Pisa and the New Mathematics of the Middle Ages.* New York: Thomas Y. Crowell, 1969.

Gillings, Richard. *Mathematics in the Time of the Pharaohs.* New York: Dover Publications, [1972] 1982.

Gilson, Étienne. *The Christian Philosophy of St. Thomas Aquinas.* trans. L. K. Shook. New York: Random House, 1956.

___. *The Spirit of Medieval Philosophy.* London: Sheed and Ward, 1936.

Gimpel, Jean. *The Medieval Machine: The Industrial Revolution of the Middle Ages.* New York: Penguin Books, 1976.

Gleick, James. *Chaos: Making a New Science.* New York: Penguin Books, 1987.

Gödel, Kurt. *On Formally Undecidable Propositions of Principia Mathematica and Related Systems.* New York: Dover Publications, [1931, 1962] 1992.

Gore, Charles, ed. *Thoughts on Religion.* London: Longmans Green, 1895.

Gould, Rupert T. *The Marine Chronometer: Its History and Development.* London: J. D. Potter, 1923.

Gould, Stephen J. *Ever Since Darwin.* New York: W. W. Norton, 1977.

Grant, Edward, ed. *A Source Book in Medieval Science.* Cambridge, MA: Harvard University Press, 1974.

Grant, Edward and John F. Murdoch, eds. *Mathematics and Its Applications to Science and Natural Philosophy in the Middle Ages: Essays in Honor of Marshall Clagett.* Cambridge: Cambridge University Press, 1987.

Grant, George. *Bringing in the Sheaves: Transforming Poverty into Production.* Atlanta: American Vision Press, 1985.

Grattan-Guiness, Ivor. *The Rainbow of Mathematics: A History of the Mathematical Sciences.* New York: W. W. Norton, [1997] 2000.

Greenstein, George. *The Symbiotic Universe.* New York: William Morrow and Company, Inc., 1988.

Gregory, John Milton. *The Seven Laws of Teaching.* Grand Rapids: Baker, [1884, 1917, 1954] 1975.

Grun, Bernard. *The Timetables of History.* New York: Simon and Schuster, 1982.

Guillen, Michael. *Bridges to Infinity: The Human Side of Mathematics.* London: Rider & Company, 1983.

___. *Five Equations that Changed the World: The Power and Poetry of Mathematics.* New York: Hyperion, 1995.

Gullberg, Jan. *Mathematics: From the Birth of Numbers.* New York: W. W. Norton, 1997.

Hahn, Alexander J. *Basic Calculus from Archimedes to Newton to its Role in Science.* New York: Spring-Verlag, 1998.

Haldane, Elizabeth S. *Descartes, His Life and Times.* London: J. Murray, 1905.

Haldene, John B. S. *Everything Has a History.* London, 1951.

Hall, A. Rupert. *From Galileo to Newton*. New York: Dover Publications, [1963] 1981.

Haller, William. *The Rise of Puritanism*. New York: Harper Torchbooks, [1938] 1957.

Hallyn, Fernand. *The Poetic Structure of the World: Copernicus and Kepler*. trans. Donald M. Leslie. New York: Zone Books, 1990.

Hamming, Richard W. "The Unreasonable Effectiveness of Mathematics." *American Mathematical Monthly*, 87 (1980), 81-90.

Hansen, Bert. *Nicole Oresme and the Marvels of Nature*. Toronto: Pontifical Institute of Mediaeval Studies, 1985.

Hardy, Godfrey H. *A Mathematician's Apology*. Cambridge: Cambridge University Press, 1967.

Hartkopf, Roy. *Math Without Tears*. Boston: G. K. Hall & Co., [1965] 1985.

Haskins, Charles Homer. *The Renaissance of the Twelfth Century*. Cambridge, MA: Harvard University Press, 1927.

___. *The Rise of Universities*. Providence: Brown University Press, 1923.

___. *Studies in the History of Mediaeval Science*. Cambridge, MA: Harvard University Press, 1924.

Haycock, Ruth. *Bible Truth for School Subjects: Science/Mathematics*. vol. 3. Whittier: Association of Christian Schools International, 1981.

Hazlitt, Henry. *Economics in One Lesson*. New York: Arlington House Publishers, [1946, 1962] 1979.

Heath, Thomas L. *A History of Greek Mathematics*. 2 vol. New York: Dover Publications, [1921] 1981.

___. *A Manual of Greek Mathematics*. New York: Oxford University Press, 1931.

___. *Aristarchus of Samos: The Ancient Copernicus*. New York: Dover Publications, [1913] 1981.

___. *Euclid: The Thirteen Books of The Elements*. 3 vol. New York: Dover Publications, [1925] 1956.

Heer, Friedrich. *The Medieval World*. trans. Janet Sondheimer. New York: New American Library, 1961.

Hegeman, David Bruce. *Plowing in Hope: Toward a Biblical Theology of Culture*. Moscow, ID: Canon Press, 1999.

Heilbron, J. L. *Geometry Civilized: History, Culture, and Technique*. New York: Oxford University Press, [1998] 2000.

___. *The Sun in the Church: Cathedrals as Solar Observatories*. Cambridge, MA: Harvard University Press, 1999.

Heisenberg, Werner. *The Physical Principles of the Quantum Theory*. trans. C. Eckart and F. C. Hoyt. Chicago: University of Chicago Press, 1930.

Hersh, Reuben. *What is Mathematics, Really?* New York: Oxford University Press, 1997.

Herz-Fischler, Roger. *A Mathematical History of the Golden Number*. New York: Dover Publications, [1987] 1998.

Hewitt, Paul G. *Conceptual Physics*. Reading, MA: Addison-Wesley, 1987.

Hilbert, David. "Die Grundlagen der elementaren Zahlenlehre." *Gesammelte Ab-handlungen*, 3 (1930), 193.

____. "On the Infinite." *Philosophy of Mathematics: Selected Readings.* ed. Paul Benacerraf and Hilary Putnam. Cambridge: Cambridge University Press, [1964] 1983.

Hillis, W. Daniel. *The Pattern on the Stone: The Simple Ideas that Make Computers Work.* New York: Basic Books, 1998.

Hilton, Peter, Derek Holton, and Jean Pedersen. *Mathematical Reflections.* New York: Springer-Verlag, [1997] 1998.

Hjellström, Björn. *Be Expert with Map & Compass.* New York: Macmillan, [1955, 1967, 1975, 1976] 1994.

Hodge, Ian. *Making Sense of Your Dollars.* Vallecito, CA: Ross House Books, 1995.

Hoffecker, W. A., ed. *Building a Christian World View.* 2 vol. Phillipsburg: Presbyterian and Reformed, 1988.

Hofstadter, Richard. *Anti-intellectualism in American Life.* New York: Alfred A. Knopf, 1969.

Hogben, Lancelot. *Mathematics for the Million.* New York: W. W. Norton, [1937, 1983] 1993.

____. *The Wonderful World of Mathematics.* New York: Doubleday, 1968.

Hoggatt, Verner E. *Fibonacci and Lucas Numbers.* Boston: Houghton Mifflin Company, 1969.

Holl, Karl. *The Cultural Significance of the Reformation.* New York, 1959.

Holmes, Arthur F. *All Truth is God's Truth.* Grand Rapids: Eerdmans, 1977.

____. *Contours of a World View.* Grand Rapids: Eerdmans, 1983.

____. *Shaping Character: Moral Education in the Christian College.* Grand Rapids: Eerdmans, 1990.

____. *The Idea of a Christian College.* Grand Rapids: Eerdmans, [1975] 1987.

____., ed. *The Making of a Christian Mind.* Downers Grove: InterVarsity Press, 1985.

Hooykaas, Reijer. *Religion and the Rise of Modern Science.* Grand Rapids: Eerdmans, 1972.

Horsburgh, E. M., ed. *Handbook of the Napier Tercentenary Celebration, or Modern Instruments and Methods of Calculation.* Los Angeles: Tomash Publishers, [1914] 1982.

Hoyle, Fred. *Highlights in Astronomy.* San Francisco: W. H. Freeman, 1975.

Huber, Richard M. *How Professors Play the Cat Guarding the Cream: Why We're Paying More and Getting Less in Higher Education.* Fairfax: George Mason University Press, 1992.

Huff, Darrell. *How to Lie with Statistics.* New York: W. W. Norton, [1954, 1982] 1993.

Hughes, Tom. *Chemistry Connections.* Dubuque, IA: Kendall/Hunt Publishing, [1975] 1983.

Huizinga, Johan. *The Autumn of the Middle Ages.* Chicago: University of Chicago Press, [1921] 1996.

Humphreys, D. Russell. *Starlight and Time: Solving the Puzzle of Distant Starlight in a Young Universe.* Colorado Springs: Master Books, 1994.

Huntley, H. E. *The Divine Proportion: A Study in Mathematical Beauty.* New York: Dover Publications, 1970.

Ivins, William M. Jr. *Art and Geometry: A Study in Space Intuitions.* New York: Dover Publications, 1964.

Jackson, Jeremy. *No Other Foundation: The Church Through Twenty Centuries.* Westchester, IL: Cornerstone Books, 1980.

Jacobs, Harold R. *Elementary Algebra.* San Francisco: W. H. Freeman, 1979.

___. *Geometry.* San Francisco: W. H. Freeman, [1974] 1987.

___. *Mathematics: A Human Endeavor.* San Francisco: W. H. Freeman, 1970.

Jaeger, Werner. *Paideia: The Ideals of Greek Culture.* 3 vol. trans. Gilbert Highet. New York: Oxford University Press, [1943] 1971.

Jaffe, Bernard. *Crucibles: The Story of Chemistry.* New York: Dover Publications, 1976.

Jaki, Stanley L. *Advent and Science.* Royal Oak, MI: Real View Books, 2000.

___. *Brain, Mind and Computers.* Washington: Regnery Gateway, [1969] 1989.

___. *Christ and Science.* Royal Oak, MI: Real View Books, 2000.

___. *Cosmos and Creator.* Edinburgh: Scottish Academic Press, 1980.

___. *Galileo Lessons.* Royal Oak, MI: Real View Books, 2001.

___. *Genesis 1 Through the Ages.* Royal Oak, MI: Real View Books, [1992] 1998.

___. *Giordano Bruno: A Martyr of Science?* Royal Oak, MI: Real View Books, 2000.

___. *God and the Cosmologists.* Royal Oak, MI: Real View Books, [1989] 1998.

___. *Is There a Universe?* New York: Wethersfield Institute, 1993.

___. *Jesus, Islam, Science.* Royal Oak, MI: Real View Books, 2001.

___. *Maybe Alone in the Universe, After All.* Royal Oak, MI: Real View Books, 2000.

___. *Planets and Planetarians.* Edinburgh: Scottish Academic Press, 1978.

___. *Science and Creation: From Eternal Cycles to an Oscillating Universe.* Edinburgh: Scottish Academic Press, 1974.

___. *The Milky Way: An Elusive Road for Science.* Devon: David & Charles, 1973.

___. *The Origin of Science and the Science of Its Origins.* Edinburgh: Scottish Academic Press, 1978.

___. *The Paradox of Olbers' Paradox.* Royal Oak, MI: Real View Books, [1969] 2000.

___. *The Relevance of Physics.* Edinburgh: Scottish Academic Press, [1966] 1992.

___. *The Road of Science and the Ways to God.* Edinburgh: Scottish Academic Press, 1978.

___. *The Savior of Science.* Grand Rapids: Eerdmans, [1988] 2000.

James, J. *The Contractors of Chartres.* 2 vol. London: Croom Helm, [1979] 1981.

Jeans, James. *Physics and Philosophy.* New York: Dover Publications, [1943] 1981.

___. *Science and Music.* New York: Dover Publications, [1937] 1968.

___. *The Growth of Physical Science.* Cambridge: Cambridge University Press, 1951.

___. *The Mysterious Universe.* New York: Macmillan, 1930.

Jeffrey, Richard C. *Formal Logic: Its Scope and Limits.* New York: McGraw-Hill Book Company, 1967.

Jehle, Paul. *Go Ye Therefore and Teach.* Plymouth: Plymouth Rock Foundation, 1982.

Jespersen, James and Jane Fitz-Randolph. *From Sundials to Atomic Clocks.* New York: Dover Publications, [1977] 1982.

Johnson, Paul. *Intellectuals*. New York: Harper & Row, 1988.

___. *Modern Times: From the Twenties to the Nineties*. New York: HarperCollins, [1983] 1991.

___. *The Birth of the Modern: World Society 1815-1830*. New York: HarperCollins, 1991.

Johnson, Phillip E. *Darwin on Trial*. Downers Grove: InterVarsity Press, [1991] 1993.

___. *Defeating Darwinism by Opening Minds*. Downers Grove: InterVarsity Press, 1997.

___. *Objections Sustained: Subversive Essays on Evolution, Law & Culture*. Downers Grove: InterVarsity Press, 1998.

___. *Reason in Balance: The Case Against Naturalism in Science, Law & Education*. Downers Grove: InterVarsity Press, 1995.

___. *The Wedge of Truth: Splitting the Foundations of Naturalism*. Downers Grove: InterVarsity Press, 2000.

Jones, Douglas and Douglas Wilson. *Angels in the Architecture: A Protestant Vision of Middle Earth*. Moscow, ID: Canon Press, 1998.

Jones, Richard F. *Ancients and Moderns: A Study of the Rise of the Scientific Movement in Seventeenth Century England*. New York: Dover Publications, [1936, 1961] 1982.

Jordan, James. *Creation in Six Days: A Defense of the Traditional Reading of Genesis One*. Moscow, ID: Canon Press, 1999.

___. "Stanley Jaki on Genesis One." *Biblical Chronology*, 10:3 (1988).

___. *Through New Eyes: Developing a Biblical View of the World*. Brentwood: Wolgemuth & Hyatt, 1988.

Kac, Mark and Stanislaw M. Ulam. *Mathematics and Logic*. New York: Dover Publications, [1968] 1992.

Kaiser, Christopher B. *Creation and the History of Science*. Grand Rapids: Eerdmans, 1991.

Kant, Immanuel. *Critique of Pure Reason*. Chicago: Encyclopedia Britannica, 1952.

___. *Metaphysical Foundations of Natural Science*. Indianapolis: Bobbs-Merrill, 1970.

Kaplan, Robert. *The Nothing That Is: A Natural History of Zero*. Oxford: Oxford University Press, 1999.

Kasner, Edward and James R. Newman. *Mathematics and the Imagination*. Redmond, WA: Microsoft Press, [1940] 1989.

Kelly, Douglas F. *Creation and Change: Genesis 1.1-2.4 in the Light of Changing Scientific Paradigms*. Ross-shire, England: Christian Focus Publications, [1997] 1999.

Kelly, Gerard W. *Short-Cut Math*. New York: Dover Publications, [1969] 1984.

Kennedy, D. James and Jerry Newcombe. *What If Jesus Had Never Been Born?* Nashville: Thomas Nelson, 1994.

Kepler, Johannes. *Epitome of Copernican Astronomy & Harmonies of the World*. trans. Charles Glenn Wallis. Amherst: Prometheus Books, [1618-1621, 1939] 1995.

___. *Johannes Kepler Gesammelte Werke*. ed. Walther von Dyck, Max Caspar, Franz Hammer, Volker Bialas. Munich: C. H. Beck, 1937-.

___. *Mysterium Cosmographicum – The Secret of the Universe.* trans. A. M. Duncan. New York: Abaris Books, 1981.

___. *New Astronomy.* trans. William H. Donahue. Cambridge: Cambridge University Press, 1992.

___. *The Six-Cornered Snowflake.* trans. Colin Hardie. Oxford: Clarendon Press, 1966.

Klein, Jacob. *Greek Mathematical Thought and the Origin of Algebra.* New York: Dover Publications, [1968] 1992.

Klein, Felix. *Elementary Mathematics from an Advanced Standpoint.* vol. 1. New York: The Macmillan Company, 1932.

Kline, Morris. *Calculus: An Intuitive and Physical Approach.* New York: John Wiley and Sons, [1967] 1977.

___. *Mathematical Thought from Ancient to Modern Times.* New York: Oxford University Press, 1972.

___. *Mathematics: A Cultural Approach.* Reading, MA: Addison-Wesley Publishing Company, 1962.

___. *Mathematics and the Physical World.* New York: Dover Publications, [1959] 1980.

___. *Mathematics and the Search for Knowledge.* New York: Oxford University Press, 1985.

___. *Mathematics for the Nonmathematician.* New York: Dover Publications, [1967] 1985.

___. *Mathematics in Western Culture.* New York: Oxford University Press, 1953.

___. *Mathematics: The Loss of Certainty.* New York: Oxford University Press, 1980.

___. *Why Johnny Can't Add: The Failure of the New Mathematics.* New York: Vintage, 1974.

___. *Why the Professor Can't Teach: The Dilemma of University Education.* New York: St. Martin's Press, 1977.

___., ed. *Mathematics: An Introduction to Its Spirit and Use.* San Francisco: W. H. Freeman, 1978.

Kneller, Karl A. *Christianity and the Leaders of Modern Science.* Royal Oak, MI: Real View Books, [1911] 1995.

Knight, David C. *Johannes Kepler and Planetary Motion.* London: Chatto & Windus, 1965.

Knott, Cargill Gilston, ed. *Napier Tercentenary Memorial Volume.* London: Longmans, Green, and Company, 1915.

Koestler, Arthur. *The Watershed: a Biography of Johannes Kepler.* Garden City: Doubleday, 1960.

Kolb, Rocky. *Blind Watchers of the Sky.* Reading, MA: Addison-Wesley Publishing Company, 1996.

Körner, Stephan. *The Philosophy of Mathematics: An Introductory Essay.* New York: Dover Publications, [1960] 1986.

Koyré, Alexandre. *From the Closed World to the Infinite Universe.* Baltimore: Johns Hopkins University Press, 1957.

____. *Metaphysics and Measurement: Essays in the Scientific Revolution.* London: Chapman & Hall, 1968.

Kozhamthdam, Job. *The Discovery of Kepler's Laws: The Interaction of Science, Philosophy and Religion.* Notre Dame: University of Notre Dame Press, 1993.

Kuhn, Thomas S. *The Copernican Revolution.* New York: Modern Library, 1957.

____. *The Structure of Scientific Revolutions.* Chicago: The University of Chicago Press, [1962] 1970.

Kurtz, Paul, ed. *Humanist Manifestos I and II.* Amherst: Prometheus Books, 1973.

Kuyk, Willem. "The Irreducibility of the Number Concept." *Philosophia Reformata,* 31 (1966), 37-50.

Kuyper, Abraham. *Christianity: A Total World and Life System.* Marlborough, NH: Plymouth Rock Foundation, [1956] 1996.

____. *Lectures on Calvinism.* Grand Rapids: Eerdmans, [1931] 1987.

Langdon, Nigel and Charles Snape. *A Way with Maths.* Cambridge: Cambridge University Press, 1984.

Latourette, Kenneth Scott. *A History of Christianity.* 2 vol. New York: Harper and Row, [1953] 1975.

Lawlor, Robert. *Sacred Geometry.* New York: Thames and Hudson, [1982] 1994.

Lee, Nigel. *The Central Significance of Culture.* Philadelphia: Presbyterian and Reformed, 1976.

Leibniz, Gottfried Wilhelm. *Leibniz Selections.* ed. Philip P. Wiener. New York: Charles Scribner's Sons, 1951.

____. *New Essays Concerning Human Understanding.* trans. Alfred Gideon Langley. New York: Macmillan, 1896.

Lewis, Clive S. *The Abolition of Man.* New York: Macmillan Publishing, [1947] 1955.

Lial, Margaret L. and Charles D. Miller. *Essential Calculus with Applications.* Glenview, IL: Scott, Foresman and Company, [1975] 1980.

Lindberg, David C. *The Beginnings of Western Science.* Chicago: University of Chicago Press, 1992.

Lindberg, David C. and Ronald L. Numbers, eds. *God and Nature: Historical Essays on the Encounter between Christianity and Science.* Berkeley and Los Angeles: University of California Press, 1986.

Lindberg, David C. and Robert S. Westman, eds. *Reappraisals of the Scientific Revolution.* Cambridge: Cambridge University Press, 1990.

Lovejoy, Arthur. *The Great Chain of Being: A Study of the History of an Idea.* Cambridge, MA: Harvard University Press, [1936] 1964.

Lowe, Ivan. "Christian Mathematician, Where Are You?" *Translation,* (January-March 1971), 6, 7, 14.

Lucas, Jerry. *Becoming a Mental Math Wizard.* White Hall, VA: Shoe Tree Press, 1991.

Lucretius. *On the Nature of Things.* trans. Charles E. Bennett. Roslyn: Walter J. Black, 1946.

Maatman, Russell. *The Unity in Creation.* Sioux Center, IA: Dordt College Press, 1978.

Macaulay, David. *Castle*. Boston: Houghton Mifflin, 1977.

___. *Cathedral: The Story of Its Construction*. Boston: Houghton Mifflin, 1973.

___. *Pyramid*. Boston: Houghton Mifflin, 1975.

___. *The Way Things Work*. Boston: Houghton Mifflin, 1988.

Machen, J. Gresham. *Education, Christianity, and the State*, ed. J. W. Robbins. Jefferson, MD: The Trinity Foundation, 1987.

MacDonald, D. B. *The Hebrew Philosophical Genius: A Vindication*. New York: Russel and Russel, [1936] 1965.

Magee, Bryan. *The Story of Thought: The Essential Guide to the History of Western Philosophy*. London: DK Publishing, 1998.

Maier, Anneliese. *On the Threshold of Exact Science: Selected Writings of Anneliese Maier on Later Medieval Natural Philosophy*. trans. Steven D. Sargent. Philadelphia: University of Pennsylvania Press, 1982.

Mandelbrot, Benoit. *The Fractal Geometry of Nature*. New York: W. H. Freeman, 1983.

Manuel, Frank. *The Religion of Isaac Newton*. London: Oxford University Press, 1974.

Maor, Eli. *e: The Story of a Number*. Princeton: Princeton University Press, 1994.

___. *To Infinity and Beyond: A Cultural History of the Infinite*. Princeton: Princeton University Press, [1987] 1991.

___. *Trigonometric Delights*. Princeton: Princeton University Press, 1998.

Marsden, George. *The Outrageous Idea of Christian Scholarship*. New York: Oxford University Press, 1997.

___. *The Soul of the American University: From Protestant Establishment to Established Nonbelief*. Oxford: Oxford University Press, 1994.

Martin, Glenn R. "Biblical Christian Education: Liberation for Leadership – An Address Given to the Marion College Faculty," *Marion College*, (August 30, 1983).

Maybury, Richard J. *Whatever Happened to Penny Candy?* Placerville, CA: Bluestocking Press, [1989, 1991] 1993.

McKay, H. *The World of Numbers*. New York: The Macmillan Company, 1946.

Menninger, Karl. *Number Words and Number Symbols: A Cultural History of Numbers*. New York: Dover Publications, [1969] 1992.

Merton, Robert K. *Science, Technology & Society in Seventeenth Century England*. New York: Howard Fertig, [1970, 1990] 2001.

Miller, Perry and Thomas Johnson. *The Puritans*. New York: Harper Torchbooks, 1963.

Minnaert, M. *The Nature of Light & Color in the Open Air*. trans. H. M. Kremer-Priest. New York: Dover Publications, 1954.

Misner, Charles, et. al. *Gravitation*. New York: W. H. Freeman, 1973.

Moore, John N. *How to Teach Origins*. Milford: Mott Media, 1983.

Morison, Samuel Eliot. *Admiral of the Ocean Sea: The Life of Christopher Columbus*. New York: MJF Books, [1942] 1970.

___. *The Great Explorers*. New York: Oxford University Press, 1978.

Moritz, Robert Edouard. *On Mathematics and Mathematicians*. New York: Dover Publications, 1958.

Morris, Henry M. *Education for the Real World*. San Diego: Master Books, [1977] 1983.

___. *Men of Science: Men of God*. San Diego: Master Books, 1984.

___. *The Biblical Basis for Modern Science*. Grand Rapids: Baker Book House, 1984.

Morton, Jean S. *Science in the Bible*. Chicago: Moody Press, 1978.

Motz, Lloyd and Jefferson Hane Weaver. *Conquering Mathematics*. New York: Plenum, 1991.

___. *The Story of Mathematics*. New York: Avon, 1993.

Muir, Jane. *Of Men and Numbers: The Story of Great Mathematicians*. New York: Dover Publications, [1961] 1996.

Murray, Charles. *The Bell Curve*. New York: The Free Press, 1994.

Murray, Iain H. *Revival & Revivalism: The Making and Marring of American Evangelicalism 1750-1858*. Edinburgh: The Banner of Truth Trust, 1994.

___. *The Puritan Hope: Revival and the Interpretation of Prophecy*. Edinburgh: The Banner of Truth Trust, [1971] 1991.

Nagel, Ernest and James R. Newman. *Gödel's Proof*. London: Routledge and Kegal Paul, 1958.

Nahin, Paul J. *An Imaginary Tale: The Story of $\sqrt{-1}$*. Princeton: Princeton University Press, 1998.

Narlikar, Jayant V. *The Lighter Side of Gravity*. New York: W. H. Freeman, 1982.

Needham, Joseph. *Science and Civilization in China*. 13 vol. Cambridge: Cambridge University Press, 1954.

___. *The Grand Titration: Science and Society in East and West*. London: George Allen and Unwin, 1969.

Neill, Stephen. *A History of Christian Missions*. London: Penguin Books, [1964] 1986.

Neugebauer, Otto. *The Exact Sciences in Antiquity*. New York: Dover Publications, [1957] 1969.

Newman, James R., ed. *The World of Mathematics*. 4 vol. New York: Simon and Schuster, 1956.

Newman, John Henry. *The Idea of a University*. ed. Frank M. Turner. New Haven, CT: Yale University Press, 1966.

Newman, Rochelle and Martha Boles. *Universal Patterns*. Bradford, MA: Pythagorean Press, [1983] 1992.

Newton, Isaac. *Opticks*. New York: Dover Publications, [1704, 1931, 1952], 1979.

___. *The Principia* or *The Mathematical Principles of Natural Philosophy*. trans. Andrew Motte. Amherst: Prometheus Books, [1687, 1848] 1995.

Newton, Richard. *Nature's Mighty Wonders*. London: S. W. Partridge, 1871.

Noebel, David A. *Understanding the Times*. Eugene: Harvest House, 1991.

Nussenzveig, H. Moyses. "The Theory of the Rainbow." *Scientific American*, 236 (1977), 116-127.

Ore, Oystein. *Number Theory and Its History*. New York: Dover Publications, [1948] 1976.

Oresme, Nicole. *Le Livre du ciel et du monde*. ed. Albert D. Menut and Alexander J. Denomy. trans. Albert D. Menut. Madison: University of Wisconsin Press, 1968.

Osserman, Robert. *Poetry of the Universe: A Mathematical Exploration of the Cosmos*. New York: Anchor Books Doubleday, 1995.

Packer, James I. *Knowing God*. Downers Grove: InterVarsity Press, 1973.

Panek, Richard. *Seeing and Believing: How the Telescope Opened Our Eyes and Minds to the Heavens*. New York: Viking, 1998.

Pappas, Theoni. *Mathematical Footprints*. San Carlos: World Wide Publishing/Tetra, 1999.

____. *More Joy of Mathematics*. San Carlos: World Wide Publishing/Tetra, 1991.

____. *The Joy of Mathematics*. San Carlos: World Wide Publishing/Tetra, 1989.

____. *The Magic of Mathematics: Discovering the Spell of Mathematics*. San Carlos: World Wide Publishing/Tetra, 1994.

Parsons, Edward A. *The Alexandrian Library*. Amsterdam: The Elsevier Press, 1952.

Pascal, Blaise. *Pensées*. trans. W. F. Trotter. New York: E. P. Dutton, 1958.

Paulos, John A. *Beyond Numeracy: Ruminations of a Numbers Man*. New York: Alfred A. Knopf, 1991.

____. *Innumeracy: Mathematical Illiteracy and Its Consequences*. New York: Hill and Wang, 1988.

Pearcey, Nancy R. and Charles B. Thaxton. *The Soul of Science: Christian Faith and Natural Philosophy*. Wheaton: Crossways Books, 1994.

Perks, Stephen C. *The Christian Philosophy of Education Explained*. Whitby: Avant Books, 1992.

Péter, Rózsa. *Playing with Infinity: Mathematical Explorations and Excursions*. New York: Dover Publications, [1961] 1976.

Pines, S. "What was Original in Arabic Science?" *Scientific Change*. ed. A. C. Crombie. New York: Basic Books, 1963.

Planck, Max. "Wissenschaftliche Selbstbiographie." *Physikalische Abhandlungen*, 3 (1948), 374.

Plantinga, Alvin and Nicholas Wolterstroff, ed. *Faith and Rationality: Reason and Belief in God*. Notre Dame: University of Notre Dame Press, 1983.

Plato. *The Collected Dialogues of Plato*. ed. Edith Hamilton and Huntington Cairns. Princeton: Princeton University Press, 1961.

Poincaré, Henri. *The Foundations of Science*. trans. George Bruce Halsted. New York and Garrison: The Science Press, 1913.

Pólya, George. *How to Solve It*. Princeton: Princeton University Press, [1945, 1957] 1971.

Pough, Frederick H. *A Field Guide to Rocks and Minerals*. Boston: Houghton Mifflin, [1953, 1955] 1960.

Poythress, Vern S. "A Biblical View of Mathematics." *Foundations of Christian Scholarship*. ed. Gary North. Vallecito, CA: Ross House, 1976.

____. "Creation and Mathematics." *The Journal of Christian Reconstruction: Symposium on Creation* (1:1). ed. Gary North. Vallecito, CA: Chalcedon, 1974.

_____. *Philosophy, Science and the Sovereignty of God.* Philadelphia: Presbyterian and Reformed, 1976.

_____. "Science as Allegory-Mathematics as Rhyme." *A Third Conference on Mathematics from a Christian Perspective.* ed. Robert L. Brabenec. Wheaton: Wheaton College, 1981.

Pratt, Richard L., Jr. *Every Thought Captive.* Phillipsburg: Presbyterian and Reformed, 1979.

Ptolemy, Claudius. *Almagest.* trans. G. J. Tooner. London: Gerald Duckworth, 1984.

Rademacher, Hans and Otto Toeplitz. *The Enjoyment of Mathematics.* New York: Dover Publications, [1966] 1990.

Ramm, Bernard. *A Christian View of Science and Scripture.* Grand Rapids: Eerdmans, 1954.

_____. *The Christian College in the Twentieth Century.* Grand Rapids: Eerdmans, 1963.

Randall, John. *The Making of the Modern Mind.* New York: Columbia University Press, [1926] 1940.

Ray, John. *The Wisdom of God Manifested in the Works of the Creation.* London: R. Harbin, [1691] 1717.

Reader, John. *Missing Links.* London: Collins, 1981.

Redondi, Pietro. *Galileo: Heretic.* Princeton: Princeton University Press, 1987.

Rehwinkel, Alfred M. *The Wonders of Creation.* Grand Rapids: Baker, 1974.

Reichenbach, Hans. *From Copernicus to Einstein.* New York: Dover Publications, [1942] 1980.

Reid, Constance. *From Zero to Infinity.* New York: Thomas Y. Crowell, 1955.

Reimer, Luetta and Wilbert. *Mathematicians are People, Too.* 2 vol. Palo Alto: Dale Seymour Publications, 1990, 1995.

_____. *Historical Connections in Mathematics: Resources for Using History of Mathematics in the Classroom.* 3 vol. Fresno: AIMS Education Foundation, 1992, 1993, 1995.

Resnikoff, H. L. and R. O. Wells, Jr. *Mathematics in Civilization.* New York: Dover Publications, [1973] 1984.

Roche, George. *Education in America.* Hillsdale: Hillsdale College Press, 1977.

_____. *The Fall of the Ivory Tower: Government Funding, Corruption, and the Bankrupting of American Higher Education,* Washington: Regnery, 1994.

Romer, Alfred. *The Restless Atom: The Awakening of Nuclear Physics.* New York: Dover Publications, [1960] 1982.

Rose, Tom. *Economics: Principles and Policy from a Christian Perspective.* Mercer, PA: American Enterprise Publications, 1986.

_____. *Economics: The American Economy from a Christian Perspective.* Mercer, PA: American Enterprise Publications, 1985.

Rosen, Edward, trans. *Kepler's Conversation with Galileo's Sidereal Messenger.* New York: Johnson Reprint, 1965.

Rothman, Tony. *Instant Physics: From Aristotle to Einstein, and Beyond.* New York: Byron Press, 1995.

Rucker, Rudy. *The Fourth Dimension.* Boston: Houghton Mifflin Company, 1984.

Ruffini, Remo. "The Princeton Galaxy." Interviews by Florence Heltizer. *Intellectual Digest*, 3 (1973), 27.

Runion, Garth E. *The Golden Section and Related Curiosa.* Glenview, IL: Scott, Foresman and Company, 1972.

Runner, H. Evan. *The Relation of the Bible to Learning.* Toronto: Wedge Publishing Foundation, 1970.

Rushdoony, Rousas J. *By What Standard?* Tyler, TX: Thoburn Press, [1958] 1983.

___. *Intellectual Schizophrenia: Culture, Crisis and Education.* Phillipsburg: Presbyterian and Reformed, [1961] 1980.

___. *Systematic Theology.* 2 vol. Vallecito, CA: Ross House, 1994.

___. *The Biblical Philosophy of History.* Nutley, NJ: The Craig Press, 1969.

___. *The Messianic Character of American Education.* Nutley, NJ: The Craig Press, 1963.

___. *The Mythology of Science.* Nutley. NJ: The Craig Press, [1967] 1979.

___. *The One and the Many: Studies in the Philosophy of Order and Ultimacy.* Fairfax: Thoburn Press, 1978.

___. *The Philosophy of the Christian Curriculum.* Vallecito, CA: Ross House, 1981.

___. *The World of Flux.* Fairfax: Thoburn Press, 1975

___. *World History Notes.* Fairfax: Thoburn Press, 1974.

Russell, Bertrand. "Recent Work on the Principles of Mathematics." *The International Monthly*, 4 (1901), 84.

___. *The Autobiography of Bertrand Russell.* 3 vol. London: George Allen and Unwin, 1969.

___. *The Impact of Science on Society.* New York: Columbia University Press, 1951.

___. *Why I am Not a Christian.* New York: Simon and Schuster, 1966.

Russell, Colin. *Cross-Currents: Interactions Between Science and Faith.* Grand Rapids: Eerdmans, 1985.

Ryken, Leland. *Worldly Saints: The Puritans As They Really Were.* Grand Rapids: Zondervan, 1986.

Saccheri, Girolamo. *Euclides Vindicatus.* trans. and ed. George Bruce Halsted. New York: Chelsea Publishing, 1986.

Sahakian, William S. *History of Philosophy.* New York: Barnes and Noble, 1968.

Sawyer, Walter W. *A Path to Modern Mathematics.* Harmondsworth Middlesex: Penguin, 1966.

___. *Prelude to Mathematics.* Harmondsworth Middlesex: Penguin, 1955.

___. *Mathematician's Delight.* Harmondsworth Middlesex: Penguin, [1943] 1957.

___. *Vision in Elementary Mathematics.* Harmondsworth Middlesex: Penguin, 1964.

___. *What is Calculus About?* New York: Random House, 1961.

Sayers, Dorothy. *The Lost Tools of Learning.* Canberra: Light Educational Ministries, [1947] 1996.

Schaeffer, Francis. *He is There and He is Not Silent.* Wheaton: Tyndale House Publishers, 1972.

___. *How Should We Then Live? The Rise and Decline of Western Thought and Culture.* Old Tappan: Revell, 1976.

___. *The Complete Works of Francis A. Schaeffer: A Christian Worldview.* 5 vol. Westchester, IL: Crossway Books, 1982.

Schaaf, Fred. *40 Nights to Knowing the Sky.* New York: Henry Holt, 1998.

___. *Wonders of the Sky.* New York: Dover Publications, 1983.

Schlect, Chris. *The Christian Worldview and Apologetics.* Moscow, ID: Logos School, 1998.

Schlossberg, Herbert. *Idols for Destruction.* Nashville: Thomas Nelson, 1983.

Schlossberg, Herbert and Marvin Olasky. *Turning Point: A Christian Worldview Declaration.* Westchester, IL: Crossway Books, 1987.

Schoenflies, A. "Die Krisis in Cantor's Mathematischen Schaffen." *Acta Mathematica,* 50 (1927), 2.

Schrödinger, Erwin. *What is Life? The Physical Aspects of the Living Cell.* Cambridge: Cambridge University Press, 1945.

Scott, Otto. *The Great Christian Revolution.* Windsor, NY: The Reformer Library, 1994.

Seidenberg, A. "The Ritual Origin of Counting." *Archive for the History of Exact Sciences,* 2 (1962-1966), 1-40.

___. "The Ritual Origin of Geometry." *Archive for the History of Exact Sciences,* 1 (1960-1962), 488-527.

Senior, John. *The Death of Christian Culture.* Harrison, NY: RC Books, [1978] 1994.

Sennholz, Hans F., ed. *Public Education and Indoctrination.* Irvington-on-Hudson: The Foundation for Economic Education, 1993.

Singer, C. Gregg. *From Rationalism to Irrationality: The Decline of the Western Mind from the Renaissance to the Present.* Phillipsburg: Presbyterian and Reformed, 1979.

Singh, Jagjit. *Great Ideas of Modern Mathematics: Their Nature and Use.* New York: Dover Publications, 1959.

Sire, James W. *Discipleship of the Mind: Learning to Love God in the Way We Think.* Downers Grove: InterVarsity Press, 1990.

___. *Habits of the Mind: Intellectual Life as a Christian Calling.* Downers Grove: InterVarsity Press, 2000.

___. *The Universe Next Door.* Downers Grove: InterVarsity Press, [1976] 1988.

Smeltzer, Donald. *Man and Number.* Emerson Books, 1958.

Smith, David Eugene, ed. *A Source Book in Mathematics.* New York: Dover Publications, [1929] 1959.

___. *History of Mathematics.* 2 vol. New York: Dover Publications, [1923] 1958.

Smith, Karl J. *College Mathematics and Calculus: With Applications to Management, Life and Social Sciences.* Pacific Grove, CA: Brooks/Cole Publishing, 1988.

Smith, Sanderson. *Agnesi to Zeno: Over 100 Vignettes from the History of Math.* Berkeley: Key Curriculum Press, 1996.

Sobel, Dava. *Longitude.* New York: Walker and Company, 1995.

Solzhenitsyn, Aleksandr. *A World Split Apart.* New York: Farrar, Straus and Giroux, 1976.

___. *Solzhenitsyn: The Voice of Freedom.* Washington: American Federation of Labor and Congress of Industrial Organizations, 1975.

Sorabji, Richard, ed. *Philoponus and the Rejection of Aristotelian Science*. London: Duckworth, 1987.

Sowell, Thomas. *Inside American Education: The Decline, The Deception, The Dogmas*. New York: The Free Press, 1993.

Sproul, R. C. *Lifeviews*. Old Tappan: Fleming H. Revell, 1986.

___. *Not a Chance: The Myth of Chance in Modern Science & Cosmology*. Grand Rapids: Baker, 1994.

Staguhn, Gerhard. *God's Laughter: Physics, Religion, and the Cosmos*. New York: Kodansha International, 1994.

Stauffer, Ethelbert. *Christ and the Caesars*. Philadelphia: The Westminster Press, 1955.

Steele, Robert. "Roger Bacon and the State of Science in the Thirteenth Century." *Studies in the History and Method of Science*. ed. Charles Singer. New York: Oxford University Press, 1921.

Stein, Sherman K. *Mathematics: The Man-Made Universe*. San Francisco: W. H. Freeman, [1963, 1969] 1976.

___. *Strength in Numbers*. New York: John Wiley & Sons, 1996.

Stein, Sherman K. and Calvin D. Crabill. *Algebra II/Trigonometry*. San Francisco: W. H. Freeman, [1970] 1976.

Steinhaus, Hugo. *Mathematical Snapshots*. New York: Oxford University Press, 1969.

Steneck, Nicholas H. *Science and Creation in the Middle Ages*. South Bend: University of Notre Dame Press, 1976.

Stephenson, Bruce. *Kepler's Physical Astronomy*. New York: Springer-Verlag, 1987.

Stevens, Peter S. *Patterns in Nature*. Boston: Little, Brown and Company, 1974.

Stewart, Ian. *Nature's Numbers*. New York: Basic Books, 1995.

___. *The Magical Maze: Seeing the World through Mathematical Eyes*. New York: John Wiley & Sons, 1997.

Stickler, Henry. *How to Calculate Quickly*. New York: Dover Publications, [1945] 1955.

Stigler, Stephen M. *Statistics on the Table: The History of Statistical Concepts and Methods*. Cambridge, MA: Harvard University Press, 1999.

Stone, Marshall. "The Revolution in Mathematics." *American Mathematical Monthly*, 68 (1961), 715-734.

Strohmeier, John and Peter Westbrook. *Divine Harmony: The Life and Teachings of Pythagoras*. Berkeley: Berkeley Hills Books, 1999.

Struik, Dirk J. *A Concise History of Mathematics*. New York: Dover Publications, 1948.

___. *A Source Book in Mathematics 1200-1800*. Princeton: Princeton University Press, 1969.

Sutton, Oliver G. *Mathematics in Action*. New York: Dover Publications, [1954, 1957] 1984.

Synge, John L. "Focal Properties of Optical and Electromagnetic Systems." *American Mathematical Monthly*, 51 (1944), 185-187.

Tannery, Paul. *La géométrie grecque*. Paris: Gauthier-Villars, 1887.

Taylor, E. L. Hebden. "The Role of Puritan-Calvinism in the Rise of Modern Science." *The Journal of Christian Reconstruction: Symposium on Puritanism and Progress* (6:1). ed. Gary North. Vallecito, CA: Chalcedon, 1979.

Taylor, Gordon Rattray, ed. *The Inventions that Changed the World.* London: Reader's Digest Association, 1982.

Taylor, Iain T. *In the Minds of Men: Darwin and the New World Order.* Toronto: TFE Publishing, 1984.

Taylor, Lloyd W. *Physics: The Pioneer Science.* New York: Dover Publications, 1959.

The Christian Teaching of Mathematics. Greenville: Bob Jones University, 1982.

The Westminster Confession of Faith. Norcross, GA: Great Commission Publications, 1994.

Thompson, D'Arcy. *On Growth and Form.* Cambridge: Cambridge University Press, [1948] 1971.

Thompson, James Edgar. *Algebra for the Practical Worker.* New York: Van Nostrand Reinhold, [1931, 1946, 1962] 1982.

___. *Arithmetic for the Practical Worker.* New York: Van Nostrand Reinhold, [1931, 1946, 1962] 1982.

___. *Calculus for the Practical Worker.* New York: Van Nostrand Reinhold, [1931, 1946, 1962] 1982.

___. *Geometry for the Practical Worker.* New York: Van Nostrand Reinhold, [1934, 1946, 1962] 1982.

___. *Trigonometry for the Practical Worker.* New York: Van Nostrand Reinhold, [1931, 1946, 1962] 1982.

Thompson, James Westphal and Edgar Nathaniel Johnson. *An Introduction to Medieval Europe: 300-1500.* London: George Allen and Unwin, 1938.

Thompson, Silvanus P. *Calculus Made Easy.* New York: St. Martin's Press, [1910, 1914] 1946.

Thorndike, Lynn. *A History of Magic and Experimental Science: During the First Thirteen Centuries of Our Era.* 2 vol. New York: Columbia University Press, 1923.

Torrance, Thomas F. *Divine and Contingent Order.* Oxford: Oxford University Press, 1981.

___. *Theology of Reconstruction.* Grand Rapids: Eerdmans, 1965.

Tozer, Aiden W. *The Knowledge of the Holy.* New York: Harper & Row, [1961] 1975.

___. *The Pursuit of God.* Harrisburg, PA: Christian Publications, [1948] 1976.

Trachtenburg, Jakow. *The Trachtenburg Speed System of Basic Mathematics.* trans. Ann Cutler and Rudolph McShane. Westport: Greenwood Press, [1960] 1981.

Transnational College of LEX. *Who is Fourier: A Mathematical Adventure.* trans. Alan Gleason. Boston: Language Research Foundation, 1995.

Turnbull, Herbert Westren. *The Great Mathematicians.* New York: Barnes & Noble, [1929] 1993.

Turner, Dorothy M. *Makers of Science: Electricity and Magnetism.* New York: Oxford University Press, 1927.

Upgren, Arthur. *Night Has a Thousand Eyes.* Cambridge, MA: Perseus Publishing, 1998.

Usher, Abbott Payson. *A History of Mechanical Inventions*. New York: Dover Publications, [1954] 1988.

Van Der Klok, Don. "A Christian Mathematics Education?" *The Christian Educators Journal*, (February 1983), 9, 10, 27.

Van Til, Cornelius. *A Christian Theory of Knowledge*. Philadelphia: Presbyterian and Reformed, 1969.

___. *A Survey of Christian Epistemology*. Phillipsburg: Presbyterian and Reformed, 1969.

___. *Apologetics*. Phillipsburg: Presbyterian and Reformed, 1976.

___. *Essays on Christian Education*. Phillipsburg: Presbyterian and Reformed, [1971] 1979.

___. *The Case for Calvinism*. Nutley, NJ: The Craig Press, 1964.

___. *The Defense of the Faith*. Phillipsburg: Presbyterian and Reformed, [1955, 1963] 1967.

___. *Why I Believe in God*. Phillipsburg: Presbyterian and Reformed, n.d.

Van Til, Henry R. *The Calvinistic Concept of Culture*. Philadelphia: Presbyterian and Reformed, 1959.

Varghese, Roy Abraham, ed. *The Intellectuals Speak Out About God*. Chicago: Regnery Gateway, 1984.

Vergara, William C. *Electronics in Everyday Life*. New York: Dover Publications, [1961] 1984.

___. *Mathematics in Everyday Things*. New York: Harper & Brothers, 1959.

___. *Science in Everyday Things*. New York: Harper & Brothers, 1958.

___. *Science: The Never-Ending Quest*. New York: Harper & Row, 1965.

Von Baeyer, Hans C. *Rainbows, Snowflakes, and Quarks: Physics and the World Around Us*. New York: McGraw-Hill, 1984.

Wallace, James and Jim Erickson. *Hard Drive: Bill Gates and the Making of the Microsoft Empire*. New York: John Wiley & Sons, 1992.

Watkins, William D. *The New Absolutes*. Minneapolis: Bethany House Publishers, 1996.

Watson, James D. *The Double Helix*. ed. Gunther S. Stent. New York: W. W. Norton, 1980.

Weaver, Jefferson Hane. *Conquering Calculus: The Easy Road to Understanding Mathematics*. New York: Plenum, 1998.

Webster, Noah. *American Dictionary of the English Language*. San Francisco: Foundation for American Christian Education, [1828, 1967, 1980, 1983] 1985.

Westfall, Richard S. *Science and Religion in Seventeenth-Century England*. Ann Arbor: The University of Michigan Press, 1973.

Weston, Anthony. *A Rulebook for Arguments*. Indianapolis: Hackett Publishing, 1992.

Weyl, Hermann. "A Half-Century of Mathematics." *American Mathematical Monthly*, 58 (1951), 523-553.

___. *Philosophy of Mathematics and Natural Science*. Princeton: Princeton University Press, 1948.

____. *Space, Time and Matter.* trans. Henry L. Brose. London: Methuen, 1922.

____. *Symmetry.* Princeton: Princeton University Press, 1952.

____. *The Theory of Groups and Quantum Mechanics.* trans. H. P. Robertson. London: Methuen, 1931.

Whitcomb, John C. and Henry M. Morris. *The Genesis Flood.* Philadelphia: Presbyterian and Reformed, 1961.

White, Andrew Dickson. *A History of the Warfare of Science with Theology in Christendom.* Albany, OR: Sage Digital Library, [1896] 1996.

White, Lynn T. *Medieval Technology and Social Change.* Oxford: Clarendon Press, 1962.

____. "Technology and Invention in the Middle Ages." *Speculum,* 15 (1940), 151.

Whitehead, Alfred North. *Adventures of Ideas.* New York: The Free Press, 1967.

____. *An Introduction to Mathematics.* New York: Henry Holt and Company, 1939.

____. *Essays in Science and Philosophy.* London: Rider and Company, 1948.

____. *Science and the Modern World.* London: Free Association Books, [1926] 1985.

____. *The Aims of Education and Other Essays.* London: Williams and Norgate, 1929.

____. *The Function of Reason.* Princeton: Princeton University Press, 1929.

Wigner, Eugene. *Symmetries and Reflections: Scientific Essays.* Cambridge and London: The MIT Press, 1970.

Wilder, Raymond. *Mathematics as a Cultural System.* Oxford: Pergamon Press, 1981.

Wilder-Smith, A. E. *He Who Thinks Has to Believe.* Minneapolis: Bethany House Publishers, 1981.

Wilson, Alistair Macintosh. *The Infinite in the Finite.* New York: Oxford University Press, 1995.

Wilson, Douglas. *Persuasions: A Dream of Reason Meeting Unbelief.* Moscow, ID: Oakcross, 1989.

____. *The Paideia of God.* Moscow, ID: Canon Press, 1999.

____., ed. *Repairing the Ruins.* Moscow, ID: Canon Press, 1996.

____., ed. *The Forgotten Heavens.* Moscow, ID: Canon Press, 1989.

Wilson, Edward O. *Consilience: The Unity of Knowledge.* New York: Alfred A. Knopf, 1998.

Windschuttle, Keith. *The Killing of History: How Literary Critics and Social Theorists are Murdering our Past.* New York: The Free Press, 1996.

Wood, Elizabeth A. *Science from Your Airplane Window.* New York: Dover Publications, [1968] 1975.

Wood, Nathan R. *The Trinity in the Universe.* Grand Rapids: Kregel, [1955] 1978.

Zacharias, Ravi. *Deliver Us From Evil: Restoring the Soul in a Disintegrating Culture.* Dallas: Word Publishing, 1996.

Zebrowski, Ernest, Jr. *A History of the Circle: Mathematical Reasoning and the Physical Universe.* New Brunswick: Rutgers University Press, 1999.

Zimmerman, Larry L. "Mathematics: Is God Silent?" *The Biblical Educator,* 2:1-3 (1980).

CREDITS

CHAPTER 2

Otto Neugebauer (page 15), courtesy of University of St. Andrews

Thales of Miletus (page 20), courtesy of University of St. Andrews

Pythagoras (page 22), courtesy of Culver Pictures, Inc.

Zeno of Elea (page 25), courtesy of University of St. Andrews

Plato (page 31), courtesy of Culver Pictures, Inc.

Aristotle (page 32), courtesy of PAR

Euclid (page 35), courtesy of The Bettmann Archive

Archimedes (page 40), courtesy of The Bettmann Archive

Hipparchus (page 46), courtesy of University of St. Andrews

CHAPTER 3

Claudius Ptolemy (page 58), courtesy of Ronald Sheridan Photo Library

Aurelius Augustinus (page 64), courtesy of Scala/Art Resource, NY

Hypatia (page 70), courtesy of University of St. Andrews

Mohammed ibn Musa al-Khowarizmi (page 83), courtesy of University of St. Andrews

Roger Bacon (page 87), courtesy of PAR

Thomas Aquinas (page 88), courtesy of The Bettmann Archive

Leonardo of Pisa (page 89), courtesy of David Smith Collection

John Wycliffe (page 96), courtesy of Bridgeman/Art Resource

CHAPTER 4

Johann Gutenberg (page 102), courtesy of Culver Pictures, Inc.

Martin Luther (page 103), courtesy of The Bettmann Archive

François Viète (page 107), courtesy of University of St. Andrews

John Napier (page 108), courtesy of Culver Pictures, Inc.

Nicholaus Copernicus (page 113), courtesy of Thomas Fisher Rare Book Library

Francis Bacon (page 114), courtesy of National Portrait Gallery

Johannes Kepler (page 115), courtesy of Culver Pictures, Inc.

Galileo Galilei (page 118), courtesy of Library of Congress

Sir Isaac Newton (page 123), courtesy of PAR

Gottfried Wilhelm Leibniz (page 128), courtesy of The Bettmann Archive

John Calvin (page 132), courtesy of The Bettmann Archive

Jakob Bernoulli (page 148), courtesy of University of St. Andrews

Leonhard Euler (page 150), courtesy of The Bettmann Archive

Voltaire (page 151), courtesy of Culver Pictures, Inc.

CHAPTER 5

René Descartes (page 155), courtesy of PAR

Pierre de Fermat (page 157), courtesy of David Smith Collection

Blaise Pascal (page 158), courtesy of PAR

Thomas Hobbes (page 159), courtesy of National Portrait Gallery, London

George Berkeley (page 160), courtesy of National Portrait Gallery, London

Edmund Halley (page 160), courtesy of The Bettmann Archive

John von Neumann (page 165), courtesy of The Bettmann Archive

Joseph-Louis Lagrange (page 166), courtesy of The Bettmann Archive

Pierre-Simon de Laplace (page 167), courtesy of University of St. Andrews

David Hume (page 168), courtesy of PAR

Immanuel Kant (page 169), courtesy of The Bettmann Archive

Niels Bohr (page 171), courtesy of Culver Pictures, Inc.

Carl Friedrich Gauss (page 173), courtesy of The Bettmann Archive

Nikolai Ivanovich Lobachevsky (page 177), courtesy of New York Public Library Collection

Janos Bolyai (page 178), courtesy of University of St. Andrews

George Friedrich Bernhard Riemann (page 179), courtesy of David Smith Collection

Felix Klein (page 180), courtesy of David Smith Collection

Georg Cantor (page 183), courtesy of David Smith Collection

Leopold Kronecker (page 185), courtesy of David Smith Collection

Henri Poincaré (page 186), courtesy of The Bettmann Archive

Alfred North Whitehead (page 187), courtesy of University of St. Andrews

Luitzen Brouwer (page 188), courtesy of University of St. Andrews

David Hilbert (page 189), courtesy of Don Miller

Kurt Gödel (page 190), courtesy of Alfred Eisenstaedt

Hermann Weyl (page 193), courtesy of Constance Reid

CHAPTER 6

Bertrand Russell (page 195), courtesy of The Bettmann Archive

Aleksandr Solzhenitsyn (page 198), courtesy of Mastro/Archive Photos

Sir James Jeans (page 200), courtesy of University of St. Andrews

Max Planck (page 200), courtesy of University of St. Andrews

Sir Arthur Stanley Eddington (page 203), courtesy of University of St. Andrews

Werner Karl Heisenberg (page 204), courtesy of University of St. Andrews

Morris Kline (page 205), courtesy of Oxford University Press

Eugene Wigner (page 207), courtesy of University of St. Andrews

Erwin Schrödinger (page 209), courtesy of The Bettmann Archive

Albert Einstein (page 210), courtesy of Culver Pictures, Inc.

James Clerk Maxwell (page 219), courtesy of University of St. Andrews

Godfrey H. Hardy (page 220), courtesy of University of St. Andrews

Edward Everett (page 223), courtesy of Library of Congress

Herbert Westren Turnbull (page 224), courtesy of University of St. Andrews

Paul A. M. Dirac (page 224), courtesy of University of St. Andrews

Charles Hermite (page 225), courtesy of University of St. Andrews

CHAPTER 7

Richard P. Feynman (page 233), courtesy of Archive Photos

Joseph Fourier (page 239), courtesy of The Bettmann Archive

SCRIPTURE INDEX

361

GENERAL INDEX

Elementary Algebra; 304
Elements of Euclid; 34, 35, 70, 175, 297, 312, 317
elements, four primal; 27, 33, 311
elephant tusks; 249
Eliphaz the Temanite; 46
ellipse; 43, 44, 116, 117, 250, 270
 definition; 44
 distribution of stress in machinery; 270
 gears and springs; 270
 planetary motion; 250
 pool table; 270
 rings of Saturn; 250
 whispering chambers; 270
ellipsis; 44
elliptic geometry; 180
elliptic integrals; 322
Elm tree; 242
Elysian fields; *303*
emanationism; 62, *91*
Empedocles; 27, 54, 311
 cyclical view of time; 28
 four primal elements; 27, 311
 survival of the fittest; 28
Emperor Augustus; 312
Emperor Constantine; 66, 313
Emperor Diocletian; 313
Emperor Julius; 39, 59, 312
Emperor Justinian; 314
Emperor Marcus Aurelius; *60*, 313
Emperor Nero; 66, 313
empiricism; xxiv, 20, 21, 31, 34, *68*, 87, 89, 93, 98, 113, 128, 131, 165, 168, 187, *194*, 221, 230, 236, *255*, *284*, 291, 298, 315, 316, 318, 321
 definition; *20*
empyrean; 68
England; 75, 86, 96, 113, 130, 131, 148, 156, 159, 275, 315, 317, 319–322, 324, 325
ENIAC; 330
Enlightenment; 67, 72, 320
Environmental Protection Agency; *233*
Ephesus; 24
epicureanism; 59, 312
Epicurus; 59, 312
epicycles; *58*, *113*
Epimenides; 185, 311

epistemology; xxiv, xxv, 7, 9, 13, 14, 17–21, 28, 30, 31, 33, 34, 54, 59, 61, 62, 65, *68*, 86, 87, 89, 91–93, 98, 106, 112, 113, *119*, *120*, 128, 131, 144, *154*, 158, 159, *162*, 165, 168–170, 172, 174, 177, 182, 187, 188, 191, 192, *194*, 198, 199, 210, 211, 218, 221, 222, 224, 226, 230–232, 235, 236, 252, *255*, *284*, 291, 294, 296, 298, 311, 315, 316, 318, 321
 a posteriori; 8, *20*, 21, 28, 89, *119*, *120*, 128, *172*, 231, 232, *255*, 291
 a priori; 8, 21, 28, 33, 89, 91, *92*, *119*, *120*, 128, 169, 172, 177, 188, 210, 231, 232, *255*, 291, 311
 analogous reasoning; 13, 18, *65*, 144, 224, 226
 definition; 7
 empiricism; xxiv, 20, 21, 31, 34, *68*, 87, 89, 93, 98, 113, 128, 131, 165, 168, 187, *194*, 221, 230, 236, *255*, *284*, 291, 298, 315, 316, 318, 321
 rationalism; xxv, 21, 30, 86, 128, 159, *162*, 191, 218
 univocal reasoning; 18, 226
Epitome astronomiae Copernicanae; 115, 319
equal sign, symbol for; 156, 318
equation(s); 14, 40, 69, 84, *107*, 141, 149, 157, 164, 166, 224, *245*, 300, 305, 317, 320–322, 324, 325
 cubic; 14, 40, *107*, 317
 determinate and indeterminate; 141
 differential; 149, 164, 166, 300–322, 325
 indeterminate; 69, 157
 integral; 325
 parametric; 305
 quadratic; 14, *245*
 quartic; *107*, 317
 theory of; 224, 320, 324
equator; 178
equiangular spiral; 149
equinoctial noon; 42
equinox; 21, 45
 precision of; 45
equivocation; 18, 205, 226, 265
Erasmus, Desiderius; 317
Eratosthenes; 39, 41–43, 52, 312
 circumference of the earth; 41
 mapmaking; 41
 poetry; 41

F

O

ABOUT THE AUTHOR

James Nickel holds B.A. (Mathematics), B.Th. (Theology and Missions), and M.A. (Education) degrees. He has worked as a professional mathematician in the United States Navy's test flight program and has over twenty-five years of experience as a computer professional in both scientific and business settings. Married with three children, he has been involved in the Christian school movement since 1978 serving as a teacher, home school parent, researcher, curriculum developer, lecturer, and writer.

He endeavors to ground all of his work upon the Trinitarian, Covenantal, and Reformed distinctives of historic, creedal Christianity. His mission is to press the crown rights of the Lord Jesus Christ in every sphere of life, expecting eventual triumph for, in the words of Cornelius Van Til, "unless we press the crown rights of our King in every realm we shall not long retain them in any realm." He labors for the cause of biblical Christian reformation heeding the injunction that true reformation begins with biblical scholarship read and applied.

He is currently Dean of the School of Mathematics, Christian Heritage Academy, International – a division of Patria Institute, LLC where he is endeavoring to develop and market a year 7-12 distance learning mathematics program.

For up-to-date information and resources about world view analysis, mathematics, and other topics, see the author's web site:

www.biblicalchristianworldview.net

For information about the work of Patria Institute, LLC, contact:

Patria Institute, LLC
Christian Heritage Academy, International
Post Office Box 211
Leavenworth, Washington 98826
509-548-7575
www.patriainstitute.com

Johannes Kepler
Courtesy of Sternwarte Kremsmünster, Linz